POCKETMEC

Mercedes-
E-Class
E200 D, E220D, E250D, E250 TD
E290TD, E300D, E300 TD
2.0, 2.2, 2.5, 2.9 and 3.0 Litre
from 1995 to 2000

COVERING:

BY PETER RUSSEK

ORDER NO.: 625

Published by
Peter Russek Publications Ltd.
3rd Floor, Block C,
Commercial Square,
High Wycombe, Bucks, HP11 2RH
Tel.: High Wycombe (01494) 440829
(01494) 526882
Fax: (01494) 473980
E-mail: russek@globalnet.co.uk
www.russek-manuals.co.uk

ISBN. NO. 1-898780-62-5

WITH FAULT FINDING SECTION AT END OF MANUAL

The publisher would like to thank Daimler Chrysler U.K. Ltd. for their invaluable support in producing this manual.

No liability can be accepted for any inaccuracies or Omissions in this workshop manual, although every possible care has been taken to make it as complete and accurate as possible. Every care has also been taken to prevent personal injury or damage to equipment when working on the vehicle We have tried to cover all models produced to the day of publication, but are unable to refer to all modifications and changes for certain markets or up-dating of models.

 Printed in England

PREFACE

Small though this Workshop Manual is in size, it lacks no detail in covering most of the servicing and repair of the Mercedes-Benz vehicles, known as "E-Class" and Series 210, introduced during 1995 with injection pump and turbo charger when applicable. The engines have 2 valves for each cylinder or 4 valves, as in the case of the 2.5 litre engines in models E250 D and E250 TD and 3.0 litre engines in models E300D and E300 TD.

Brief, easy-to-follow instructions are given, free from all necessary complications and repetitions, yet containing all the required technical detail and information, and many diagrams and illustrations.

Compiled and illustrated by experts, this manual provides a concise source of helpful information, all of which has been cross-checked for accuracy to the manufacturer's official service and repair procedures, but many instructions have derived from actual practice to facilitate your work. Where special tools are required, these are identified in the text if absolutely necessary and we do not hesitate to advise you if we feel that the operation cannot be properly undertaken without the use of such tools.

The readers own judgement must ultimately decide just what work he will feel able to undertake, but there is no doubt, that with this manual to assist him, there will be many more occasions where the delay, inconvenience and the cost of having the van off the road can be avoided or minimised.

The manual is called "Pocket Mechanic" and is produced in a handy glove pocket size with the aim that it should be kept in the vehicle whilst you are travelling. Many garage mechanics themselves use these publications in their work and if you have the manual with you in the car you will have an invaluable source of reference which will quickly repay its modest initial cost.

Finally – We do not claim that we know everything. If you have any suggestions to facilitate certain operations, please do not hesitate to pass them onto us, so other fellow motorists can benefit later on.

0. INTRODUCTION

Our "Pocket Mechanics" are based on easy-to-follow step-by-step instructions and advice, which enables you to carry out many, jobs yourself. Moreover, now you have the means to avoid these frustrating delays and inconveniences which so often result from not knowing the right approach to carry out repairs which are often of a comparatively simple nature.

Whilst special tools are required to carry out certain operations, we show you in this manual the essential design and construction of such equipment, whenever possible, to enable you in many cases to improvise or use alternative tools. Experience shows that it is advantageous to use only genuine parts since these give you the assurance of a first class job. You will find that many parts are identical in the various makes covered, but our advice is to find out before purchasing new parts - **Always buy your spare parts from an officially appointed dealer.**

0.0. General Information

Four-cylinder, five-cylinder and six-cylinder engines are covered in the manual, having in common a electronically controlled injection pump.

In the category "Four-cylinder Engines" with 8 valves you will find two engines within the model years in question: A 2.0 litre (1997 c.c.) engine with a performance of 75 BHP (55 kW) at 4600 rpm is fitted to model E 200 D with model designation 210.003 and with engine type 604.917. An engine with the same capacity but with a performance of 95 BHP (70 kW) at 5000 rpm is fitted to model E220 D with model designation 210.004 and with engine type 604.912. Both engines belong to the engine series "604", denoting a four-cylinder engine. Both engines have 8 valves.

Three five-cylinder engines are covered in the manual. The E250 D is fitted with an engine with a capacity of 2.5 litre (2497 c.c.) and a performance of 113 BHP (83 kW) at 5000 rpm. The engine type 605.912 is fitted to model 210.015. The same engine is also fitted to model 210.210.

The E 250 TD is fitted with a similar engine, but with a performance of 150 BHP (110 kW) at 4400 rpm. The engine designation is 605.962. The same engine is also fitted to models 210.215 and 210.217, depending on the model year.

The five-cylinder engine with a capacity of 2.9 litres (2874 c.c.) and with 2 per cylinder and a performance of 129 BHP (95 kW) at 4000 rpm is fitted to model E290 TD with model designation 210.217 and with engine type 602.982. The same engine is also fitted to model 210.617.

The six-cylinder engine, pre-fix "606", is fitted to model E300D (model with a capacity of 3.0 litres (2996 c.c.) with 24 valves and develops a performance of 139 BHP (100 kW) at 5000 rpm. Model identification is 210.020, engine identification 606.912.

The E300 with turbo charger (E300 TD) has a performance of 177 BHP (130 kW) at 4400 rpm and is fitted to model 210.025 or 210.225, depending on the model year.

All models belong to the model series W210, but it should be noted that the performance figures given above can vary, depending on the country.

The advantages of a multi-valve technology – 6-cyl. Engines

One of the basis problems of a four stroke engine is the filling of the cylinders during the induction stroke with the necessary amount of the fuel/air mixture. The problem is increased with increasing engine speed, as the opening period of the valves is

shortened. Technicians refer to filling loss. To compensate the valve diameters are selected as large as possible so that more fuel/air mixture can enter, but the disadvantage is, of course, the larger diameter of the compression chamber.
This is the main reason for the introduction of the multi-valve technology. Four valve heads make up a lager opening area than two large valve heads and the size of the compression chamber can remain the same.
The advantages of the four valves can therefore given as follows:

- Four valves enable larger opening diameters for inlet and exhaust gases. This helps the engine performance and the fuel consumption. This is one of the reasons that an engine with four valves per cylinder has a better consumption than a valve with two valves.
- Engines with four valves per cylinder have smaller valves and thereby less weight to be moved. A quicker response of the valve gear is therefore possible. One more advantage of this construction that valve springs with a lower pressure is necessary to close the valves.
- Smaller valves are able to cool down quicker during the closing period.
- Engines with four valves per cylinder also enable to obtain a higher compression ratio.

The vehicles covered in this manual are fitted with a five-speed manual transmission or an automatic transmission with four or five speeds.
The front axle of the new series W210 has a double wishbone.
The so-called multi-link independent rear suspension comprises the rear axle carrier, provided with double-row angular ball bearings. The wheel carriers are guides by 5 specially located links, referred to as camber strut, pulling strut, pushing strut, track rod and spring links, the latter being the actual suspension arms. The hydraulic shock absorbers are fitted between the spring links and the body. A torsion bar is fitted the spring links and the frame floor by means of a connecting link. The well-known level control system in other Mercedes Benz models is either fitted as standard or as an optional extra throughout the range.
Disc brakes on all four wheels, with dual-line brake system and brake servo is fitted. The handbrake acts on the rear wheels.
New on the Series W210 is the installation of a rack and pinion steering, replacing the re-circulating ball-type steering used on earlier models. The servo-assistance, however, has been retained.

0.1. Vehicle Identification

The type identification plate is located at the R.H. side on the upper face of the radiator frame and contains the vehicle type, chassis number, permissible maximum weight and the permissible axle load on front and rear axle. The paint code is located in the plate opposite.
The chassis number can also be found in the engine compartment bulkhead. The first six numbers refer to the vehicle type, the 7th number refers to the steering and the 8th number to the transmission. The following six numbers are the actual serial number.
The engine number is stamped into the cylinder block above the injection pump flange at the front end of the engine.
The code numbers and letters must always be quoted when parts are ordered. Copy the numbers on a piece of paper and take it to your parts supplier. You will save yourself and your parts department delays and prevents you from ordering the wrong parts.

0.2. General Servicing Notes

The servicing and overhaul instructions in this Workshop Manual are laid out in an easy-to-follow step-by-step fashion and no difficulty should be encountered, If the text and diagrams are followed carefully and methodically. The "Technical Data" sections form an important part of the repair procedures and should always be referred to during work on the vehicle.

In order that we can include as much data as possible, you will find that we do not generally repeat in the text the values already given under the technical data headings. Again, to make the best use of the space available, we do not repeat at each operation the more obvious steps necessary - we feel it to be far more helpful to concentrate on the difficult or awkward procedures in greater detail. However, we summarise below a few of the more important procedures and draw your attention to various points of general interest that apply to all operations.

Always use the torque settings given in the various main sections of the manual. These are grouped together in separate sub-sections for convenient reference.

Bolts and nuts should be assembled in a clean and very lightly oiled condition and faces and threads should always be inspected to make sure that they are free from damage burrs or scoring. DO NOT degrease bolts or nuts.

All joint washers, gaskets, tabs and lock washers, split pins and "O" rings must be replaced on assembly. Seals will, in the majority of cases, also need to be replaced, if the shaft and seal have been separated. Always lubricate the lip of the seal before assembly and take care that the seal lip is facing the correct direction.

References to the left-hand and right-hand sides are always to be taken as if the observer is at the rear of the vehicle, facing forwards, unless otherwise stated.

Always make sure that the vehicle is adequately supported, and on firm ground, before commencing any work on the underside of the car. A small jack or a make shift prop can be highly dangerous and proper axle stands are an essential requirement for your own safety.

Dirt, grease and mineral oil will rapidly destroy the seals of the hydraulic system and even the smallest amounts must be prevented from entering the system or coming into contact with the components. Use clean brake fluid or one of the proprietary cleaners to wash the hydraulic system parts. An acceptable alternative cleaner is methylated spirit, but it this is used, it should not be allowed to remain in contact with the rubber parts for longer than necessary. It is also important that all traces of the fluid should be removed from the system before final assembly.

Always use genuine manufacturer's spares and replacements for the best results.

Since the manufacturer uses metric units when building the cars it is recommended that, these are used for all precise units. Inch conversions are given in most cases but these are not necessarily precise conversions, being rounded off for the unimportant values.

Removal and installation instructions, in this Workshop Manual, cover the steps to take away or put back the unit or part in question. Other instructions, usually headed "Servicing", will cover the dismantling and repair of the unit once it has been stripped from the vehicle it is pointed out that the major instructions cover a complete overhaul of all parts but, obviously, this will not always be either necessary and should not be carried out needlessly.

There are a number of variations in unit parts on the range of vehicles covered in this Workshop Manual. We strongly recommend that you take care to identify the precise model, and the year of manufacture, before obtaining any spares or replacement parts.

Std.: To indicate sizes and limits of components as supplied by the manufacturer. Also to indicate the production tolerances of new unused parts.

O/S Paris supplied as Oversize or Undersize or recommended limits for such parts, to enable them to be used with worn or re-machined mating parts.

U/S O/S indicates a part that is larger than Std. size U/S may indicate a bore of a bushing or female part that is smaller than Std.

Max.: Where given against a clearance or dimension indicates the maximum allowable If in excess of the value given it is recommended that the appropriate part is fitted.

TIR: Indicates the Total Indicator Reading as shown by a dial indicator (dial gauge).

TDC: Top Dead Centre (No. 1 piston on firing stroke).

MP: Multi-Purpose grease.

0.3. Recommended Tools

To carry out some of the operations described in the manual we will need some of the tools listed below:

Fig. 0.1 – A double open-ended spanner in the upper view and an open-ended/ring spanner in the lower view. Always make sure that the spanner size is suitable for the nut or bolt to be removed and tightened.

As basic equipment in your tool box you will need a set of open-ended spanners (wrenches) to reach most of the nuts and bolts. A set of ring spanners is also of advantage. To keep the costs as low as possible we recommend a set of combined spanners, open-ended on one side and a ring spanner on the other side. Fig. 0.1 shows a view of the spanners in question. Sockets are also a useful addition to your tool set.

A set of cross-head screwdrivers, pliers and hammers or mallets may also be essential. You will find that many bolts now have a "Torx" head. In case you have never seen a "Torx" head bolt, refer to Fig. 0.2. A socket set with special t "Torx" head inserts is used to slacken and tighten these screws. The size of the bolts are specified by the letter "T" before the across-flat size.

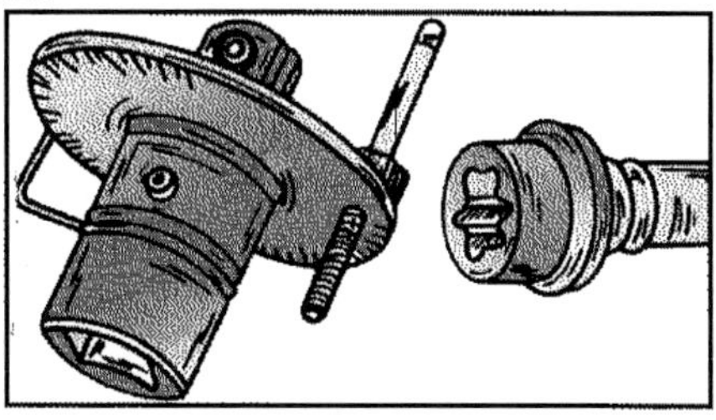

Fig. 0.2 – A graduated disc is used to "angle-tighten" nuts and bolts. "Torx" head bolts are shown on the R.H. side.

Circlip pliers may also be needed for certain operations. Two types of circlip pliers are available, one type for external circlips, one type for internal circlips. The ends of the pliers can either be straight or angled. Fig. 0.3 shows a view of the circlip pliers. Apart from the circlip pliers you may also need the pliers shown in Fig. 0.4, i.e. side cutters, combination pliers and water pump pliers.

Every part of the vehicle is tightened to a certain torque value and you will therefore need a torque wrench which can be adjusted to a certain torque setting. In this connection we will also mention a graduated disc, shown in Fig. 0.4, as many parts of the vehicle must be angle-tightened after having been tightened to a specific torque. As some of the angles are not straight-forward (for example 30 or 60 degrees), you will either have to estimate the angle or use the disc.

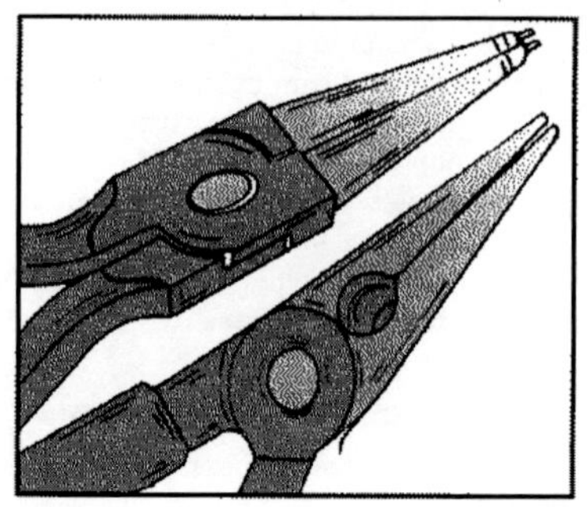

Fig. 0.3 – Circlip pliers are shown in the upper view. The type shown in suitable for outside circlips. The lower view shows a pair of pointed pliers.

Finally you may consider the tool equipment shown in Fig. 0.4 which will be necessary from time to time, mainly if you intend to carry out most maintenance and repair jobs yourself.

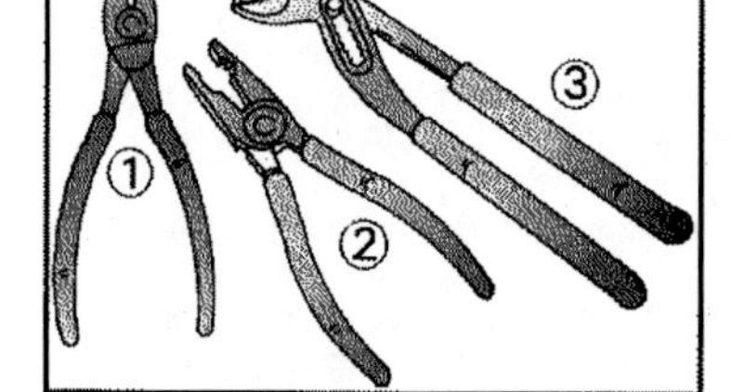

Fig. 0.4 – Assortment of pliers suitable for many operations.
1 Side cutter
2 Combination pliers
3 Water pump pliers

0.4. Before you start

Before you carry out any operations on your vehicle it may be of advantage to read the following notes to prevent injuries and damage to the vehicle:

- Never carry out operations underneath the vehicle when the front or rear is only supported on the jack. Always place chassis stands in position (refer to next section). If no chassis stands are available and if the wheels are removed place on wheel on top of the other one and place them under the side of the vehicle where you work. If the jack fails the vehicle will drop onto the two wheels, preventing injury.
- Never slacken or tighten the axle shaft nuts or wheel bolts when the vehicle in resting on chassis stands.
- Never open the cooling system when the engine is hot. Sometimes it may, however, be necessary. In this case place a thick rag around the cap and open it very slowly until all steam has been released.
- Never allow brake fluid or anti-freeze to come in contact with painted areas.
- Never inhale brake shoe or brake pad dust. If compressed air is available, blow off the dust whilst turning the head away. A mask should be worn for reasons of safety.
- Remove oil or grease patches from the floor before you or other people slip on it.
- Do not work on the vehicle wearing a shirt with long sleeves. Rings and watches should be removed before carrying out any work.
- If possible never work by yourself. If unavoidable ask a friend or a member of the family to have a quick look to check that's everything is OK.
- Never hurry up your work. Many wheel bolts have been left untightened to get the vehicle quickly back on the road.
- Never smoke near the vehicle or allow persons with a cigarette near you. A fire extinguisher should be handy, just in case.

- Never place a hand lamp directly onto the engine to obtain a better view. Even though that the metal cage will avoid direct heat it is far better if you attach such a lamp to the open engine bonnet.
- Never drain the engine oil when the engine is hot. Drained engine oil must be disposed of in accordance with local regulation.

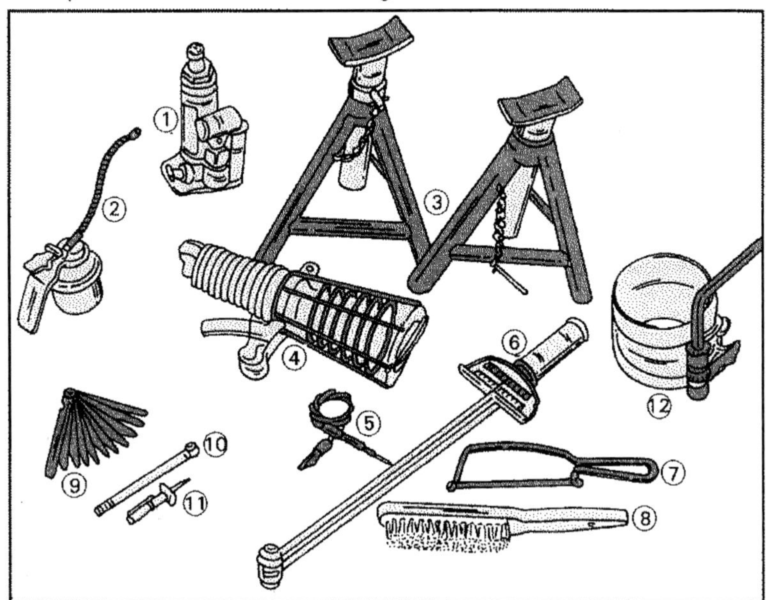

Fig. 0.5 – Recommended tools to service and repair your vehicle.

1 Hydraulic jack
2 Oil can
3 Chassis stands
4 Electric hand lamp
5 Test lamp (12 volts)
6 Torque wrench
7 Small hand saw
8 Wire brush
9 Feeler gauges
10 Tyre pressure gauge
11 Tyre profile depth checker
12 Piston ring clamp band

0.5. Dimensions and Weights (typical)

Overall length:..4800/4740 mm (Estate 4765 mm)
Overall width:..1740/1800 mm
Overall height:..1443/1448 mm
Wheelbase:..2800 mm, 2830 mm, 3600 mm
Front track:..1500 or 155 mm
Rear track:..1490 or 1540 mm
Width:..1740, 180 mm (Estate 1740 mm)
Ground clearance:..160 mm
Kerb Weights (as available):
Typical examples, refer to your Owners' Manual for details:
- E200D:..1340 or 1365 kg
- E220D:..1390 kg
- E250D:..1420 kg

- E250 D Estate: ....1530 kg
- E250 TD: ....1535 kg
- E290D:...1465 kg
- E300 D :...1470 kg (USA 1580 kg
- E300 TD: .... 1490 or 1535 kg, depending on model year
- E300 TD Estate:..1600 kg

Permissible total weight with trailer:
- E200D:... 1910 or 1960 kg, depending on model year
- E220D:...1960 kg
- E250D:...1990 kg
- E250 D Estate: ....2180 kg
- E250 TD: ....1990 kg
- E290D:...2040 kg
- E300 D :..2030 kg
- E300 TD: ....2030 kg
- E300 TD Estate:..2235 kg

0.6. Capacities

Engines:
- Oil and filter change – 604 and 605 engines, with and without ASSYST: .....6.5 litres
- Oil and filter change – 606-engine, with and without ASSYST: ....7.5 litres
- Difference between Max/Min: ....2.0 litres

Cooling System (with heater, approx.):
- 604 engine – without A/C system: ....6.0 litres
- 604 engine – with A/C system: ....7.0 litres
- 605 engine:..7.5 litres
- 606 engine, E300D: ....7.5 litres
- 606 engine, E300 TD: ....9.0 litres

Manual transmission: ....1.5 litres
Brake fluid:...0.45 to 0.6 litres
Steering system:...1.0 litre
Fuel tank: .... Depending on model 65, 70 or 80 litres
Fuel reserve capacity: .... Depending on model 8, 9 or 10 litres

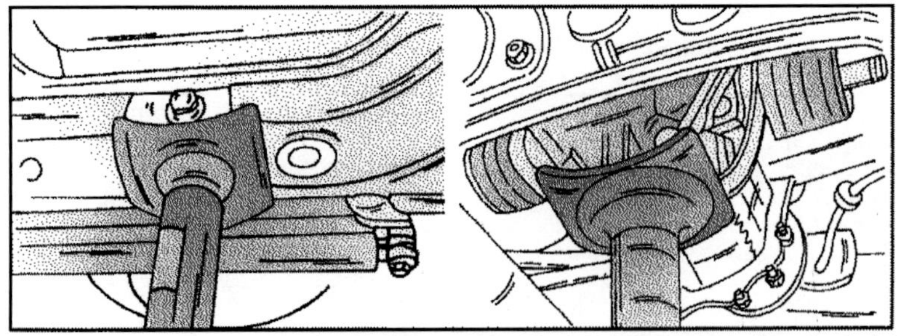

Fig. 0.6 – Jacking up the front end of the vehicle. The L.H. view shows where the jack is placed underneath the front crossmember. The R.H. view shows the jacking up of the rear of the vehicle. The jack is placed underneath the centre piece of the rear axle.

0.7. Jacking up the Vehicle

To prevent damage to the underside of the vehicle, apply a jack or chassis stands only to the points specified below:

Fig. 0.7 – Jacking up one side of the vehicle. Place the jack (2) underneath the side of the body as shown. Chassis stands are placed at the position shown.

The front end of the vehicle should be lifted up by placing a jack underneath the transverse crossmember of the front suspension as shown in Fig. 0.6, taking care not to damage the undercover for the engine compartment. To lift the rear end of the vehicle, place the jack underneath the rear axle centre piece, as shown in Fig. 0.6 on the R.H. side. Make sure the jack is sufficient to take the weight of the vehicle. The vehicle can also be jacked up on one side. In this case place the jack underneath the hard rubber inserts near the wheels, as shown in Fig. 0.7 on one side of the vehicle. Never place a jack underneath the oil sump or the gearbox to lift the vehicle.

Chassis stands should only be placed on the L.H. and R.H. sides under the side of the body without damage to the paint work. Use chassis stands of the construction shown in Fig. 0.8, should be used, but again make sure that they are strong enough to carry the weight of the vehicle. Make sure the vehicle cannot slip off the stands.

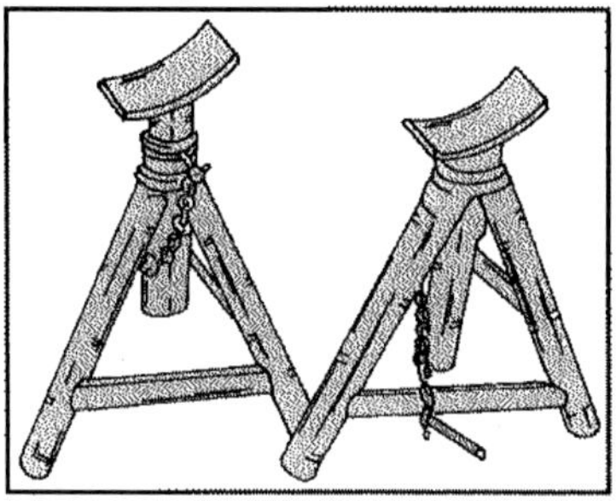

Fig. 0.8 – Three-legged chassis stands are the safest method to support the vehicle when work has to be carried out on the underside of the vehicle.

Before lifting the front of the vehicle engage first or reverse gear when a manual transmission is fitted or place the gear selector lever into the "P" (park) position when an automatic transmission is fitted. Use suitable chocks and secure the front wheels when the rear end of the vehicle is jacked up.
Always make sure that the ground on which the vehicle is to be jacked up is solid enough to carry the weight of the vehicle.

Note: *It is always difficult to raise a vehicle first on one side and then on the other. Take care that the vehicle cannot tip-over when the first side is lifted. Ask a helper to support the vehicle from the other side. Never work underneath the vehicle without adequate support.*

1. ENGINES

1.0 Main Features

Fitted Engines:
- E200 Diesel (4-cyl.) – in model 210.003: 604.917
- E220 Diesel (4-cyl.) – in model 210.004: 604.912
- E250 Diesel (5-cyl.) in model 210.010: 605.912
- E250 Turbo diesel (5-cyl.) in model 210.015 and 210.210: 605.962
- E290 Turbo diesel (6-cyl.) – in model 210.017: 602.982
- E300 Diesel (6-cyl.) – in model 210.020: 606.912
- E300 Turbo diesel (6-cyl.) – in model 210.025: 606.962

Injection Order:
- Four-cylinder.: .... 1 – 3 – 4 - 2
- Five-cylinder: .... 1 – 2 – 4 – 5 – 3
- Six-cylinder: .... 1 – 5 – 3 – 6 – 3 – 4

Arrangement of cylinders: .... In line

Camshafts:
- Except E290 TD: Two overhead camshafts, No. marked in end face
- E290 TD: 1 camshaft

Arrangement of valves: .... Overhead

Cylinder bore: .... 87.00 mm or 89.00 mm (E 290 TD)

Piston stroke:
- E200 D: .... 86.6 mm
- E290 TD: .... 92.40 mm
- All other engines: .... 84.00 mm

Capacity:
- 604 engine: .... 2155 c.c.
- 605 engine: .... 2497 c.c.
- 602 engine: .... 2874 c.c.
- 606 engine: .... 2996 c.c.

Compression Ratio:
- 602 engine: .... 19.5 : 1
- 604, 605, 606 engines: .... 22.0 : 1

Max. kW/B.H.P. (DIN) (variations are possible):
- E200 D: .... 55 (75 BHP at 4600 rpm
- E220 D: .... 70 kW (95 BHP) at 5000 rpm
- E250 D: .... 83 kW (113 BHP) at 5500 rpm
- E250 TD: .... 150 BHP (110 kW) at 4400 rpm
- E290 TDI: .... 95 kW (129 BHP) at 4000 rpm
- E300 D: .... 100 kW (136 BHP) at 5000 rpm
- E300 TD to 1996: .... 108 kW (147 BHP) at 4600 rpm
- E300 TD from 1997: .... 130 kW (177 BHP) at 4400 rpm

Max. Torque (variations are possible):
- E200 D: .... 13.0 kgm at 3600 rpm
- E220 D: .... 15.3 kgm at 3100 rpm

- E250 D: .... 17.3 kgm at 2800 – 4600 rpm
- E250 TD: .... 28.0 kgm at 2800 rpm
- E290 TD: .... 30.6 kgm at 1800 rpm
- E300 D: .... 21.4 kgm at 2200 - 4600 rpm
- E300 TD to 1996: .... 27.8 kgm at 2400 rpm
- E300 TD from 1997: .... 33.6 kgm at 1600 rpm

Crankshaft bearings: .... 5 (4-cyl.), 6 (5-cyl.), 7 (6 cyl.)

Cooling system	Thermo system with water pump, thermostat, cooling fan, tube-type radiator
Lubrication	Pressure-feed lubrication with gear-type oil pump, driven with chain from crankshaft. With full-flow and by-pass oil filter
Air cleaner	Dry paper element air cleaner

General Information

The diesel engines fitted to the E-Class vehicles are different in many ways. We assume that you may have changed from a type W124 model to a W210 model and you may therefore know something about the engine series 601, 602, and 603 fitted to the earlier models. The following information will tell you something about the new engines of type "604", "605", "606" and "602", the latter fitted to the E290 TD.

- The cylinder head, as already mentioned, has four valves per cylinder (either 16 valves in the case of a four-cylinder or 24 valves in the case of a six-cylinder) or as previously 8 valves. The cylinder head is made of light alloy metal. The valve seats, made of hardened steel, are pressed into the cylinder head. The valves are „gliding" in brass valve guides and are arranged as „overhead" valves, i.e. they are inserted vertically, valve head down, into the combustion chambers.
- With the exception of the 602 engine two camshafts are fitted. The shaft on the R.H. side, seen in the direction of drive, is responsible for the opening and closing of the exhaust valves, the L.H. shaft is the inlet valve camshaft. The bearings for the camshafts are not machined directly into the cylinder head.
- The adjustment of the valves is no longer necessary on these engines. Hydraulic compensating elements are fitted which will ensure the correct valve clearance at all times. The function of the hydraulic valve clearance compensating elements is to eliminate valve clearance, i.e. the dimensional changes in the valve train (valve lash) due to heat expansion and wear are compensated by the elements. The hydraulic valve compensating element are fitted into the rocker levers and operate the valves directly via a ball socket: The rocker arm is in constant contact with the cam. The compensating elements cannot be repaired, but can be checked for correct functioning.
- The valve tappets are inserted between camshafts and valve ends. The cams push against the ends of the tappets to operate the valves. The tappets are known as bucket tappets.

1.1. Engine – Removal and Installation

1.1.0. ENGINE WIRING HARNESS – DISCONNECTION AND CONNECTION

Before the removal of the engine we will show you where the engine wiring harness must be disconnected. The components are shown in Figs. 1.1 and 1.2. You will have to refer to these illustrations for any other jobs, so remember where the connections can be found. The illustrations are shown with the component numbers as given in the wiring diagrams and can therefore be cross-referenced accordingly.

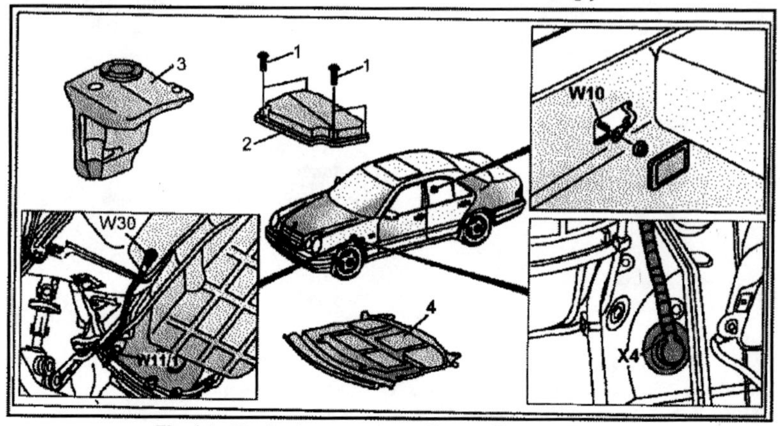

Fig. 1.1 – Engine harness connections – Also see Fig. 1.2.

1 Component compartment cover screws
2 Component compartment cover
3 Fluid container
4 Undercover(sound proofing)
W10 Battery earth lead below rear seat
W11/1 Battery earth cable
W30 Earth cable
X4 Connector, terminal 30, foot well

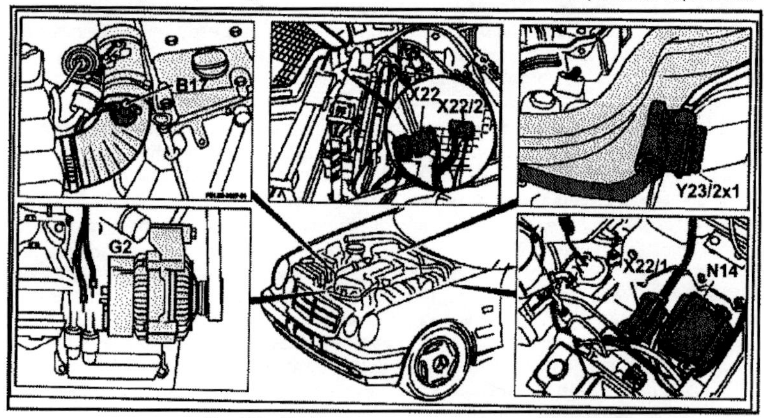

Fig. 1.2 – Engine harness connections – Also see Fig. 1.1.

B17 Intake air temperature sensor
G Alternator
N14 Glow-plug relay
X22 Engine compartment/transmisson connector
X22/1 As X22 for automatic
X22/2 Connector, terminal 50
Y23/2x1 Injection pump connector

1.1.1. SOUND PROOFING UNDERNEATH THE VEHICLE

Again the panels underneath the vehicles must be removed in order to remove the engine and transmission and will also be necessary for other work to be carried out on the underside of the vehicle. The sound proofing consists of three parts, front section, centre section and rear section. The attachment of the front and rear sections are shown in Fig. 1.3. The centre section is attached at the four corners of the square panel.

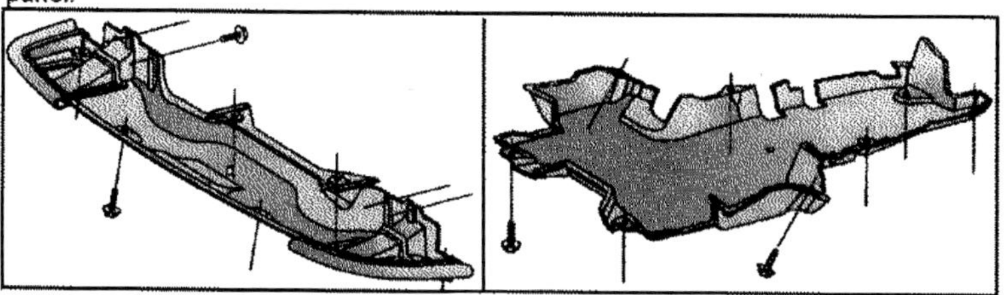

Fig. 1.3 – The front (left) and rear (right) sound proofing bottom sections.

1.1.2 REMOVAL OF THE ENGINE

The engine and gearbox should be removed as a single unit out of the engine compartment. The gearbox can then be removed from the engine. We would like to point out that the total weight of the power unit is approx. 200 kg and a suitable hoist or crane is required to lift out the assembly. The following instructions are given in general for all engines, but is aimed specifically at engines 604, 605 and 606. The engine bonnet must be placed in vertical position or can be removed to prevent damage to the paint work. Proceed as follows for the engine in question:

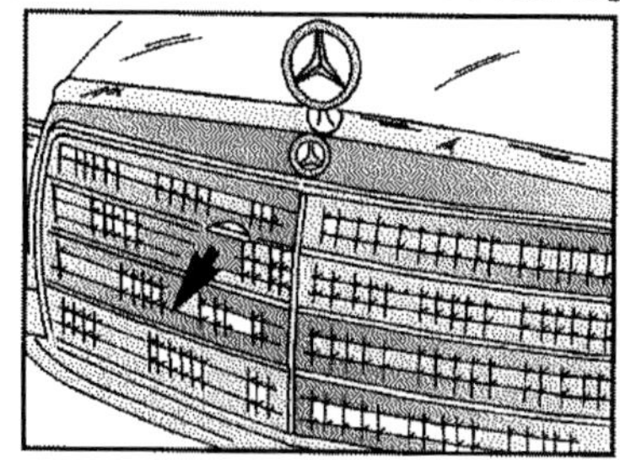

Fig. 1.4 – The arrow points to the locking lever for the engine bonnet in the centre of the radiator grille.

- Place suitable covers over both wings to prevent damage to the paint work. Bring the engine bonnet into vertical position. To do this, lift the bonnet until the bonnet stay on the L.H. side engages with the lock. The safety hook of the bonnet is located in the centre of the radiator grille at the position shown in Fig. 1.4.

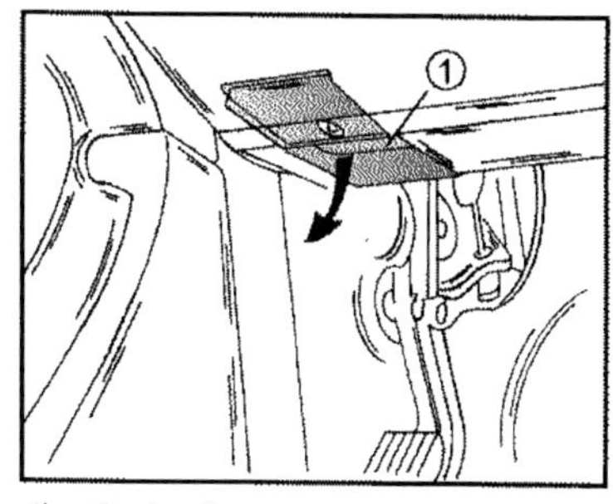

Fig. 1.5 – The locking lever (1) must be moved as described to lock the engine bonnet in the vertical position.

- Your Operators Manual will give you instructions how to lock the bonnet. Otherwise push the locking lever (1) in Fig. 1.5 in the direction of the arrow and list the bonnet upwards, without allowing the lever to lock. Carry out the same operation on the R.H. side and bring the bonnet in the vertical position. .Some of the operations are only referred to, i.e. "remove the radiator" will be described in section "Cooling System".

Figs. 1.6 and 1.7 show the parts to be removed in the case of the 604/605/606 engines. The operations are similar as described for the 602 engine (E290 TD), with the difference that some additional parts must be removed/disconnected. Numbers given below refer to the two illustrations.

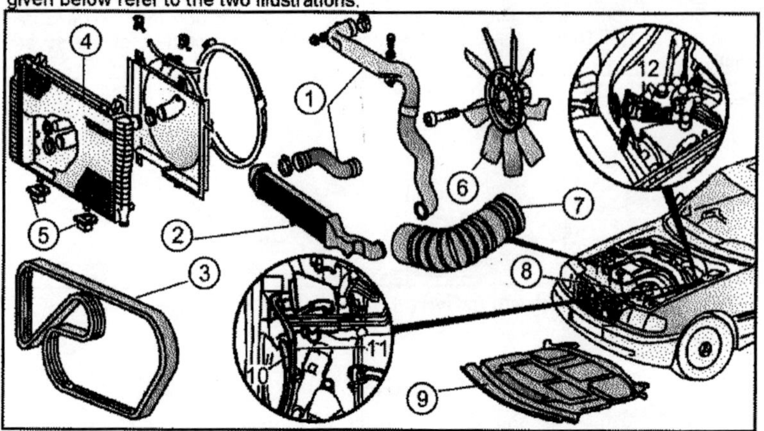

Fig. 1.6 – Removal and installation of the engines. Continued in Fig. 1.7.

- Disconnect the positive battery lead from the battery and move the cable well away
- Remove the sound-proofing bottom section in accordance with Fig. 1.3.
- Remove the charge air pipes (1) on the left and right of the air charge cooler (2) in Fig. 1.6. The removal and installation of the pipes is described under a separate heading.

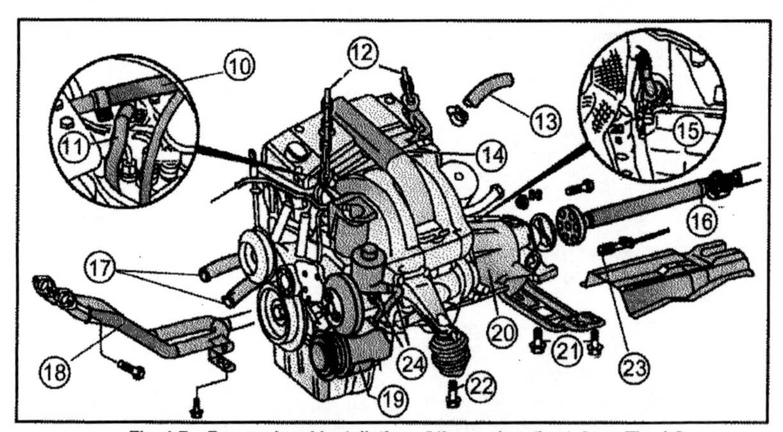

Fig. 1.7 – Removal and installation of the engine. Cont. from Fig. 1.6.

- Drain the cooling system and remove the radiator (4) as described in section "Cooling System". Remove the poly V-belt (3) as described later on in the same section.
- If A/C is fitted insert a guard plate with dimensions 400 x 680 x 1 mm, made of sheet metal or plastic and insert the plate at position (10) to protect the radiator/condenser.

- Disconnect all vacuum lines. These include the vacuum pump, intake manifold, brake servo unit, vacuum unit at vacuum control valve in the case of an automatic transmission, the vacuum units of the exhaust gas re-circulation system and in the case of a turbo charged engine at the vacuum control unit for the turbo charger. Make sure you know where to connect the lines during installation.
- Detach the coolant hoses (17) and (13) from the cylinder head and the coolant thermostat housing. Note that in the case of a 606 engine (E300D and TD) the coolant hose (13) can only be disconnected after the engine is at an angle. Inspect the hoses and hose clips before re-using them.
- Disconnect the fluid lines (24) from the power steering pump. The reservoir can be emptied to avoid loss of fluid.
- Detach the control pressure cable (12) at the position shown if an automatic transmission is fitted.
- Disconnect the engine wiring harness (14) as described earlier on, referring to Figs. 1.1 and 1.2.
- Disconnect the two fuel lines (10) and (11). The fuel filler cap should be opened. Fuel may run out. Take the necessary precautions.
- Unbolt the exhaust system (18) at the turbo charger or exhaust manifold and the transmission intermediate bearing. Section "Exhaust System" will give further details.
- Disconnect the gearchange rods from the transmission (MT, A/T).
- In the case of a manual transmission disconnect the clutch fluid line from the transmission).
- Slacken the clamp nut
- Disconnect the propeller shaft (16) from the transmission (also see section "Propeller Shaft"). Push the shaft towards the rear.
- Remove the guard plate near the plug connection (15) for the electronic control unit of the automatic transmission.
- Remove the A/C compressor (19). Attach the compressor with a piece of wire to the bottom at the side of the engine compartment. All pipes must remain connected.
- Remove the bolts (21) and (22) of the engine mountings and engine support. Tighten the bolts (22) to 3.5 kgm, the bolts (21) to 3.0 kgm during installation. The bolt and nut securing the engine rear crossmember is tightened with 4.0 kgm to the body.
- Attach suitable ropes or chains as shown by (12) to the engine lifting eyes and attach the ends to a hoist or crane. Operate the lifting equipment until the ropes/chains are just tight.
- Carefully lift the engine and transmission from the engine compartment, trying to obtain an angle of 45°. Continuously check that none of the cables, leads, etc. can get caught in the engine. Stop the lifting operation immediately as soon as problems can be seen. Under no circumstances force the assembly out of the engine compartment.

602.982 Engine (E290 Turbodiesel) – Removal and Installation

Figs. 1.8 and 1.9 show details for the removal and installation of this engine. The vehicle must be jacked up and placed on chassis stands as necessary. Proceed as follows, referring to the two illustrations as necessary:

- Disconnect the positive battery lead from the battery and move the cable well away
- Remove the sound-proofing bottom section (8) in accordance with Fig. 1.3.

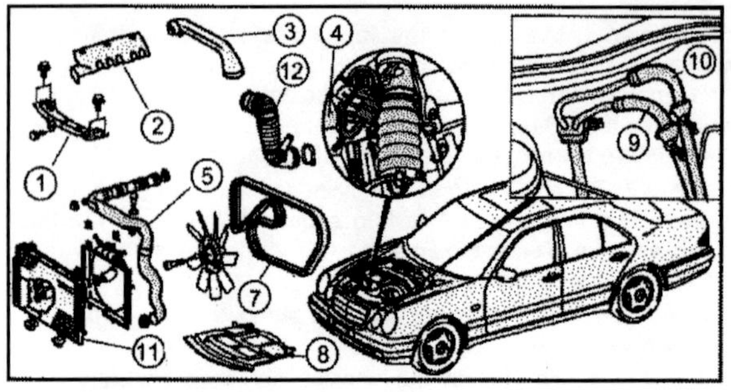

Fig. 1.8 – Removal and installation of the engine (E290 TD).

- Remove the charge air pipes (5) at the top at the front of the engine.
- Remove the viscous fan as described in section "Cooling System".
- Remove the charge air pipe (3), the engine trim panel (2) and the upper frame cross-member (1).
- Remove the radiator as described in section "Cooling System".

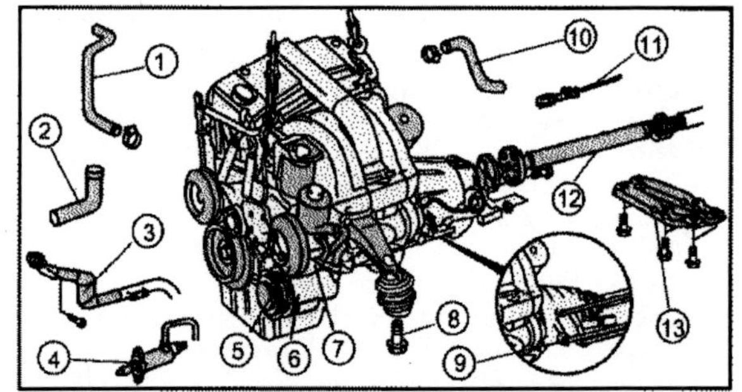

Fig. 1.9 – Removal and installation of the engine (E290 TD).

- If an A/C system is fitted remove the poly V-belt (7) as described as in section "Cooling System".
- If A/C is fitted insert a guard plate with dimensions 400 x 680 x 1 mm, made of sheet metal or plastic and insert the plate at position (10) to protect the radiator/condenser.
- Disconnect the engine wiring harness as described earlier on, referring to Figs. 1.1 and 1.2.
- Remove the air intake hose (12).
- Detach the vacuum line connections (4) at the vacuum units of exhaust gas re-circulation system, charge pressure and throttle valve control.
- Open the fuel filler cap, wait for the pressure to release and clamp the fuel pipes (9) and (10) with a suitable G-clamp. Then disconnect the two fuel hoses. Take care as fuel will be running out. .
- Detach a vacuum pipe at the charge air distribution pipe.

The following number references can be followed in Fig. 1.9:

- Detach the coolant hoses (1) and (10) at the rear of the cylinder head and at the coolant thermostat housing.
- Slacken the bolts securing the pulley of the power steering pump (7). Tighten the bolts to 3.0 kgm. Then remove the power steering pump with all pipes connected.
- Remove the A/C compressor (6). Attach the compressor with a piece of wire to the bottom at the side of the engine compartment. All pipes must remain connected.
- Remove the charge air hose (2) at the bottom.
- Unbolt the exhaust system (3) from the turbo charger and release the exhaust system from the transmission intermediate bearing.

- Disconnect the propeller shaft (12) from the transmission flange and push the shaft to the rear.
- Detach the park lock interlock (11) at the transmission with the selector lever in position "P" (with A/T).
- Remove the rear engine support (13).
- Detach the gearchange linkages (9) at the transmission end.
- Remove the clutch slave cylinder (4) (with manual transmission) and place it to one side with the fluid pipe attached.
- Unscrew the bolts (8) from the engine mountings on each side of the engine.
- Attach suitable ropes or chains as shown by (12) in Fig. 1.7 to the engine lifting eyes and attach the ends to a hoist or crane. Operate the lifting equipment until the ropes/chains are just tight.
- Carefully lift the engine and transmission from the engine compartment, trying to obtain an angle of 45°. Continuously check that none of the cables, leads, etc. can get caught in the engine. Stop the lifting operation immediately as soon as problems can be seen. Under no circumstances force the assembly out of the engine compartment.

1.1.3 INSTALLATION OF THE ENGINE

The installation is carried out in reverse order to the removal, but the following points should be noted during the installation:

- Do not connect or refit any of the disconnected parts until the engine and transmission are refitted to their mountings and the engine is free from the ropes or chains.
- Remove the engine mounts if their condition requires it. Keep oil or grease away from the rubber parts.
- If work has been carried out on engine or transmission make sure that the oil drain plug(s) have been tightened and check the oil level after installation.

Note: As the vehicles are fitted with a fault diagnostic system it is possible that some faults are stored in the memory after certain cables have been disconnected. We recommend that you have the fault memory checked at a dealer.

- Connect the propeller shaft in accordance with the instructions in section 4.
- Tighten the nuts and bolts to the correct tightening torque values.
- Refill the cooling system with anti-freeze of the correct strength for the temperatures to be expected. If the original anti-freeze is re-used, check its strength before filling it in.
- Check the air cleaner element before re-use. Dirty elements should be replaced, mainly if overhaul work has been undertaken on the engine.
- After starting the engine and allowing it to warm up, check the cooling system for leaks. Drive the vehicle a few miles to check, for example, for exhaust rattle.

1.2 Engine - Dismantling

Before commencing dismantling of the engine, all exterior surfaces should be cleaned, as far as possible, to remove dirt or grease. Plug the engine openings with clean cloth first to prevent any foreign matter entering the cavities and openings. Detailed information on engine dismantling and assembly is given in the sections dealing with servicing and overhaul (sections commencing at 1.4.) and these should be followed for each of the sub-assemblies or units to be dealt with.

Follow the general dismantling instructions given below.

- Dismantling must be carried out in an orderly fashion to ensure that parts, such as valves, pistons, bearing caps, shells, tappets and so on, are replaced in the same positions as they occupied originally. Mark them clearly, but take care not to scratch or stamp on any rotating or bearing surfaces. A good way to keep the valves in order is by piercing them through an upside-down cardboard box and writing the number against each valve. Segregate together the tappets, the springs and retainers with collets for each valve, if possible in small plastic bags for each individual valve.
- If a proper engine dismantling stand is not available, it will be useful to make up wooden support blocks to allow access to both the top and bottom faces of the engine. The cylinder head, once removed from the block, should be supported by a metal strap, screwed to the manifold face and secured by two nuts onto the manifold studs.

1.3 Engine - Assembling

The assembly of the engine is described in the following section for the component parts in question.

1.4. Engine – Servicing and Overhaul

1.4.0 CYLINDER HEAD AND VALVES

1.4.4.0 Technical Data

Cylinder Head (all dimensions in mm):

Height of cylinder head – 604, 605, 606 engine:............................126.85 – 127.15 mm
Min. height after machining – 604, 605, 606 engine:.................................. 126.80 mm
Distortion of cylinder head face:
Max. Distortion of cylinder head face:
- In longitudinal direction:.. 0.04 mm
- In transverse direction:.. 0.00 mm

Max. deviation of faces between upper and
lower head sealing faces:... 0.14 mm
Distance from end of valve stem to
base of camshaft bearing:...23.7 – 24.2 mm
Projection of pre-chamber:..3-2 – 3.7 mm

Valves

Valve Head Diameter – 604, 605, 606 engine to 10/94:
- Inlet valves:..28.90 – 29.10 mm
- Exhaust valves:...25.90 – 26.10 mm

Valve Head Diameter – 605, 605, 606 engine from 11/94:
- Inlet valves:..28.90 – 29.10 mm
- Exhaust valves:...25.90 – 26.10 mm

Valve seat angle – All valves:.. 45° 15'
Height of valve head – All valves:.. 1.0 – 1.2 mm

Valve Stem Diameter 604, 605, 606 engine to 10/94:
- Inlet valves:..5.960 – 5.975 mm
- Exhaust valves:...5.955 – 5.970 mm

Valve Stem Diameter 604, 605, 606 engine from 11/94:
- Inlet valves:..5.960 – 5.975 mm
- Exhaust valves:...6.955– 6.970 mm

Valve identification:
Engines 604, 605, 606 engine to 10/94:
- Inlet valves:.. E606 11
- Exhaust valves: A606 05/06
Engines 604, 605, 606 engine from 11/94:
- Inlet valves:.. E606 11
- Exhaust valves: A606 05/07
Valve seat width:
- Inlet valves:..1.9 mm
- Exhaust valves: ....1.9 mm
Valve Length:
- All valves, all model years:..105.50 – 105.90 mm

Valve seats
Valve seat width – Inlet/Exhaust:...1.9 mm
Valve seat angle:.. 45° - 15°
- Upper correction angle:.. 15°
- Lower correction angle:... 60°

Valve Guides
Inlet Valve Guides (604, 605, 606 engines):
- Length:.. 37.5 mm
- Inner diameter ....6.000 – 6.015 mm
- Interference fit in guides:...+ 0.12 – 0.031 mm
Exhaust Valve Guides:
- Length:.. 37.5 mm
- Inner diameter – to 10/94...6.000 – 6.015 mm
- Inner diameter – from 11/94 ....7.000 – 6.015 mm
- Max. inner diameter ....up to 0.015 mm more
- Interference fit in guides bores: ....+ 0.029 – 0.051 mm
Valve Guide Bore in Cylinder Head:
- Standard size: ....12.500 – 12.510 mm
- Standard size 1: .... 12.53 mm
- Repair size 1:.. 12.700 mm
Valve Guide Outer Diameter:
- Standard size: ....12.540 – 12.550 mm
- Standard size 1 (grey colour):...12.560 – 12.570 mm
- Repair size 0 (red colour):...12.740 – 12.750 mm

Valve Timing

	602 Engine	***Other Engines***
Inlet valve opens:	20° after TDC	12° after TDC
Inlet valve closes:	13° after BDC	13° after BDC
Exhaust valve opens:	21.5° before BDC	25° before BDC
Exhaust valve closes:	20.5° before TDC	14° before TDC

Camshaft
Camshaft end float:: .... 0.07 – 0.15 mm
Camshaft end float: wear limit:.. 0.18 mm
Camshaft Bearing Diameter:
- Standard diameter:...30.944 – 30.950 mm
Camshaft bearing clearance:
- New/wear limit:...0.050 – 0.091 mm/0.11mm

1.4.0.1. Cylinder Head – Removal and Installation

605, 605, 606 Engine

The following information should be noted when work is carried out on a cylinder head:

- The cylinder head is made of light-alloy. Engine coolant, engine oil, the air required to ignite the fuel and the exhaust gases are directed through the cylinder head. Glow plugs, injectors, pre-combustion chambers and valve tappets are fitted to the cylinder head. Also in the cylinder heads you will find the camshaft. The exhaust manifold and the inlet manifold are bolted to the outside of the head. The fuel enters the head on one side and exits on the other side, i.e. the head is of the well-known "crossflow" type.
- The cylinder head is fitted with various sender units, sensors and switching valves, responsible for certain functions of the temperature control.
- As the cylinder head is made of light alloy, it is prone to distortion if, for example, the order of slackening or tightening of the cylinder head bolts is not observed. For the same reason never remove the cylinder head from a hot engine,
- A cylinder head cannot be checked in fitted position. Sometimes the cylinder head gasket will "blow", allowing air into the cooling system. A quick check is possible after opening the coolant reservoir cap (engine fairly cold). Allow the engine to warm-up and observe the coolant. Visible air bubbles point in most cases to a "blown" gasket. Further evidence is white exhaust smoke, oil in the coolant or coolant in the engine oil. The latter can be checked at the oil dipstick. A white, grey emulsion on the dipstick is more or less a confirmation of a damaged cylinder gasket.
- If you are convinced that water has entered the engine and you want to get home or to the nearest garage, unscrew the injectors and crank the engine with the starter motor for a while to eject the water. Refit the injectors, start the engine and drive to your destination without switching off the engine. This is the only method to avoid serious engine damage (bent connecting rods for example).
- Cylinder heads for the non-turbo and turbo diesel are not the same (ignore this remark in countries without turbo diesel).

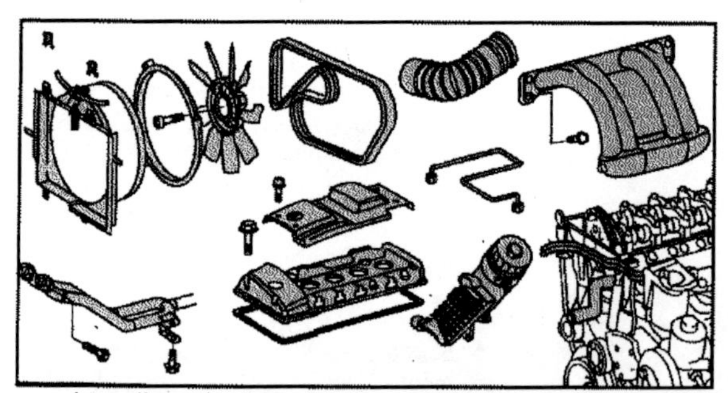

Fig. 1.10 – Removal and installation of the engine (604, 605, 606). The illustrated parts must be removed. See also Fig. 1.11.

The cylinder head must only be removed when the engine is cold. The head is removed together with the exhaust manifold, but the inlet manifold must be separated from the cylinder head before the head can be lifted off. This also applies to various other component parts. New cylinder head gaskets are wrapped in plastic and must only be unwrapped just before the gasket is fitted. The cylinder head can be removed with the engine fitted and these operations are described below, but note that operations may vary, depending on the equipment fitted. Figs. 1.10 and 1.11 show details of the parts to be removed during the removal of the cylinder head. Proceed as follows:

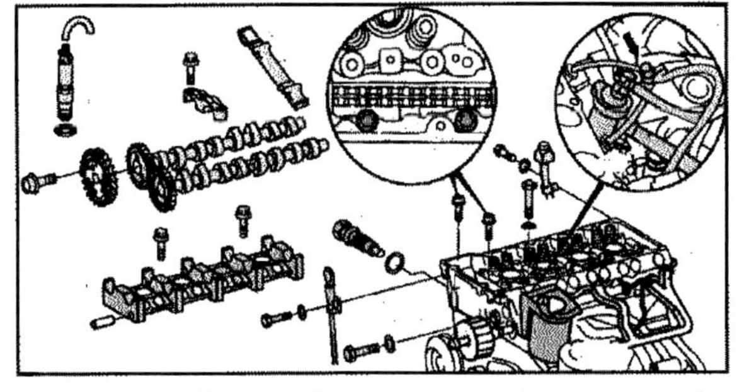

Fig. 1.11 – The illustrated parts must be removed to remove the cylinder head (605, 605, 606 engine).

- Place the engine bonnet into vertical position as described during the removal of the engine. Remove the engine compartment panels underneath the engine compartment as already described.
- Drain the cooling system. In the case of a 606 engine with turbo charger the radiator must be removed as described in section "Cooling System".
- Remove the single poly V-belt from the front of the engine as described later on.
- Remove the cylinder head cover as described following the removal of the cylinder head. This involves the removal of the intake manifold on 604 and 605 engines, the resonance intake manifold in the case of a E250D and the charge air distribution pipe in the case of a E250 TD, also described below. To remove the cylinder head cover six bolts must be removed. Two are located on each long side of the cover and two on the timing side of the engine. If an automatic transmission is fitted, there is a regulating rod fitted across the cylinder head cover which must be separated on one side at the ball joint. A sticking cylinder head cover must not be freed by tapping it with a hammer. If difficult to remove, try to unstick it by pushing it by hand to one side. Use a plastic mallet carefully, if necessary.
- Disconnect all coolant, fuel and vacuum hoses and the electrical cables connected to the cylinder head or any other unit on the head which cannot be removed together with the head.
- Disconnect the exhaust pipe flange from the exhaust manifold or the turbo charger from its connection on the engine. The exhaust manifold remains on the cylinder head.
- Remove the chain tensioner and the timing chain slide rail as described in connection with the timing mechanism. The chain tensioner plug must be unscrewed by applying a spanner to the hexagon. The plug is located above the water pump and the thermostat cover, next to the large tube.
- Disconnect the injection pipes. Immediately cover the open connections in suitable manner to prevent entry of dirt.
- Remove the camshaft housing as described under a separate heading.
- Remove the injectors (section "Diesel Fuel Injection").
- Disconnect the cables from the glow plugs.
- Disconnect the engine wiring harness from the cylinder head.
- Open the fuel filler cap to release the pressure in the system and disconnect the two fuel lines at the position shown in Fig. 1.11 (top right). On models with oil-water heat exchanger at the oil filter remove the oil filter. On other models remove the oil filter and place it to one side with the fuel lines attached.
- In the inside of the chain case remove two 8 mm socket head bolts with an Allen key. An extension and a socket is required to reach the bolts.
- Remove the oil pressure line, the oil return line and the charge air pipe from the turbo charger. Details are shown in Fig. 1.12.

Fig. 1.12 – Parts to be removed when a turbo charger is fitted.

1 Air intake tube
2 Turbo charger
3 Pressure oil pipe
4 Connection flange for 5
5 Oil return pipe
6 Charge air pipe

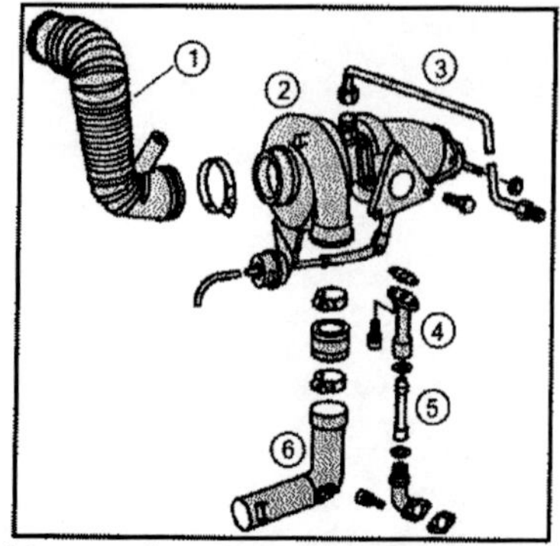

- Unscrew the cylinder head bolts in reverse order to the one shown in Figs. 1.13. or 1.14. A multi-spline bit must be used to slacken the bolts. A normal Allen key is not suitable, as it will damage the bolt heads. Immediately after removal of the bolts measure their length. If the dimension from the end of the bolt to the underside of the bolt head is more than 83.0, 105.6 or 118.5 mm, depending where the bolts are located, replace the bolt heads. New bolts have a length of 102.0 mm, i.e. bolts which nearly approach the max. length of 104 mm should be replaced.

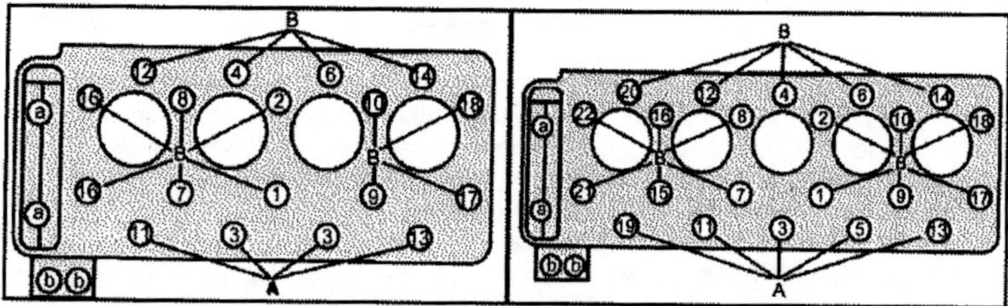

Fig. 1.13 – Tightening sequence for the cylinder head. In the L.H. view for the 604 engine, in the R.H. view for the 605 engine. The tightening sequence for the six-cylinder engine is shown below. Bolts "A" are 115 mm long, bolts B are standard bolts (102 mm).

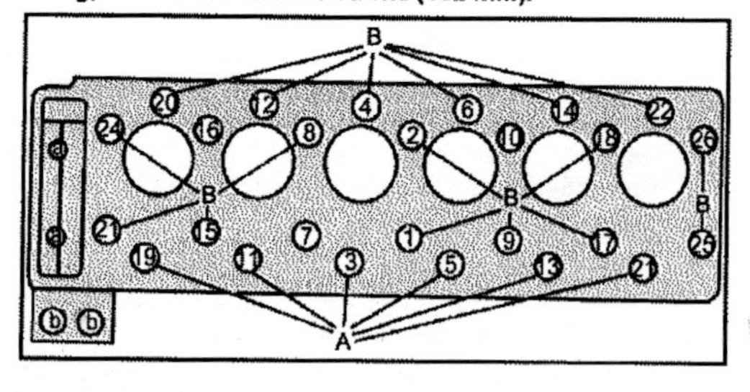

Fig. 1.14 – Tightening sequence for the cylinder head for the 606 engine (six-cylinder). Bolts "A" are 115 mm long, bolts B are standard bolts (102 mm).

- Lift off the cylinder head. If a hoist or other lifting device is available, hook a rope or chain to the two lifting eyes and lift off the head. Fig. 1.15 shows details of the final removal of the cylinder head.
- Immediately after removal clean the cylinder head and block surfaces of old gasket material.
- If necessary overhaul the cylinder head as described on the following pages.

The installation of the cylinder head is carried out as follows, taking into account some of the points already mentioned during removal.

- Measure the length of each cylinder head bolt between the end of the thread to the underside of the bolt head. The standard length is 102 mm. Any bolt longer than 104 mm must be replaced.

Fig. 1.15 – Removal of the cylinder head, shown on 4-cyl. engine.

1 Bolts, M7, secure brackets (4)
2 Cylinder head bolts, M10, 102 mm long
3 Cylinder head bolts, M10, 115 mm long
4 Brackets
5 Cylinder head
6 Cylinder head gasket
7 Dowel sleeves

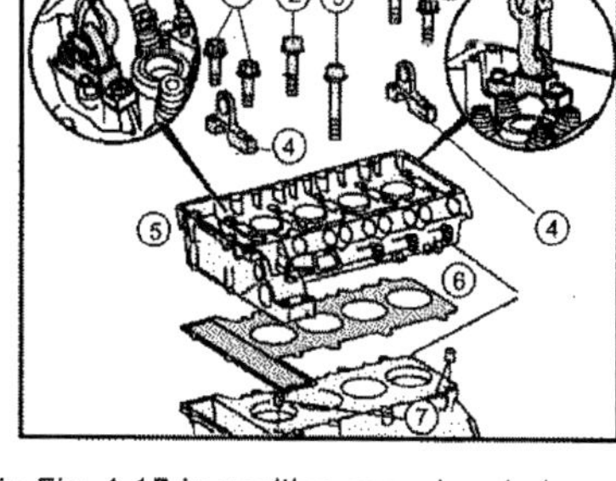

- Coat the bolt thread and the underside of the bolt heads with engine oil. The bores in the cylinder block must be free of oil.
- Place a new cylinder head gasket (6) in Fig. 1.15 in position over dowel sleeves (6). If the crankcase has been re-machined an oversize gasket (1.85 mm thick) must be used. A machined crankcase will come from the engine shop with the correct gasket. Make sure you obtained the correct one.
- Place the cylinder head (6) in position and tap it down with a mallet. Insert the cylinder head bolts and tighten them hand-tight.

The bolts are now tightened in the following order. Figs. 1.13 and 1.14 show the tightening sequence for the three engines. Bolts "a" are Allen-head bolts (M8 thread), bolt "b" (with washers) secure the fuel filter. Bolts (B) have a "Torx" head and a suitable socket is required. Bolts (a) secure the cylinder head to the timing cover. Tighten the bolts as follows:

- Tighten the cylinder head bolts to 1.5 kgm (10 ft.lb.) in the order shown.
- Tighten the cylinder head bolts to 3.5 kgm (18 ft.lb.) in the order shown.
- Tighten bolts (a) to 2.5 kgm (14.5 ft.lb.).
- Tighten cylinder head bolts a further 90° in the order shown and allow the bolts to settle for 10 minutes.
- Tighten the cylinder head bolts again by 90° in the order shown. A re-tightening of the cylinder head bolts is not necessary.
- All other operations are carried out in reverse order.

602 Engine (E290 TD)

Fig. 1.16 shows the parts to be removed. The numbers are referred to in the description.

- Place the engine bonnet into vertical position as described during the removal of the engine. Remove the engine compartment panels underneath the engine compartment as already described.
- Remove the charge air pipe (1) at the front.
- Remove the viscous fan clutch (5) together with the fan shroud (4) as described in section "Cooling System".
- Remove the charge air distribution pipe (15) as described below.
- Remove the oil filter housing (9).
- Disconnect the battery earth cable.
- Remove the cylinder head cover (12).
- Drain the cooling system.
- Detach the coolant hoses at the engine side.
- Rotate the engine until the piston of No. 1 cylinder is at the top dead centre position. Lock the crankshaft in this position or make sure that crankshaft and camshaft can not rotate during the remaining operations.

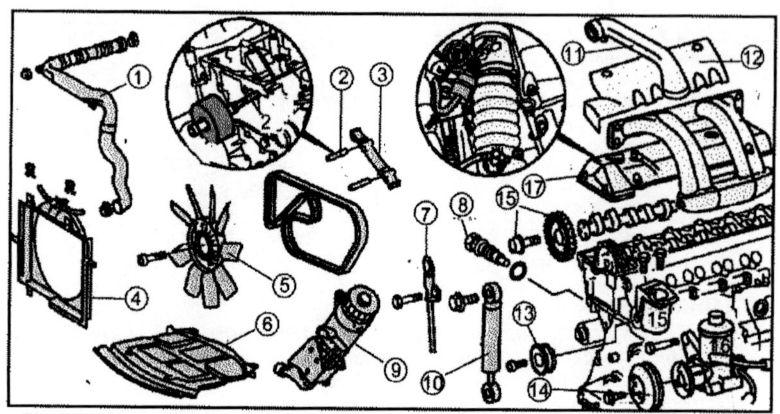

Fig. 1.16 – Removal and installation of the cylinder head if a 602 engine is fitted (E290 TD).

- Detach the exhaust pipe from the turbo charger and the bracket from the transmission.
- Remove the bracket on the turbo charger and the engine.
- Disconnect the oil return pipe from the turbo charger on the crankcase side. The attachment is similar as shown in Fig. 1.12.
- Slacken the bolts (14) securing the pulley to the steering pump (16),
- Slacken the tensioning device for the poly V-belt and unbolt the shock absorber (10) from the cylinder head. The pulley of the power steering pump (16) can now be removed.
- Remove the poly V-belt pulley (13) to gain access to the bottom guide rail pin.
- Remove the power steering pump (16) with the fluid lines connected, place it to one side and attach it to the side of the engine compartment with a piece of wire.
- Remove the fuel filter (15) together with the bracket and pipes connected and place it to one side.
- Disconnect the vacuum pipes from the vacuum pump.
- Remove the bolt securing the oil dipstick tube (7) to the cylinder head.
- Disconnect the large intake air hose from the side of the turbo charger after slackening the hose clip and also the hose at the bottom of the turbo charger.
- Remove the chain tensioner (8) as described in the section dealing with the timing mechanism.
- Disconnect the oil pressure feed pipe from the turbo charger as described for the other engines. The attachment is similar as shown in Fig. 1.12.
- Disconnect the injection pipes.
- Disconnect the engine wiring harness on the cylinder head.
- Open the fuel filler cap to release the pressure in the system and disconnect the two fuel lines. The fuel hoses can be pinched together with suitable G-clamps.
- Detach the vacuum connections on vacuum units of the exhaust re-circulation system and boost pressure and throttle valve control.
- Remove the timing chain guide rail (3). The bearing pins (2) must be removed with the special puller shown. Refer to the section dealing with the timing mechanism for details.
- In the inside of the chain case remove two 8 mm socket head bolts with an Allen key. An extension and a socket is required to reach the bolts.

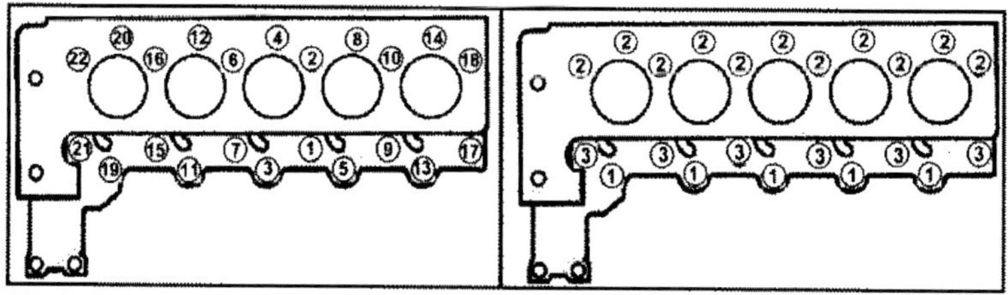

Fig. 1.17 – The cylinder head bolts are tightened in the order shown in the L.H. view. Slacken them in reverse order. The R.H. view shows the location/length of the bolt.
1 M10 x 80 mm 2 M10 x 102 mm 3 M10 x 115 mm

- Remove the camshaft sprocket (15) and the camshaft as described in the section dealing with the timing mechanism.
- Disconnect the oil leak lines on the injectors in order to gain access to the cylinder head bolts.
- Unscrew the cylinder head bolts in reverse order to the one shown in Figs. 1.17. A multi-spline bit must be used to slacken the bolts. A normal Allen key is not suitable, as it will damage the bolt heads. Immediately after removal of the bolts measure their length. If the dimension from the end of the bolt to the underside of the bolt head is more than 83.6, 105.6 or 118.6 mm, depending where the bolts are located, replace the bolt heads. New bolts have a length of 80.0, 102.0 or 115.0 mm, i.e. bolts which nearly approach the max. length should also be replaced. Fig. 1.17 shows on the R.H. side where the individual bolts are located.
- Bolt on the brackets (4) shown in Fig. 1.15 to the cylinder head. The front bracket is fitted to the first camshaft bearing cap mounting, the rear bracket to the R.H. hole of the sixth camshaft bearing cap mounting.
- Lift off the cylinder head. If a hoist or other lifting device is available, hook a rope or chain to the two lifting brackets, similar as shown in Fig. 1.15.
- Immediately after removal clean the cylinder head and block surfaces of old gasket material.
- If necessary overhaul the cylinder head as described on the following pages.

The installation of the cylinder head is carried out as follows, taking into account some of the points already mentioned during removal.

- Measure the length of each cylinder head bolt between the end of the thread to the underside of the bolt head in accordance with the dimensions given above. The location of the bolts are shown in Fig. 1.17. Replace any bolts which have been stretched.
- Coat the bolt thread and the underside of the bolt heads with engine oil. The bores in the cylinder block must be free of oil.
- Place a new cylinder head gasket in position over dowel sleeves
- Place the cylinder head in position and tap it down with a mallet. Insert the cylinder head bolts and tighten them hand-tight.

The bolts are now tightened in the following order. Fig. 1.17 shows the tightening sequence. The tightening takes place irrespective of the bolt length.

- Tighten the cylinder head bolts to 1.0 kgm (7 ft.lb.) in the order shown.
- Tighten the cylinder head bolts to 3.5 kgm (18 ft.lb.) in the order shown.
- Tighten bolts securing the Allen-head bolts inside the timing cover to the (a) to 2.5 kgm (14.5 ft.lb.).
- Tighten cylinder head bolts a further 90° in the order shown and allow the bolts to settle for 10 minutes.

- Tighten the cylinder head bolts again by 90° in the order shown. A re-tightening of the cylinder head bolts is not necessary.

Also note the following tightening torques on this engine:

Camshaft sprocket flange to camshaft: .... 2.5 kgm + 90°
Shock absorber (V-belt) to cylinder head: .... 2.5 kgm
Engine support to crankcase: .... 2.0 kgm + 90°
Pulley, power steering pump: .... 3.0 kgm
Power steering pump to support bracket: .... 2.5 kgm

- All other operations are carried out in reverse order.

1.4.0.2. Cylinder Head – Operations before Removal is possible

Cylinder Head Cover – Removal and Installation – 604, 605 and 606 Engines

The cylinder head cover is shown in Fig. 1.18. Before the cover can be taken off, remove the intake manifold (604, 605 engine), the resonance intake manifold (606 engine, E300D) or the charge air distributor pipe (606 engine, E300 TD). The operations are described on the following pages. The injection pipes must also be removed. This is described in section "Fuel Injection System").

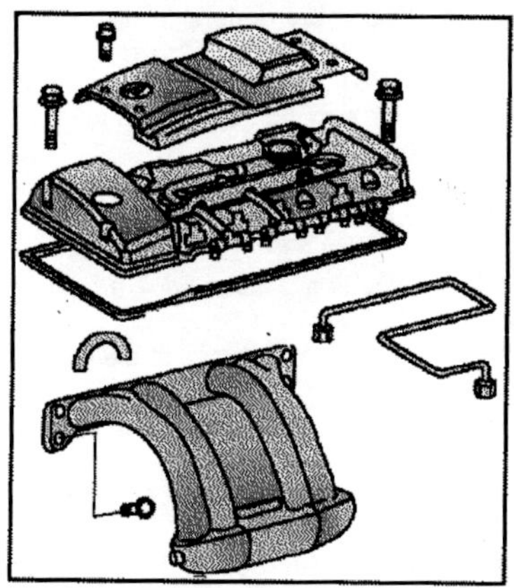

Fig. 1.18 – The parts to be removed during the removal of the cylinder head cover (engines 604, 605, 606).

Remove the cylinder head cover trim. During installation take care not to trap a leak-oil pipe between trim panel and cylinder head cover.

The cylinder head cover can now be removed. Take off the gasket and check its condition. Only replace if you think it is necessary.

The installation is a reversal of the removal procedure. The cover bolts are tightened to 1.0 kgm (7 ft.lb.). The trim cover is located on bushes which must engage correctly,

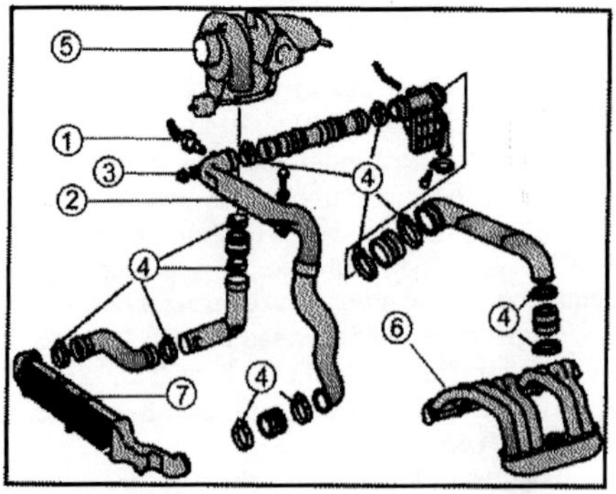

Fig. 1.19 – Removal and installation of the charge air pipes.

1 Plug connector
2 Charge air pipes
3 Rubber mount
4 Hose clamps
5 Turbo charger
6 Intake manifold
7 Charge air cooler

Charge Air Pipe – Removal and Installation – 602 (E290 TD), 606 (E300 TD) Engines

The parts to be removed are shown in Fig. 1.19. Removal and installation of the pipe (2) is fairly easy. The pipe is attached by means of the rubber mount (3) at one end. The cable connector plug must be withdrawn from the intake air temperature sensor (1) before the pipe is removed. The connecting points must be thoroughly clean before installation. Tighten the hose clamps (4) to 0.4 kgm only.

Intake Manifold – Removal and Installation – 604, 605 Engines

The parts to be removed are shown in Fig. 1.20. Removal and installation of the manifold is fairly easy. Note the differences in the case of a 606 engines. Proceed as follows:

Fig. 1.20 – Removal and installation of the inlet manifold (604, 605 engines).

1 Crankcase ventilation
2 Cylinder head cover trim panel
3 Intake manifold
4 Bolt
5 Bolt
6 Sealing rings
7 Vacuum hose
8 Bracket
9 Bracket
10 Bracket

- Remove the crankcase ventilation assembly (1).

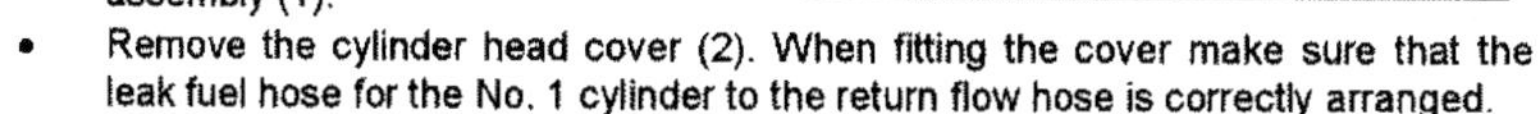

- Remove the cylinder head cover (2). When fitting the cover make sure that the leak fuel hose for the No. 1 cylinder to the return flow hose is correctly arranged.

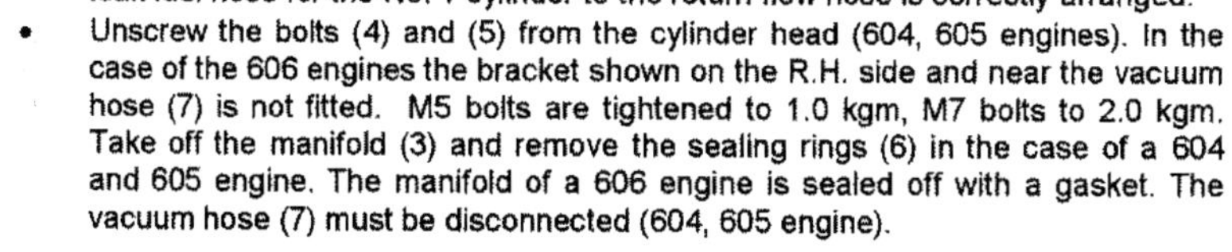

- Unscrew the bolts (4) and (5) from the cylinder head (604, 605 engines). In the case of the 606 engines the bracket shown on the R.H. side and near the vacuum hose (7) is not fitted. M5 bolts are tightened to 1.0 kgm, M7 bolts to 2.0 kgm. Take off the manifold (3) and remove the sealing rings (6) in the case of a 604 and 605 engine. The manifold of a 606 engine is sealed off with a gasket. The vacuum hose (7) must be disconnected (604, 605 engine).

Fig. 1.21 – Removal and installation of the resonance intake pipe (606 engine).

1 Intake pipe
2 "O" sealing rings
3 "O" sealing rings
4 Vacuum hose
5 Bracket
6 Securing bolt

Resonance Intake Pipe – Removal and Installation – 606 Engine only

The pipe (1) in question is shown in Fig. 1.21. Removal and installation is carried out by referring to the diagram. The sealing rings (2) and (3) must be replaced if damaged, the vacuum hose (4) must be disconnected.

1.4.0.3. Cylinder Head - Dismantling

The following description assumes that the cylinder head is to be overhauled in an engine shop. If only a top overhaul is asked for, you can remove the valves as described below and can then decide if a complete overhaul of the head is required. The cylinder head must be removed.

The valve stem oil seals can be replaced with the cylinder head fitted. Signs of worn oil seals are blue exhaust smoke when the vehicle is coasting (gear engaged), when the engine is accelerated after idling for a while or blue smoke when starting the cold engine. If the oil consumption has reached 1 litre per 600 miles, replace the valve stem oil seals as described under a separate heading. The valve stem oil seals can also be replaced with the engine fitted to the vehicle, but various special tools are required. It is therefore a job for the Mercedes dealer.

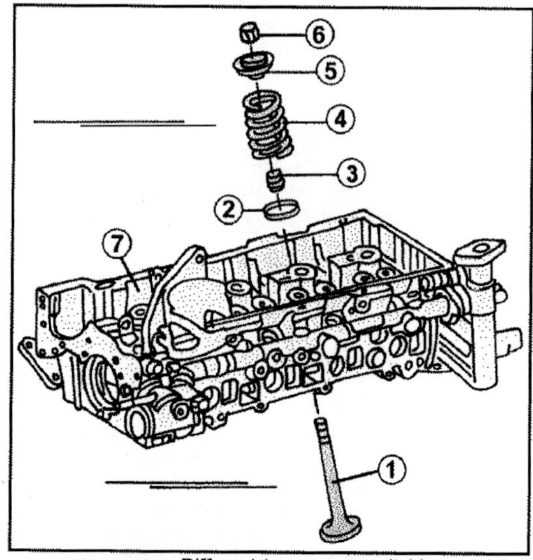

Fig. 1.22 – Valves removed from the cylinder head (4-cyl.).
1 Valve
2 Lower spring seat
3 Valve stem oil seal
4 Valve spring
5 Upper valve spring cup
6 Valve cotter halves

The valves can be removed as follows. After removal you will have the parts shown in Fig. 1.22, shown on the example of the four-cylinder engine.

- Remove all auxiliary parts from the cylinder head, including the exhaust manifold.
- A valve spring compressor is required to remove the valves. Fig. 1.23 shows such a compressor. Different types are available, must it must be possible to compress the springs. Valves are held in position by means of valve cotter halves. Compress the springs and remove the valve cotter halves (2) with a pair of pointed pliers or a small magnet (1).

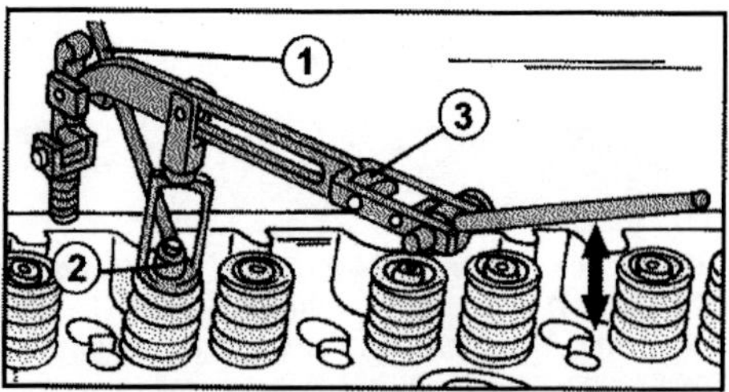

Fig. 1.23 – Removal of valves with a valve lifter (3). The valve cotter halves (2) can be removed with a small magnet (1.

- If a valve spring compressor is not available, it is possible to use a short piece of tube to remove the valve cotter halves. To do this, place the tube over the upper valve spring collar and hit the tube with a blow of a hammer. The valve cotter halves will collect in the inside of the tube and the components can be removed. The valve head must be supported from the other side of the cylinder head. Keep the hammer in close contact with the tube to prevent the cotter halves from flying out.
- Remove the valve spring collar (5) and the valve spring (4) in Fig. 1.22. The valve springs (one spring per valve) are identified with a paint spot and only a spring with a paint spot of the same colour must be fitted.
- Remove valve stem oil seals (1) in the L.H. view of Fig. 1.24 carefully with a screwdriver or a pair of pliers (2).

- Remove the valves one after the other out of the valve guides and pierce them in their fitted order through a piece of cardboard. Write the cylinder number against each valve if they are to be re-used.

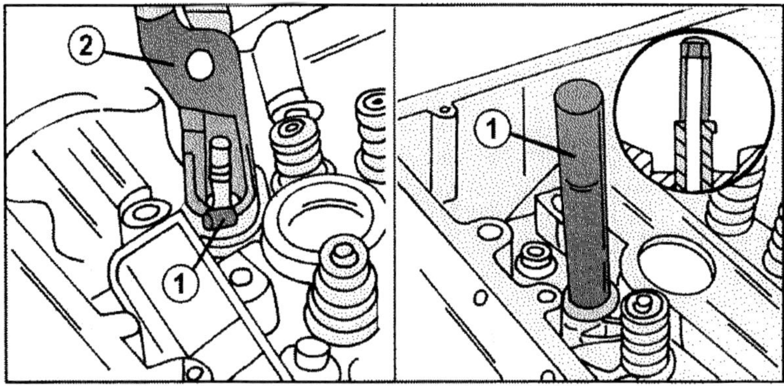

Fig. 1.24 – Removal of valve stem oil seals on the left. Use a pair of pliers (2) to remove the seal (1). The R.H. view shows the installation of a seal with the piece of tube (1).

A few words should be said about the camshafts. The camshafts of a four-cylinder engine has five bearings, the shaft of a five-cylinder engine has six bearings, the shaft of the six-cylinder engine seven bearings. The lower part of the bearing location is machined directly into the cylinder head. The removal and installation of the camshafts is described later on in connection with the timing mechanism.

The camshaft sprockets are marked with timing marks, which indicate the top dead centre position of piston No. 1 when they are aligned in the position shown during the removal instructions. Camshafts are marked with a number, referred to in the section dealing with the timing mechanism. If a shaft is replaced, only fit a shaft with the same number. The parts lists will have the correct camshaft(s) for the engine in question.

1.4.0.4. Cylinder Head - Overhaul

The cylinder head must be thoroughly cleaned and remains of old gasket material removed. The checks and inspections are to be carried out as required.

Valve Springs: If the engine has a high mileage, always replace the valve springs as a set. To check a valve spring, place the old spring and a new spring end to end over a long bolt (with washer under bolt head) and fit a nut (again with a washer). Tighten the nut until the springs are under tension and measure the length of the two springs. If the old spring is shorter by more than 10%, replace the complete spring set.

The springs must not be distorted. A spring placed with its flat coil on a surface must not deviate at the top by more than 2 mm (0.08 in.).

Valve Guides: Valve guides for inlet and exhaust valves are made of cast iron and have different diameters, Guides for exhaust valves are shorter and have a larger inner diameter.

Clean the inside of the guides by pulling a petrol-soaked cloth through the guides. Valve stems can be cleaned best by means of a rotating wire brush. Measure the inside diameter of the guides. As an inside micrometer is necessary for this operation, which is not always available, you can insert the valve into its guide and withdraw it until the valve head is approx. level with the cylinder head face. Rock the valve to and fro and check for play. Although no exact values are available, it can be assumed that the play should not exceed 1.0 - 1.2 mm (0.04 -0.047 in.). Mercedes workshops use gauges to check the guides for wear.

Guides are removed with a shouldered mandrel from the combustion chamber side of the cylinder head. If guides with nominal dimension 1 can be used, drive them in position, until the retaining is resting against the cylinder head. If repair guides are fitted, the locating bores in the cylinder head must be reamed out to take the new guides. As dry ice is required to fit the new valve guides, we recommend to have the work carried out in a workshop.

Before a valve guide is replaced, check the general condition of the cylinder head. The guides must be reamed after installation, and after the cylinder head has cooled down, if applicable, to their correct internal diameter, given in Section 1.4.0.0 (page 20), noting the different diameters for inlet and exhaust valves. Valve guides have been changed on engines manufactured after Nov. 1994. The inner diameter is now 7 mm, compared to 6 mm on earlier engines. Th new valve guides can be recognised on the circlip in Fig. 1.25. The later valve guides are made of grey cast iron.

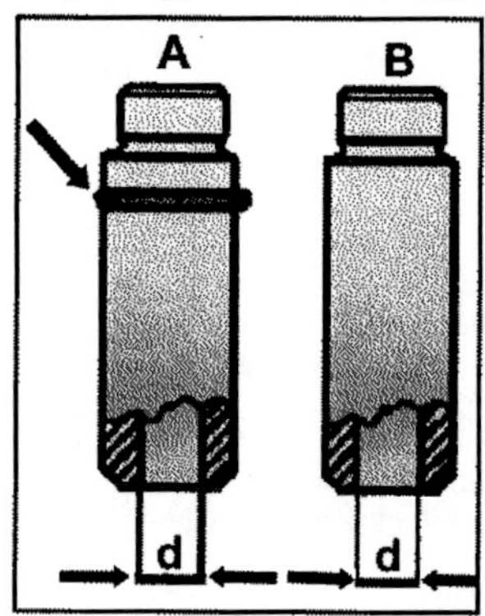

Fig. 1.25 – The circlip shown by the arrow has been discontinued on valve guides with 7 mm inner diameter (B). (A) = earlier valve guide.

Valves must always be replaced if new valve guides are fitted. The valve seats must be re-cut when a guide has been replaced. If it is obvious that seats cannot be re-ground in the present condition, new valve seat inserts must be fitted. Again this is an operation for a specialist and the work should be carried out in a workshop.

Valve Seats: If the camshaft bearings are excessively worn, fit a new or exchange cylinder head. In this case there is no need to renovate the valve seats.

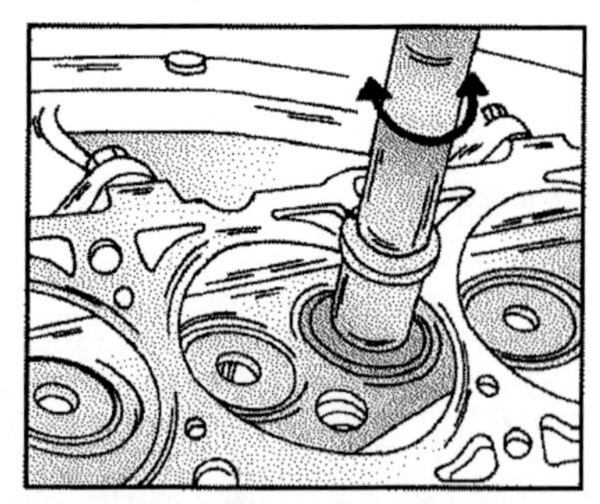

Fig. 1.26 – Grinding-in of valves, using a suitable sucker.

Check all valve seats for signs of pitting or wear. Slight indentations can be removed with a 45° cutter. If this operation is carried out properly, there should be no need to grind-in the valves. Valve seat inserts can be fitted to the cylinder head. Replacement of valve seat inserts will require that the old seat insert is removed by machining. The machining must not damage the bottom face of the head recess. As this is a critical operation,- we advise you to bring the cylinder head to your Mercedes Dealer who has the necessary equipment and experience to do the job. It may be possible to obtain a reconditioned cylinder head in exchange for the old one to avoid time delay. In this case remove all ancillary parts from the old head and refit them to the new head.

Valves can be ground into their seats in the conventional manner. Fig. 1.26 shows detail. Attach the suction tool to the valve and move it towards the left and right. Here

and then lift the valve, rotate it by a quarter of a turn and then re-grind as described and shown. If necessary apply additional grinding paste.
After completing the operation clean all affected parts from grinding paste and check the seat on valves and valve seats. An uninterrupted seat must be visible on valves and seats. Otherwise have the seats re-machined in an engine shop or in severe cases replace the cylinder head.

Valves: The stems of inlet valves and exhaust valves have been specially treated, i.e. the ends of the valve stems must not be ground off.
Valves can be cleaned best with a rotating wire brush. Check the valve faces for wear or grooving. If the wear is only slight, valves can be re-ground to their original angle in a valve grinding machine, but make sure that there is enough material left to have an edge on the valve head. The valve head thickness must be 0.5 - 0.7 mm in the case of the inlet valves and 0,5 - 0.6 mm in the case of the exhaust valves. Again we recommend an engine shop.

Check the valve stem diameters and in this connection the inside diameters of the valve guides. If there is a deviation from the nominal values, it may be necessary to replace the valve guides (see above). Also check the end of the valve stems. There should be no visible wear in this area.
Always quote the model year and the engine number when ordering new valves, as different valves are used. These are marked by means of a number in the end of the valve stem.
Sometimes it is only required to replace the exhaust valves, if these for example are burnt out at their valve head edges.

Fig. 1.27 – Checking a camshaft for run-out at the centre bearing journal.

Cylinder Head: Thoroughly clean the cylinder head and cylinder block surfaces of old gasket material and check the faces for distortion. To do this, place a steel ruler with a sharp edge over the cylinder head face and measure the gap between ruler and face with feeler gauges. Checks must be carried out in longitudinal and diagonal direction and across the face. If a feeler gauge of more than 0,10 mm (0.004 in.) can be inserted, when the ruler is placed along or across the cylinder head, have the cylinder head face re-ground. The dimension between the face of the valve heads and the cylinder head face will change after regrinding, but the workshop will correct it accordingly.

Camshaft: Place the camshaft with both end journals into "V" blocks or clamp the shaft between the centres of a lathe and apply a dial gauge to the centre journal, as shown in Fig. 1.27. Slowly rotate the shaft to check for run-out. If the dial gauge reading exceeds 0.01 mm, fit a new shaft. Make sure to fit the correct shaft if the shaft is to be replaced. Check the identification number when ordering a new shaft.

Replacing Valve Stem Oil Seals (Cylinder Head fitted): Valve stem oil seals are available in repair kits. Included in the repair kits are protective sleeves which must be pushed over the valves during installation of the seals.

Normally a special tool is used to fit the seals, as shown in Fig. 1.24 on the R.H. side. but a well fitting piece of tube of suitable diameter can be used. Take care not to damage the sealing lip and the spring.

The valve cotter halves and the valve springs must be removed to replace the seals. To prevent the valves from dropping into the combustion chambers, the pistons must be at the top dead centre position. This is easy in the case of a four-cylinder engine, as two pistons are always at top dead centre. The operation is more difficult on a five-cylinder engine, as the crankshaft must be rotated by a certain angle to move the next piston to TD.C. Only attempt this operation if you are competent. Start by setting the piston of the No. 1 cylinder to the top dead centre position. Valve stems seals of cylinders Nos. 1 and 4 can be replaced in this position.

- Remove the camshafts (see description later on).
- Remove the valve cotter halves of the first cylinder as described in Section 1.4.0.3. Remove the valve springs in the case of a four-cylinder engine.
- Use a pair of pliers (Fig. 1.24, L.H. view) or a screwdriver to remove the valve stem seals, without damaging the valve stems.
- Coat the new seals with engine oil and carefully push them over the valve stems. The protective sleeve must be used for the inlet valves. Push the seals over the valve guides until properly in position.
- Fit the valve springs with the paint spot towards the bottom, fit the upper spring cup and compress the springs until the valve cotter halves can be inserted. Make absolutely sure that the cotter halves have properly engaged before the crankshaft is rotated.
- Replace the valve stem seals of cylinder No. 4 in the same manner.
- Lift the camshaft sprocket slightly to prevent disengagement of the timing chain and rotate the crankshaft of a four-cylinder engine by half a turn. The valve stem seals of cylinders Nos. 2 and 3 can be replaced in the same manner. In the case of the other engines, rotate the crankshaft until both valves of the next cylinder are closed. As already mentioned, take care when rotating the crankshaft. Both valves of a cylinder must be at the same height, before the valve cotter halves are removed.
- Refit the camshaft and associated parts.

Note: Operate the valve spring compressor very slowly, as valve cotter halves sometimes stick to the valve stems. Observe the valve spring during compression. Only the spring should move, not the valve. Prevent by all means that the valve can press against the piston.

1.4.0.5. Cylinder Head - Assembly

The assembly of the cylinder head is a reversal of the dismantling procedure. Note the following points:

- Lubricate the valve stems with engine oil and insert the valves into the correct valve guides.
- Valve stem seals are different for inlet and exhaust valves. Make sure to order the correct seals. The repair kit contains fitting sleeves and these must be used to fit the seals (see last Section).
- The sleeves are fitted over the valve stem before the seal is pushed in position

- Fit the valve spring and valve spring collar over the valve and use the valve lifter to compress the spring. Insert the valve cotter halves and release the valve spring lifter. Make sure that the cotter halves are in position by tapping the end of the valve stem with a plastic mallet. Place a rag over the valve end - just in case.
- Fit the camshafts as described later on and carry out all other operations in reverse order to the dismantling procedure.

1.4.0.6. Hydraulic Valve Clearance Compensation

The function of the hydraulic valve clearance compensating elements is to eliminate valve clearance, i.e. the dimensional changes in the valve train (valve lash) due to heat expansion and wear are compensated by the elements. The rocker arm is in constant contact with the cam.

The compensating elements cannot be repaired, but can be checked for correct functioning as described below. Figs. 1.28 and 1.29 show sectional views of a valve with clearance compensation. We will give a short description of the operation. All references refer to Fig. 1.28.

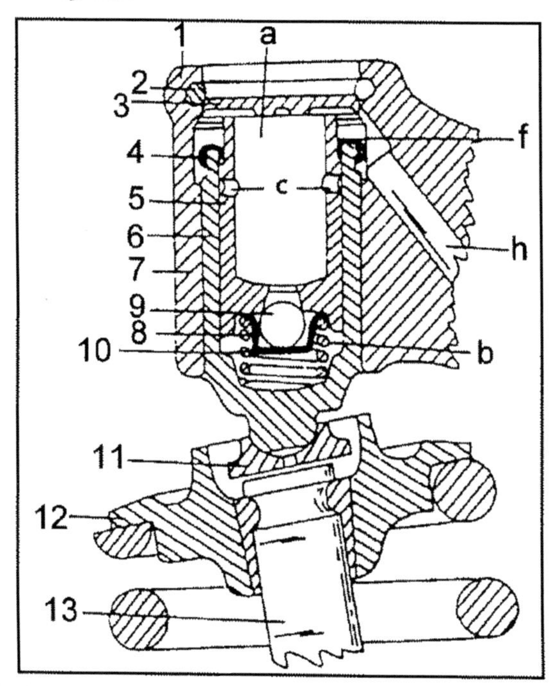

Fig. 1.27 - Sectional view of a valve clearance compensating element

1 Rocker lever
2 Lock ring
3 Washer
4 Closing cap
5 Thrust pin
6 Guide sleeve
7 Ball guide
8 Thrust spring
9 Ball, 4 mm
10 Thrust spring
11 Ball socket
12 Valve spring retainer
13 Valve
a Oil chamber
b Work chamber
c Return bores
f Ring groove
h Oil bores

The hydraulic valve compensating element are fitted into the rocker levers and operate the valves directly via a ball socket (11). Each element consists of the following components:

- The thrust pin (5) with oil supply chamber and the return bores and the ball valve (check valve), i.e. items (9), (7) and (10). The ball valve separates the supply chamber from the work chamber.
- The guide sleeve (6) with the work chamber (b), the thrust spring (8) and the closing cap (4).
- When the engine is stopped and the tappet is held under load from the cam, the element can completely retract. The oil displaced from the work chamber (b) flows through an annular gap, i.e. the clearance between the guide sleeve and the thrust pin to the oil supply chamber (a).
- When the cam lobe has moved past the valve tappet, the thrust pin (5) will be without load. The thrust spring (8) forces the thrust pin upwards until the valve tappet rests against the cam.

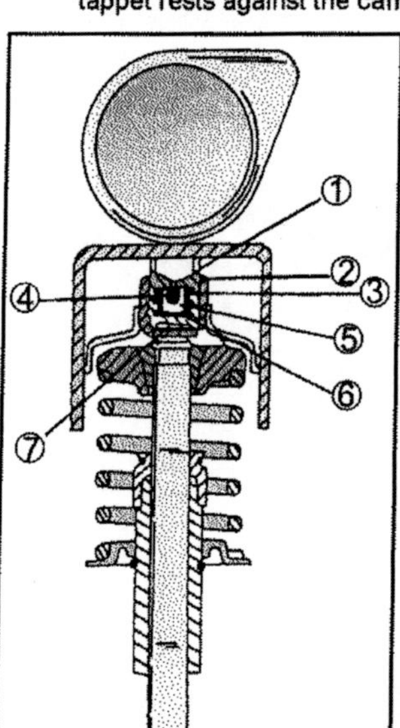

Fig. 1.29 – Sectional view of a valve and a compensating element,
1 Thrust pin
2 Lock Ring
3 Thrust spring
4 Ball guide
5 Ball
6 Thrust spring
7 Guide bush

The vacuum resulting from the upward movement of the thrust pin in the work chamber (b) opens the ball valve and the oil can flow from the supply chamber (h) into the work chamber (b). The ball valve closes when the valve tappet presses against the cam and puts the thrust pin under load. The oil in the work chamber acts as a "hydraulic rigid connection" and opens the valve in question.

When the engine is running and depending on the engine speed and the cam position, the thrust pin is only pushed down slightly.

The oil contained in the oil supply chamber (a) is sufficient to fill the work chamber (b) under all operating conditions of the engine. Oil or leak oil which is not required, as well as air are able to escape via the annular gap between the washer (3) and the rocker lever. The oil ejected from the work chamber flows via the annular gap between the guide sleeve and the thrust pin and the two return bores (c) into the oil supply chamber.

If the tappets are removed, note the following points:

- Always keep the tappets in an upright position, i.e. the open side towards the top.
- After removal of a tappet (see below), mark the cylinder number and the compensating element in suitable manner. Always fit original parts in their same locations.

Checking a hydraulic compensating element: As the elements are in continuous contact with the camshaft, you will rarely hear noises from the area of the hydraulic elements. If noises can be heard, check the elements as follows:

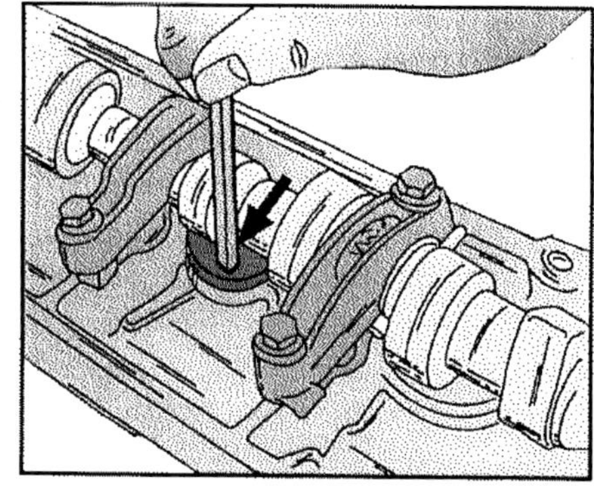

Fig. 1.30 – Checking of a hydraulic compensating element.

- Start the engine and run it approx. 5 minutes al 3000 rpm.
- Remove the cylinder head cover.
- Rotate the crankshaft until the cam for the tappet to be checked is pointing vertically towards the top.
- Use a drift, as shown in Fig. 1.30 and push the tappet towards the inside, or try to move the tappet with the fingers.
- If the tappet cannot be depressed or excessive clearance can be felt between the tappet and the back of the cam, replace the tappet. The tappet is supplied together with the hydraulic compensating element. Mercedes workshops can reset the tappet to its original position, but this operation is beyond the scope of the home mechanic.

Tappet Removal and Installation:

- Remove the camshaft as described later on.
- Use a suction tool to remove the tappets. Mark them, if they are to be refitted.
- Fit the tappets into their original bores, if re-used. Refit the camshafts as described and carry out all other operations in reverse order to the removal procedure.

1.4.1. PISTON AND CONNECTING RODS

1.4.1.0. Technical Data

All dimensions are given in metric units.

Pistons	**602/604**	**605/606 Engine**
Piston Diameter – Std.	89.000 mm	87.000 mm
Class A:	88.970 - 88.976 mm	86.000 - 86.976 mm
Class X:	88.975 – 88.983 mm	86.975 - 86.983 mm
Class B:	88.982 – 88.988 mm	86.982 - 86.988 mm

Piston Running Clearance:
- New .... 0.017 – 0.043 mm
- Wear limit .... 0.12 mm

Max. weight difference within engine:
- New .... 4 grams
- Wear limit .... 10 grams

Piston Pins:

Pin diameter .... 25.995 – 26.000 mm or 26.995 – 27.000 mm

Piston pin running clearance:
- In small end bush .... 0.018 – 0.029 mm

- In piston ....0.004 – 0.015 mm

Piston Rings:
Piston Ring Gaps:
- Upper rings............ ....0.20 – 0.45 mm
  - Wear limit............ ....1.0 mm
- Centre rings........... ....0.20 – 0.40 mm
  - Wear limit............ ....1,0 mm
- Lower rings............ ....0.20 - 0.40 mm
  - Wear limit:........... ....1.0 mm

Side Clearance in Grooves:
  - Upper rings:..0.090 - 0.120 mm
  - Wear limit:... 0.20 mm
  - Centre rings...0.050 - 0.080 mm
  - Wear limit:... 0.15 mm
- Lower rings...0.030 - 0.065 mm
  - Wear limit:... 0.10 mm

Connecting Rods
Distance from centre small end bore to
centre big end bore.. 145.0 mm
Width of con rod at big end bore.. 24.000 mm
Basic bore diameter of big end bore... 47.95 mm
Basic bore diameter of small end bore .... 29.500 mm

Small End Bush:
- Outer diameter.. . . . . 29.50 mm
- Inner diameter. . . . . 27.000 mm
Max. twist of connecting rods . . . . .0.10 mm per 100 mm
Max. bend of connecting rods . . . . .0.045 mm per 100 mm
Max. weight difference in same engine:. . ...5 gram (per set)
Connecting Rod Bolts:
- Thread: .... M9 x 1
- Diameter of stretch neck.. 7.4 mm
- Min. diameter of stretch neck:.. 7.1 mm

Connecting rod bearing details: .... See under "Crankshaft"

1.4.1.1. Piston and Connecting Rods – Removal

The pistons are made of light-alloy.

Three piston rings are fitted to each piston. The two upper rings are the compression rings, i.e. they prevent the pressure above the piston crown to return to the crankcase. The lower ring is the oil scraper ring. Its function is to remove excessive oil from the cylinder bore, thereby preventing the entry of oil into the combustion chamber. The three rings are not the same in shape. The upper ring has a rectangular section, the centre ring has a chamfer on the inside and the lower ring is chrome-plated on its outside and also has a coil spring. Only the correct fitting of the piston rings will assure the proper operation of the piston sealing.
The connecting rods connect the pistons with the crankshaft, a piston pin connects the piston with the connecting rods. Fig. 1.31 shows details of the various parts.

Pistons and connecting rods are pushed out towards the top of the cylinder bores, using a hammer handle alter connecting rod bearing caps and shells have been removed. Before removal of the assemblies note the following points:

- The engine must be removed, oil sump, oil pump and cylinder head must be removed.
- Pistons and cylinder bores are graded in three diameter classes within specified tolerance. The class number is stamped into the upper face of the cylinder block, next to the particular cylinder bore, as shown in Fig. 1.32.
- Each piston is marked with an arrow to indicate the fitting direction, as shown in the upper circle in Fig. 1.31. The arrow points in the direction of drive.
- Mark each piston and the connecting rod before removal with the cylinder number. This can be carried out by writing the cylinder number with paint onto the piston crown. Also mark an arrow, facing towards the front of the engine (the arrow in the piston crown will be covered by the carbon deposits). When removing the connecting rod, note the correct installation of the big end bearing cap. immediately after removal mark the connecting rod and the big end bearing cap on the same side (exhaust side). This is best done with a centre punch (cylinder No. 1 one punch mark, etc., see Fig. 1.31, lower circle).

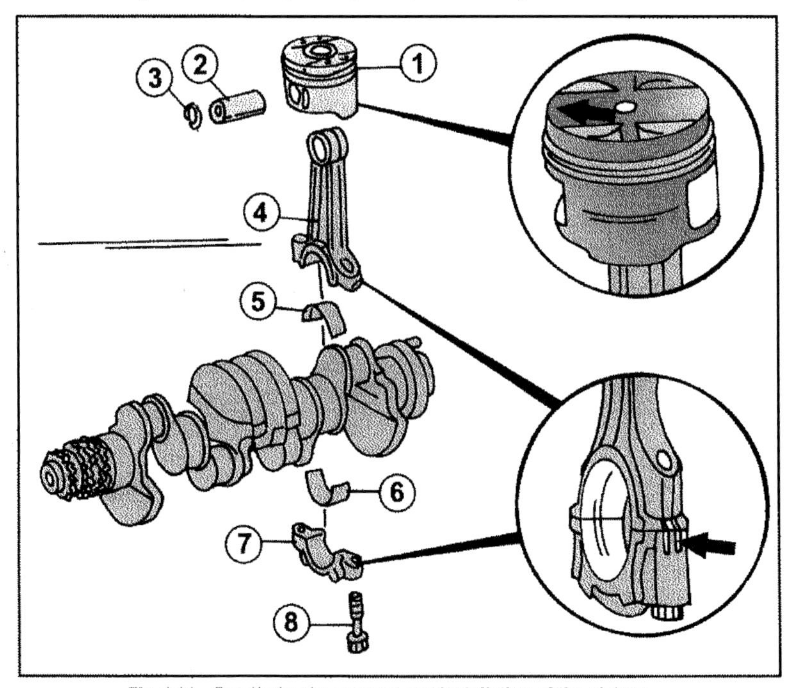

Fig. 1.31 – Details for the removal and installation of the pistons.

1 Piston
2 Piston pin
3 Circlip
4 Connecting rod
5 Upper bearing shell
6 Lower bearing shell
7 Big end bearing cap
8 Bearing cap bolt

Mark the big end bearing shells with the cylinder number. The upper shells have an oil drilling (to lubricate the piston pin).

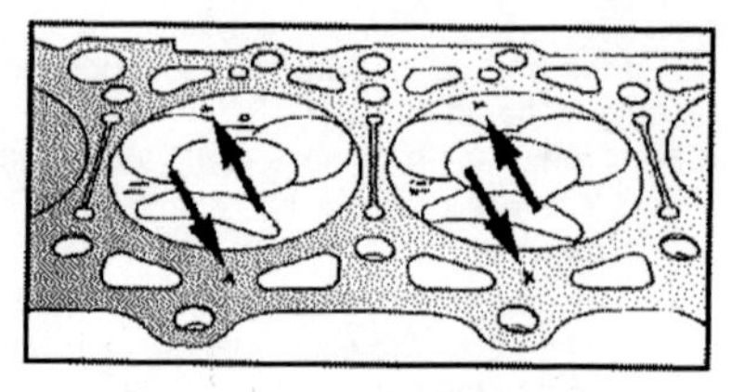

Fig. 1.32 – Identification (marking) of piston crown and cylinder block face with the group code letters.

- Big end bearing journals can be re-ground to four undersizes (in steps of 0.25 mm between sizes). Corresponding bearing shells are available.
- Remove the bearing caps and the shells and push the assemblies out of the cylinder bore. Any carbon deposits on the upper edge of the bores can be carefully removed with a scraper.
- Remove the piston pin snap rings. A notch in the piston pin bore enables a pointed drift to be inserted, as shown in Fig. 1.33, to remove the rings. Press the piston pins out of the pistons. If necessary heat the piston in boiling water.
- Remove the piston rings one after the other from the pistons, using a piston ring pliers if possible (Fig. 1.34). if the rings are to be re-used, mark them in accordance with their pistons and position.

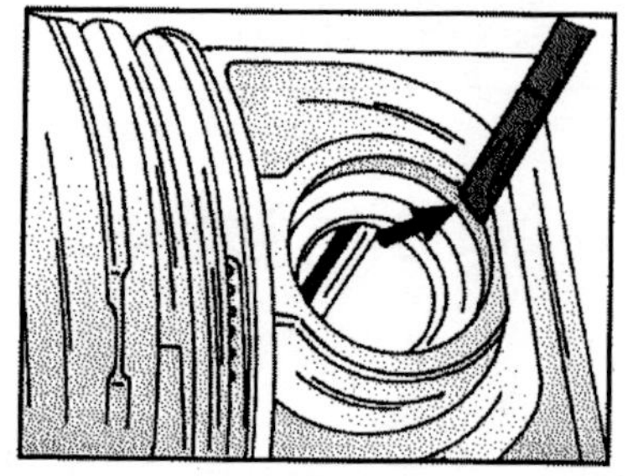

Fig. 1.33 – Removal of the securing clip for the piston pins. Apply the screwdriver blade at the ring gap.

Fig. 1.34 – Shown below is the removal or installation of piston rings with a pair of piston ring pliers. Never expand rings more than necessary to prevent breakage.

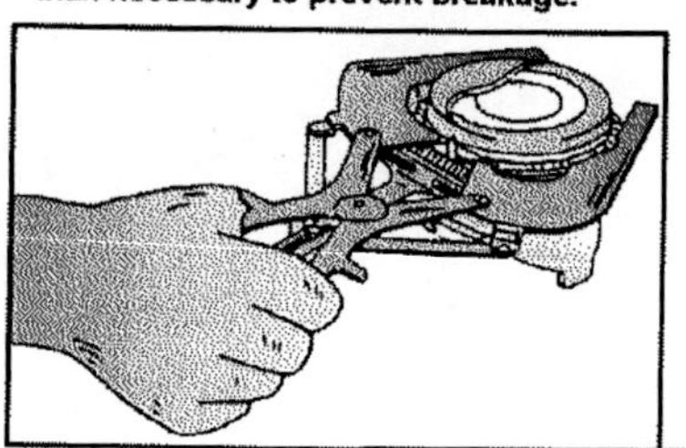

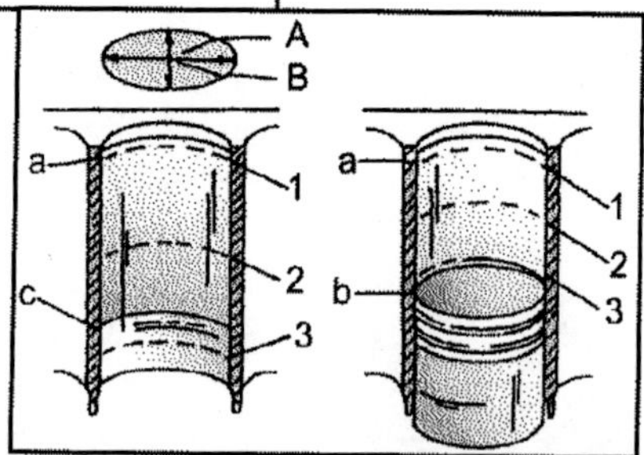

Fig. 1.35 – Measuring diagram for the cylinder bores. Numbers 1, 2 and 3 indicate the three levels where measurements should take place.

A = Measurement parallel to bore
B = Measurement across bore
a = Upper return point of upper piston ring
b = Bottom dead centre of piston
c = Lower return point of oil control ring
1-3 = Measuring points

1.4.1.2. Measuring the Cylinder Bores

An inside caliper is necessary to measure the diameter of the cylinder bores. The following operations are not possible if none is available or cannot be hired.

Cylinder bores must be measured in longitudinal and transverse direction and at three positions down the bore, i.e. 10 mm (0.4 in.) from the upper bore edge, 10 mm from the lower bore edge and once in the centre, totalling 6 measurements. The worst measurement must be taken when deciding on the size for the pistons to be fitted (Fig. 1.35). Fig. 1.36 shows how an internal micrometer is used.

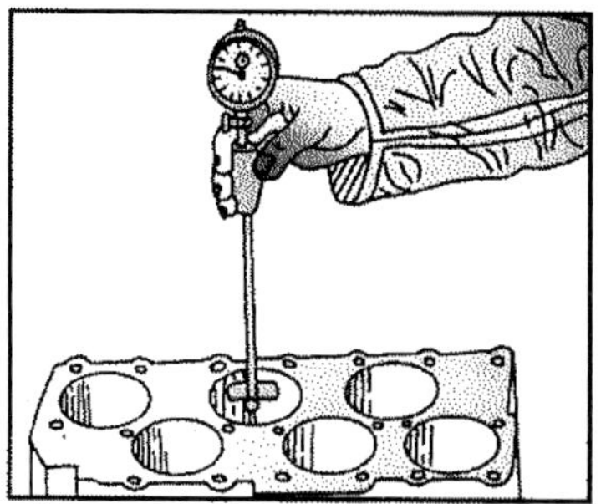

Fig. 1.36 – Measuring the diameter of a cylinder bore. The illustration shows a different engine but the principle is the same.

Note that all cylinder bores must be re-bored, even if only one of the bores is outside the diameter limit. A tolerance of 0.20 mm is permissible. If the wear is outside the limits, it is possible to have new cylinder liners fitted to the block. Your dealer will advise you what can be done.

The final cylinder bore diameter is determined after measuring the piston diameter. To measure the diameter, apply an outside micrometer 10 mm (0.4 in.) from the bottom edge and at right angle to the piston pin bore, as shown in Fig. 1.37. Add the piston running clearance to this dimension and 0.03 mm for the honing of the cylinders. The piston running clearance must not exceed 0.12 mm.

To measure the running clearance, determine the piston and cylinder bore diameters as described above and calculate the difference between the dimensions.

If the difference is more than 0.12 mm, have the cylinder bores re-bored to fit oversize pistons.

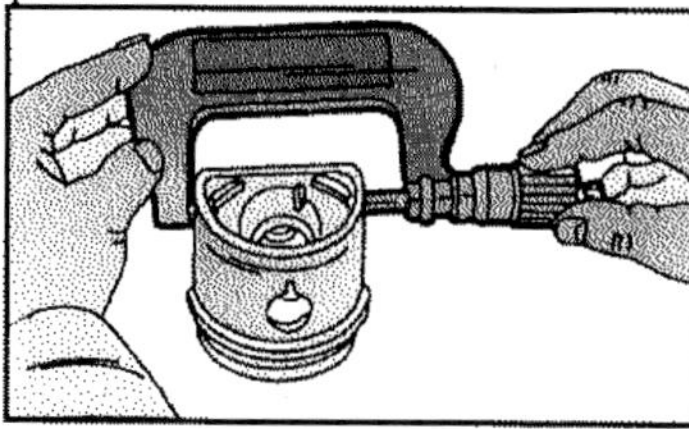

Fig. 1.37 – Measuring the diameter of a piston. Apply the micrometer at right angle to the piston pin approx. 4 mm away from the outer edge of the piston skirt. If in doubt have the pistons measured in an engine workshop.

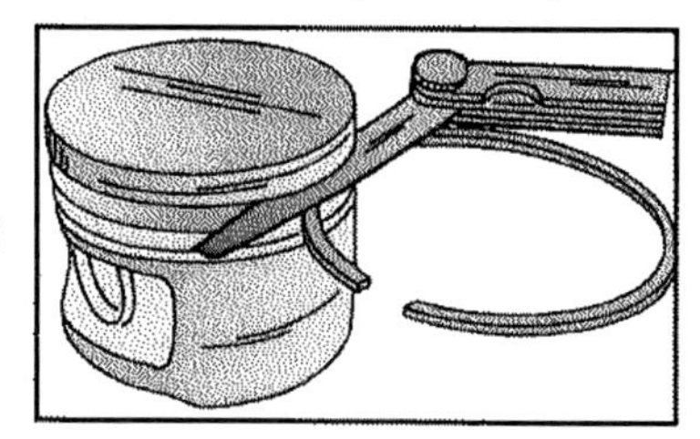

Fig. 1.38 – Checking the side clearance of piston rings.

1.4.1.3. Checking Pistons and Connecting Rods

All parts should be thoroughly inspected. Signs of seizure, grooves or excessive wear requires the part to be replaced. Check the pistons and connecting rods as follows:

- Check the side clearance of each piston ring in its groove by inserting the ring together with a feeler gauge, as shown in Fig. 1.38. The grooves must be thoroughly cleaned before the check. If the wear limit exceed the values in the technical data is reached, either the rings or the piston are worn.

- Check the piston ring gap by inserting the ring from the bottom into the cylinder bore. Use a piston and carefully push the piston ring approx. 1 in. further into the bore. This will square it up. Insert a feeler gauge between the two piston ring ends to check the ring gap, as shown in Fig. 1.39. Refer to Section 1.4.1.0. for the wear limits. Rings must be replaced, if these are exceeded. New rings should also be checked in the manner described.

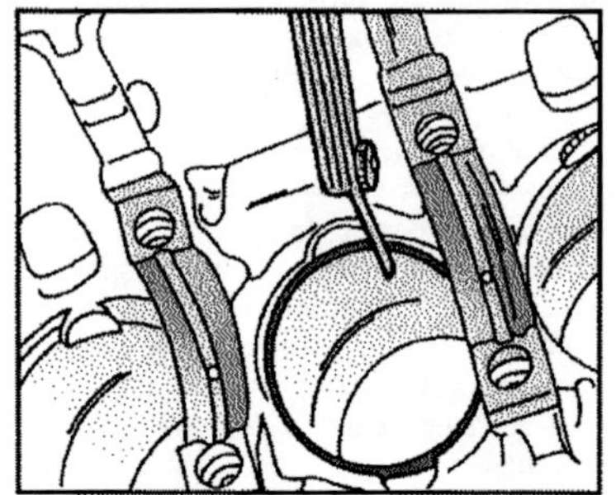

Fig. 1.39 – Checking piston ring gaps.

- Piston pins and small end bushes must be checked for wear or seizure. One individual connecting rod can be replaced, provided that a rod of the same weight group is fitted. Connecting rods are marked with either one or two punch marks and only a rod with the same mark must be fitted. The small end bushes can be replaced but we recommend to take them to an engine shop to have them replaced.
- Before re-using the connecting rod bolts check their length as shown in Fig. 1.40. Replace any bolt that is longer the 48 mm.
- Connecting rods should be checked for bend or twist, particularly when the engine has covered a high mileage. A special jig is necessary for this operation and the job should be carried out by an engine shop

The following information concern the connecting rods:

- Connecting rods which were over-heated due to bearing failure (bluish colour) must not be refitted.

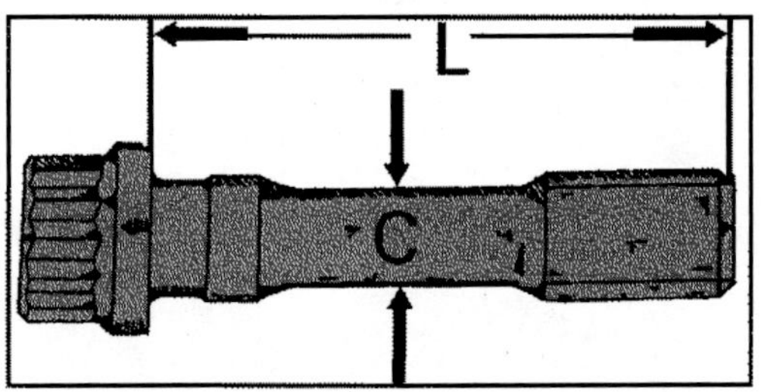

Fig. 1.40 – Connecting rod bolts can stretch in their length "L". By measuring the stretch neck "c", you will know if the bolts can be re-used.

- Connecting and bearing caps are matched to each other and must be fitted accordingly.
- New connecting rods are supplied together with the small end bearing bush and can be fitted as supplied.
- If the piston pin has excessive clearance in the small end bush, have a new bush fitted to obtain the correct piston pin running clearance and again we recommend to have the job carried out by an engine shop.

1.4.1.4. Checking the Big End Bearing Clearance

These operations are described in connection with the crankshaft (Section 1.4.3.).

1.4.1.5. Piston and Connecting Rods - Assembly

Fig. 1.31 shows a piston together with the connecting rod. The assembly can be carried out by referring to the illustration.

- Before fitting the pistons check the piston crowns to check the position of the arrow.
- Insert the connecting rod into the piston and align the two bores. Make sure that the arrow in the piston crown and locating lugs for the bearing shell location are facing the L.H. side of the engine, as shown in Fig. 1.31.
- Generously lubricate the piston pin with engine oil and insert it into the piston and connecting rod, using thumb pressure only. Never heat the piston to fit the piston pin. Fit the circlips to both sides of the piston, making sure of their engagement around the groove. Move the piston up and down to check for free movement. Fit the snap rings into the piston bores. Make sure the are located properly.

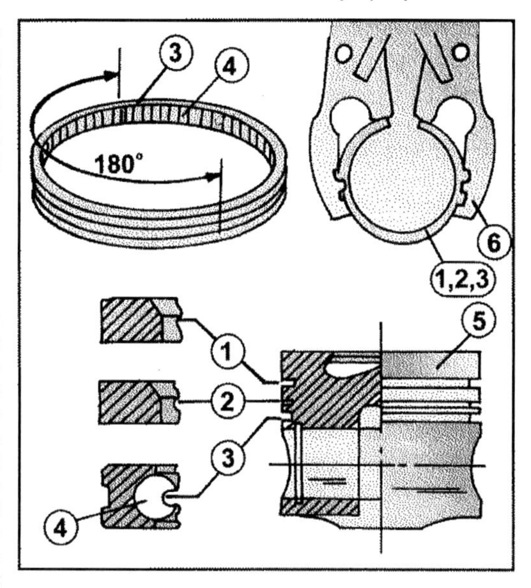

Fig. 1.41 – Sectional view of the piston. The section and arrangement of the piston ring gaps must be observed. The fitting of rings its shown on the top (R.H.). The L.H. view shows the oil control ring and the coil spring.

1 Straight compression ring
2 Tapered ring
3 Oil control ring
4 Coil spring
5 Piston

- Using a pair of piston ring pliers, fit the piston rings from the top of the piston, starting with the bottom ring. The two compression rings could be mixed up and Figs. 1.41 should be referred to avoid mistakes. Rings are either marked with "Top" or a paint mark which must point to the piston crowns. Arrange the piston ring gaps at 120° apart. The oil control ring must be fitted in accordance with the insert in Fig. 1.41. Under no circumstances mix-up the upper and lower compression rings.

1.4.1.6. Pistons and Connecting Rods – Installation

- Generously lubricate the cylinder bores with oil. Markings on connecting rods and bearing caps must be opposite each other. The arrows in the piston crowns must face towards the front of the engine, the marks in connecting rods and big end bearing caps must be opposite each other, both as shown in Fig. 1.31.
- Arrange the piston rings at equal spacings of 120° around the circumference of the piston skirt. The spring (4) in the oil control ring (3) must be arranged as shown in Fig. 1.41. Use a piston ring compressor to push the rings into their grooves, as shown in Fig. 1.42. Check that all rings are fully pushed in.
- Rotate the crankshaft until two of the crankpins are at the bottom in the case of the four-cylinder engine. On the five-cylinder engine you will have to fit the assemblies one after the other. Place the engine on its side to facilitate the

installation. Bearing shell (5) in Fig. 1.31 should be in the con rod. Shells (5) and (6) must not be interchanged.

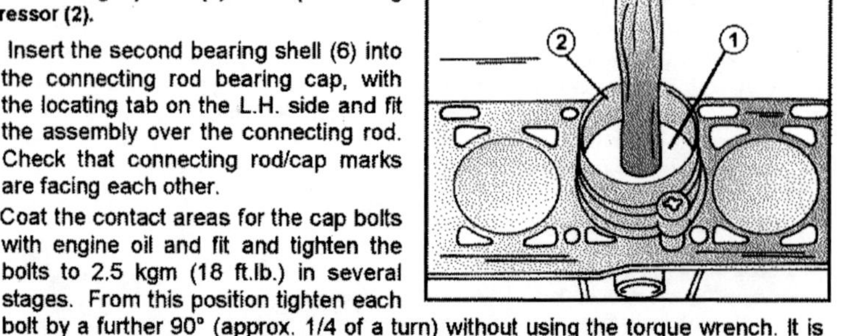

Fig. 1.42 – Fitting a piston (2) with a piston ring compressor (2).

- Insert the second bearing shell (6) into the connecting rod bearing cap, with the locating tab on the L.H. side and fit the assembly over the connecting rod. Check that connecting rod/cap marks are facing each other.
- Coat the contact areas for the cap bolts with engine oil and fit and tighten the bolts to 2.5 kgm (18 ft.lb.) in several stages. From this position tighten each bolt by a further 90° (approx. 1/4 of a turn) without using the torque wrench. It is assumed that the stretch bolts have been measured as previously described.
- Rotate the crankshaft until the two remaining crankpins are at bottom dead centre (four-cylinder) or the next crankpin is at the bottom (other engines) and fit the two other or three other piston/connecting rod assemblies in the same manner.
- Check the pistons and connecting rods once more for correct installation and that each piston is fitted to its original bore, if the same parts are refitted.
- With a feeler gauge measure the side clearance of each big end bearing cap on the crankpin. The wear limit is 0.50 mm.

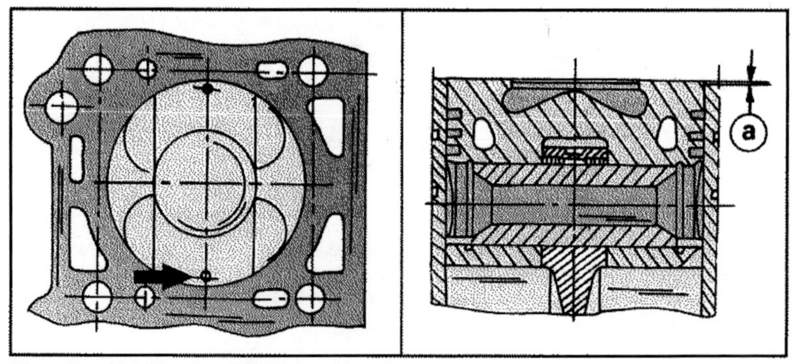

Fig. 1.43 – Checking the piston protrusion. The L.H. view shows where the measurement is carried out. The R.H. view shows with "a" where the piston protrudes above the cylinder block face.

- If pistons have been replaced measure the protrusion of the pistons above the cylinder block face. This must be between 0.8 to 0.62 mm on both engines. To check the protrusion place the piston to the top dead centre position (must be exact) and measure the protrusion at the points shown by the arrows in the L.H, view of Fig. 1.43., i.e. in the direction of the piston pin. You measure the part of the piston above the cylinder block face as shown in the R.H. view of Fig. 1.43. The protrusion will determine the thickness of the cylinder head gasket during assembly, i.e. either a standard thickness gasket or a repair thickness gasket must be used. With the values obtained you will be able to obtain the correct gasket from your parts department.

1.4.2. TIMING CASE COVER

The timing case cover can be replaced with the engine fitted to the vehicle if for example work on the timing chain is required. The front of the engine must be accessible to gain access to the timing cover. The parts to be removed are shown in Fig. 1.44 for an engine with one camshaft. Proceed as follows to remove and refit the cover. Some of the operations are only mentioned, but are described in the section in question. Note that on vehicles with A/C system can be treated in the same manner as only the compressor must be removed, with all lines/pipes attached.

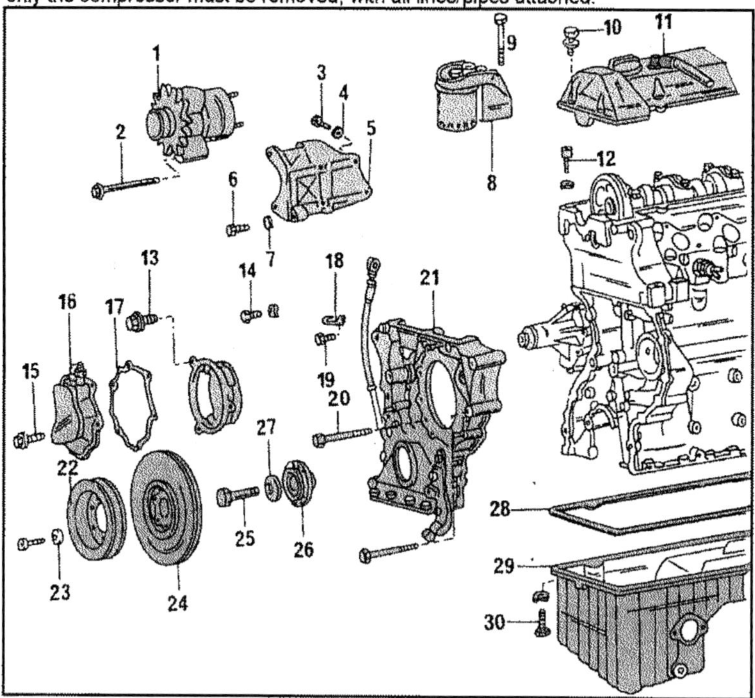

Fig. 1.44 – Parts to be removed during the removal of the timing cover if one camshaft is fitted to the engine.

1 Alternator
2 Bolt, 4.5 kgm
3 Bolt, 4.5 kgm
4 Washer
5 Alternator mounting bracket
6 Bolt, 2.5 kgm
7 Washer
8 Fuel filter
9 Bolt, 1.0 kgm
10 Bolt, 1.0 kgm
11 Cylinder head cover
12 Bolt. 2.5 kgm
13 Bolt, 1.0 kgm
14 Bolt, 1.0 kgm
15 Bolt, 1.0 kgm
16 Vacuum pump
17 Gasket
18 Bracket
19 Bolt, 2,5 kgm
20 Bolt, 2.5 kgm
21 Timing cover
22 Belt pulley
23 Washer
24 Vibration damper
25 Bolt, 32 kgm
26 Damper pulley
27 Spring washer
28 Oil sump gasket
29 Oil sump
30 Bolt, 2.5 kgm

The following items are fitted to the timing cover, shown in Fig. 1.44 for the engine in question:

Vacuum pump for brake system, water pump, fuel filter, steering pump, bearing bolt for poly V-belt tensioning device, front crankshaft oil seal, pointer for TDC indication, oil dipstick guide tube. The arrangement is, however, not the same on all engines.

Proceed as follows, noting the differences when mentioned, as it is impossible to describe the operations in detail on all engines. Fig. 1.45 shows a view of the timing cover with the location of some of the parts.

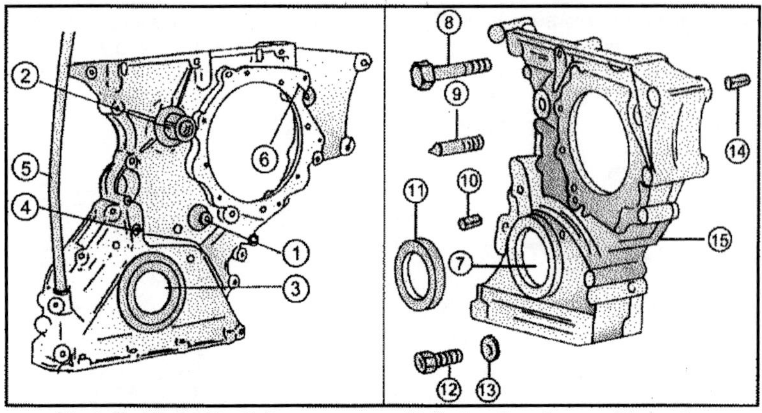

Fig. 1.45 – View of the timing case cover with the location of some of the parts.

1 Pointer for TDC sensor
2 Bore for poly V-belt tensioner
3 Crankshaft front oil seal
4 TDC pointer
5 Oil dipstick guide tube
6 Opening for vacuum pump
7 Opening for oil seal
8 Bolt, cover to crankcase
9 TDC pointer
10 Dowel pin (TDC sensor)
11 Crankshaft oil seal
12 bolt, cover to crankcase
13 Spring washer
14 Dowel pin
15 Timing case cover

- Disconnect the battery earth cable.
- Remove the noise dampening panel underneath the engine compartment as described earlier on.
- Remove the charge air pipe as described during the removal of the cylinder head, the viscous fan clutch as described in section "Cooling System" (not the same on all engines), the poly V-belt ("Cooling System") and the radiator in the case of a 606 engine (with and without turbo charger).
- Remove the pulleys from the coolant pump and the power steering pump. The pulley must be prevent from rotation.
- Remove the cylinder head cover as described earlier on.
- Remove the vacuum pump. The gasket or the "O" sealing ring must be replaced during installation.
- Rotate the crankshaft until the piston of the No. cylinder is at top dead centre (TDC) position. This will enable you to adjusting the TDC sensor during installation.
- Remove the vibration damper.
- Remove the alternator and the alternator bracket (carrier).
- Unscrew the oil pump securing bolts. The bolts securing the timing cover to the oil sump must be fully removed, the remaining oil sump bolts must be slackened, but remain in position. Note the tightening torques during installation. M6 bolts to 1.0 kgm (to crankcase), M8 bolts to 2.5 kgm (to crankcase), oil sump to transmission bell housing to 4.0 kgm.
- Remove the coolant hoses at the front of the cylinder head. Check and if necessary replace the hoses and hose clamp if not in good condition.
- Remove the power steering pump and attach it with the connected pipes/hoses at the side of the engine compartment.

- Remove the oil filter and place it to one side with the fuel lines connected.
- Remove the guide rail pins from the cylinder head as described in connection with the timing mechanism.

Fig. 1.46 – The two bolts (1) must be removed inside the timing case.

- Remove the poly V-belt tensioning device (only if the timing cover is to be replaced).
- Remove the bolts securing the injection pump to the timing case cover. Tighten the bolt to 2.0 kgm during installation.

Fig. 1.47 – The location of the bolts with different length.
1 M6 x 70 mm
2 M6 x 60 mm
3 M6 x 40 mm

- Unscrew the bracket for the TDC sensor. During installation screw on the bracket so that the pointer is aligned with the TDC marking in the vibration damper.
- Unscrew the two bolts securing the timing cover to the cylinder head in the inside of the timing case. Tighten the bolts to 2.5 kgm.
- Remove the timing case cover, but note the length of the bolts. All bolts have an M6 thread, but are 60 mm, 70 mm or 40 mm long. Refer to Fig. 1.47 or mark their position accordingly.
- With the cover removed extract the crankshaft oil seal. If a new cover is fitted transfer the tensioning device for the poly V belt from the old cover to the new cover. The same applies to the oil-water heat exchanger (7), but the gasket (9) must be replaced. The bolts are tightened to 1.4 kgm (10 ft.lb). The tensioning device is tightened to 3.0 kgm (22 ft.lb.) to the cover, the tensioning pulley, if removed, is tightened to 3.6 kgm (26 ft.lb.).

The installation if a reversal of the removal procedure, noting the tightening torques already given. Remember to fill the engine with oil and the cooling system.

1.4.3. CRANKSHAFT AND BEARINGS

1.4.3.0. Technical Data

All dimensions in metric units.

Machining tolerances:

Max. out-of-round of journals:	0.005 mm
Max. taper of main journals	0.010 mm
Max. taper of crankpins:	0.015 mm
Max. run-out of main journals*:	
Journals Nos. II and IV	0.07 mm
Journal No. III:	0.10 mm

* Crankshaft placed with Nos. I and V journals (602/604 or I and IV journals (605/606) in "V" blocks.

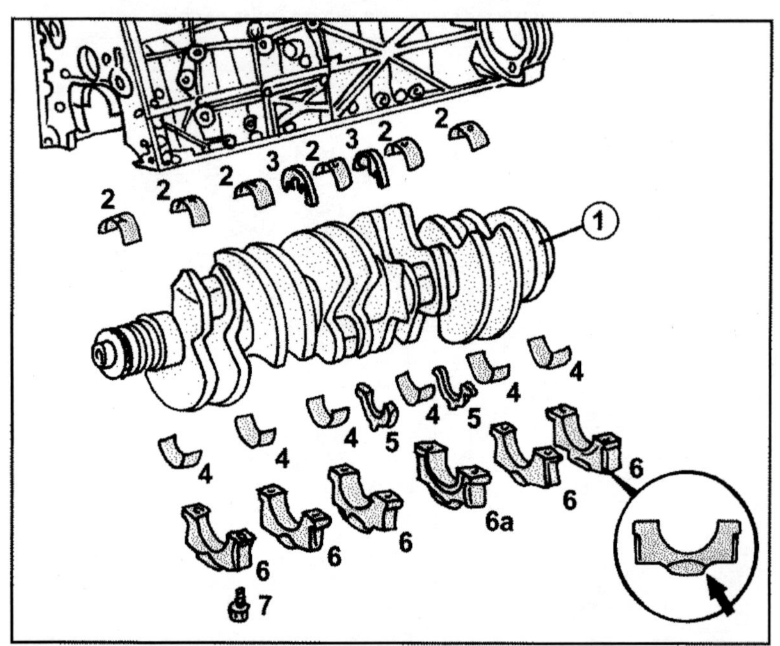

Fig. 1.48 – Crankshaft and bearings of a 602 engine. One bearing less is fitted to the 604 engine, one more to the 605/606 engine. All bearing caps are numbered (arrow).

1 Crankshaft
2 Bearing shells, crankcase
3 Thrust washers, crankcase
4 Bearing shells in bearing caps
5 Thrust washers, cap
6 Main bearing caps
6a Fitted bearing cap
7 Bearing cap bolts

Main Bearing Journal Diameter – all engines:

Nominal:	57.950 – 57.965 mm
1st repair size:	57.700 – 57.715 mm
2nd repair size:	57.450 – 57.465 mm
3rd repair size:	57.200 – 57.215 mm
4th repair size:	56.950 – 56.965 mm

Basic Bearing Bores (all engines):

For main bearings:	62.500 – 62.519 mm
For big end bearings:	51.600 - 51.619 mm

Crankpin Diameter – all engines:

Nominal dimension:	47.950 – 47.965 mm
1st repair size	47.000 – 47.715 mm
2nd repair size:	47.450 - 47.560 mm
3rd repair size:	47.200 – 47.215 mm
4th repair size:	46.950 - 46.965 mm

Width of Crankpins:

Nominal Dimension	27.960 - 28.044 mm
Repair sizes:	Up to 28.30 mm

Bearing Running Clearances (all engines):
Main bearings:..0.03 - 0.07 mm (best 0.055 mm)
Big end bearings: ....0.03 - 0.07 mm (best 0.50 mm)
Wear limit:..0.080 mm
Bearing End Float (all engines):
Main bearings: ....0.10 - 0.25 mm
Big end bearings: .... 0.12 - 0.26 mm
Wear limit - Main bearings: ....0.30 mm
Wear limit - Big end bearings:..0.50 mm

Bearing Shells (all engines):

	Main Bearings	*Big End Bearings*
Nominal Dimension:	2.25 mm	1.80 mm
1st repair size:	2.37 mm	1.92 mm
2nd repair size:	2.50 mm	2.05 mm
3rd repair size	2.62 mm	2.17 mm
4th repair size	2.75 mm	2.30 mm

Connecting Rod Bolts (all engines):
Thread: . . . . M9 x 1
Diameter of stretch neck: . . . . .7.4 mm
Min. diameter of stretch neck: . . . . 7.1 mm
Tightening torque: . . . . 0.5 + 2.5 kgm (18 ft.lb-) + 90°

1.4.3.1. Crankshaft - Removal and Installation

The engine must be removed to take out the crankshaft. The operations are similar on all engines.

- Remove the transmission from the engine. Take care not to distort the clutch shaft.
- Counterhold the flywheel in suitable manner and evenly slacken the clutch securing bolts. Use a centre punch and mark the clutch and flywheel at opposite points. Lift off the clutch plate and the driven plate. Immediately clean the inside of the flywheel and unscrew the flywheel. See also Fig. 1.49.

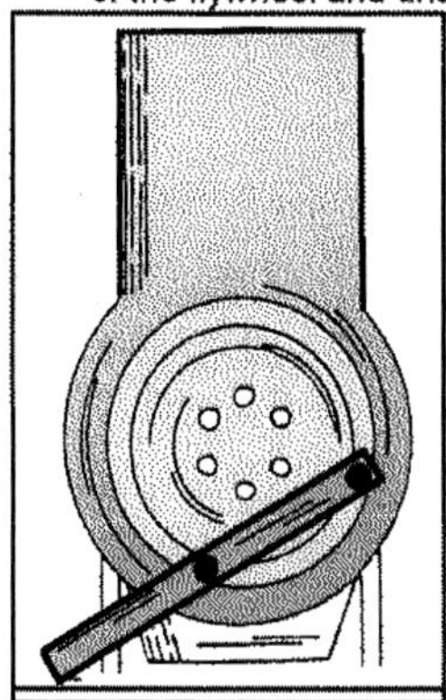

Fig. 1.49 – Flywheels and driven plates can be prevented from rotating as shown when the bolts are slackened. When tightening the bolts arrange the bar on the other side.

- Remove the drive plate for a torque converter of an automatic transmission in the same manner.
- With the flywheel still locked, remove the crankshaft pulley bolt and remove the crankshaft pulley/damper as described later on.
- Remove the cylinder head as described in Section 1.4.0.1. and the timing cover as described in Section 1.4.2. Remove the oil sump and oil pump.
- Remove the pistons and connecting rods as described in Section 1.4.1.1.
- The crankshaft end float should be checked before the crankshaft is removed. To do this, place a dial gauge with a suitable holder in front of the cylinder block and place the gauge stylus against the end flange of the crankshaft, as shown in Fig. 1.50.

Fig. 1.50 – Checking the crankshaft end float.

- Use a screwdriver to push the crankshaft all the way to one end and set the gauge to "0" ' Push the shaft to the other side and note the dial gauge reading. The resulting value is the end float. If it exceeds 0.30 mm (0.012 in.) replace the thrust washers during assembly, but make sure to fit washers of the correct width. These are located left and right at the centre bearing. Note that only two washers of the same thickness must be fitted.

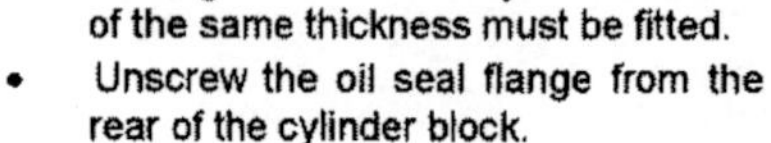

- Unscrew the oil seal flange from the rear of the cylinder block.
- Unscrew the main bearing bolts (7) in Fig. 1.48 evenly across. The bearings caps are marked with the numbers 1 to 5 or 1 to 6, depending on the engine. The numbers are stamped into the centre of the caps, as shown in Fig. 1.51. No. 1 cap is located at the crankshaft pulley side.
- Remove the bearing shells (4) from the bearing journals (they could also stick to the caps) and immediately mark them on their back faces with the bearing number. Also remove the thrust washers (5).
- Lift the crankshaft (1) out of the cylinder block and remove the remaining thrust washers from the centre bearing location and the remaining bearings shells. keep the shells together with the lower shells and the bearing caps. These shells have an oil bore and a groove and must always be fitted into the crankcase when the crankshaft is installed.

Fig. 1.51 – The arrows show the numbering of the main bearing caps (four-cylinder). Caps must be fitted in their original order.

1.4.3.2. Inspection of Crankshaft and Bearings

Main and crankpin journals must be measured with precision instruments to find their diameters. All journals can be re-ground four times and the necessary bearing shells are available, i.e. undersize shells can be fitted.

Place the crankshaft with the two end journals into "V" blocks and apply a dial gauge to the centre main journal. Rotate the crankshaft by one turn and read off the dial gauge. If the reading exceeds 0.06 mm, replace the crankshaft.

Check the main bearing and big end bearing running clearance as follows:

- Bolt the main bearing caps without shells to the crankcase, oil the bolt threads and fit each cap. Tighten the bolts to 5.5 kgm (39.6 ft.lb.) and then angle-tighten them a further 90 – 100°. Bearing caps are offset and can only be fitted in one position.
- Referring to Fig. 1.52 and measure the bearing bores in directions A, B and C and write down the results. If the basic diameter is exceeded (see Section 1.4.3.0.), the bearing cap and/or the cylinder block must be replaced.

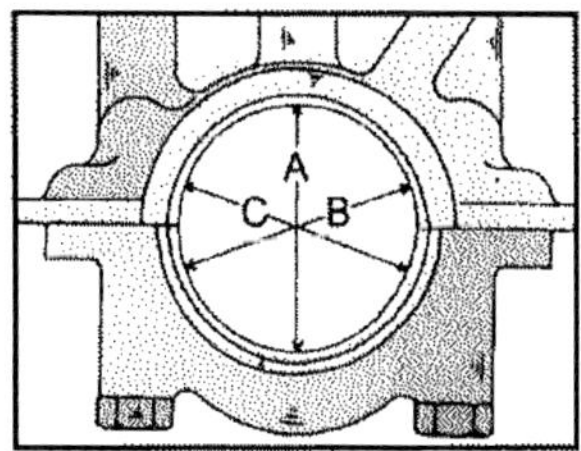

Fig. 1.52 – Measuring the inside diameter of the fitted bearing shells.

- Remove the bearing caps and refit them, this time with the well cleaned bearing shells. Re-tighten the bolts as specified.
- Measure the diameter of each bearing in accordance as shown in Fig. 1.50 and write down the results. Deduct the journal diameter from the bearing diameter. The resulting difference is the bearing running clearance, which should be between 0.031 - 0.073 mm, with a wear limit of 0.080 mm.

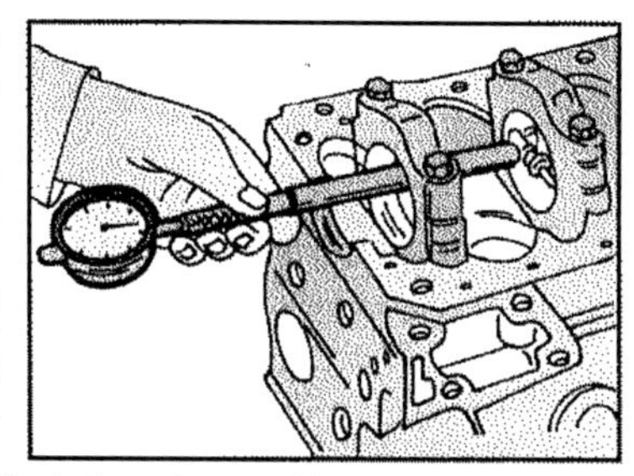

Fig. 1.53 – Measuring the inside diameter of the fitted main bearing caps together with the shells.

- Check the big end bearing clearances in a similar manner, but bolt the bearing caps to the connecting rods. Tighten the nuts to the value given in the technical data and angle-tighten as above. The bearing clearance should be between 0.031 - 0.073 mm, with the same wear limit.

Selection of bearing shells is rather complicated, and we advise you to take the cylinder block to an engine shop, if the above measurements have revealed that new bearing shells are necessary.

1.4.3.3. Crankshaft - Installation

Thoroughly clean the bearing bores in the crankcase and insert the shells (2) with the drillings into the bearing bores, with the tabs engaging the notches. Fit the thrust washers (3) to the respective bearing, with the oil grooves towards the outside, as shown in Fig. 1.54.

Fig. 1.54 – Correct fitting of the crankshaft thrust washers.

- Use the two forefingers as shown in Fig. 1.55 to hold the thrust washers against the bearing cap and fit the cap in position.
- Lift the crankshaft in position and fit the bearing caps with the inserted shells (again shells well oiled and locating tabs in notches). Fit the two thrust washers to the centre bearing cap, again with the oil groove towards the outside. Place this cap in position, guiding the two thrust washers in order not to dislodge them. Use the forefingers to hold the washers as shown in Fig. 1.55. Quote the engine number when new washers are ordered. Fig. 1.56 shows the location of the bearing shells and thrust washers. The numbers refer to Fig. 1.48.
- Check the numbering of the bearing caps and fit the well oiled bolts. A number is stamped into the bearing cap and indicates where cap No. 1 is fitted. Fit the remaining caps in accordance, always matching the number in the cap with the number in the crank-case.

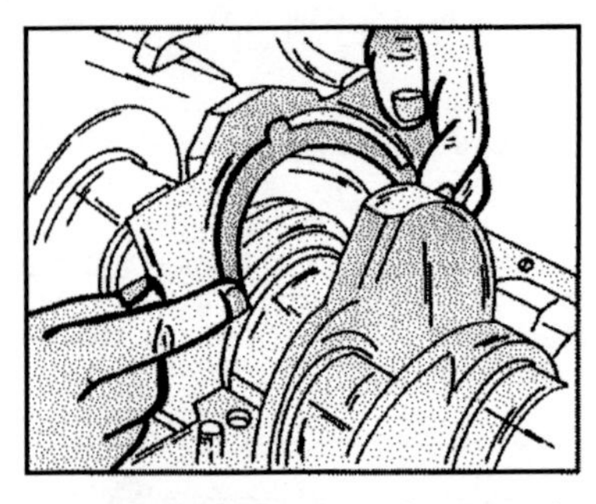

Fig. 1.55 – Fitting the main bearing cap together with the thrust washer.

- Tighten the bolts from the centre towards the outside in several steps to a torque reading of to 5.5 kgm (39.5 ft.lb.) and from this position a further 90° (a quarter of a turn).
- Rotate the crankshaft a few times to check for binding.
- Re-check the crankshaft end float as described during removal. Attach the dial gauge to the crankcase as shown in Fig. 1.50. The remaining operations are carried out in reverse order to the removal procedure. The various sections give detailed description of the relevant operations, i.e. piston and connecting rods, rear oil seal flange, timing mechanism, flywheel and clutch or drive plate, oil pump, oil sump and cylinder head.

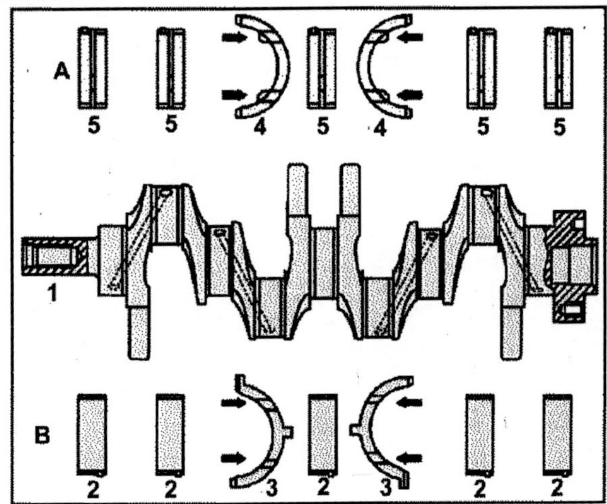

Fig. 1.56 – Main bearing shells and thrust washers as fitted to the crankcase (A) and main bearing caps (B).

Fig. 1.57 – Crankshaft main bearing caps are marked as shown. The crankcase is marked with the same number, in this case with "1".

1.4.3.4. Flywheel or Drive Plate (Automatic)

The engine can be fitted with a conventional flywheel or a dual-mass flywheel. Always check the height of the old flywheel before fitting a new one.

Both flywheel and drive plate can be

replaced with the engine fitted without re-balancing of the crankshaft. Proceed as described:

- Remove the transmission (Section 3.1.).
- Counterhold the flywheel in suitable manner and remove the clutch after having marked its relationship to the flywheel. Remove the drive plate in a similar manner. 8 bolts are used to secure the flywheel. A hole has been drilled between two of the bores and a similar hole is drilled into the crankshaft. These two bolts must be aligned when the flywheel or the drive plate is fitted. Fig. 1.61 shows the alignment bore in the case of the flywheel. The drive plate has a similar hole.
- Remove the flywheel or the drive plate. Distance washers are used in the case of the drive plate, which can also be removed. Measure the diameter of the mounting bolts at their smallest section (stretch neck). If less than 8.1 mm, replace the bolts. The measurement is carried out as shown earlier on for the connecting rod bolts.
- If the flywheel or the starter ring looks worn, take the wheel to your dealer to have the flywheel re-machined and/or the ring gear replaced.

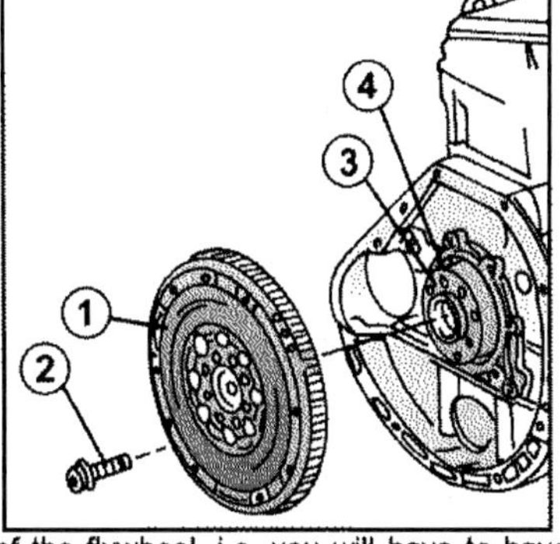

Fig. 1.58 – View of the flywheel with the location of the dowel pin.

1 Flywheel
2 Flywheel bolt, 45° + 90°
3 Cylindrical dowel pin
4 Crankshaft

Fit the flywheel or the drive plate with the alignment bores in line. Fit a distance washer underneath and on top of the drive plate. Tighten the bolts evenly across to 4.5 kgm (32 ft.lb.) and from this position a further 90°. The angle is important to give the stretch bolts their correct tension.

Engines for manual transmissions are fitted with a ball bearing inside the flywheel. The bearing must be pressed out of the flywheel, i.e. you will have to have access to a press. Place the flywheel onto a press table as shown in Fig.1.56, insert the bearing and press the bearing in position. Grease the bearing after installation.

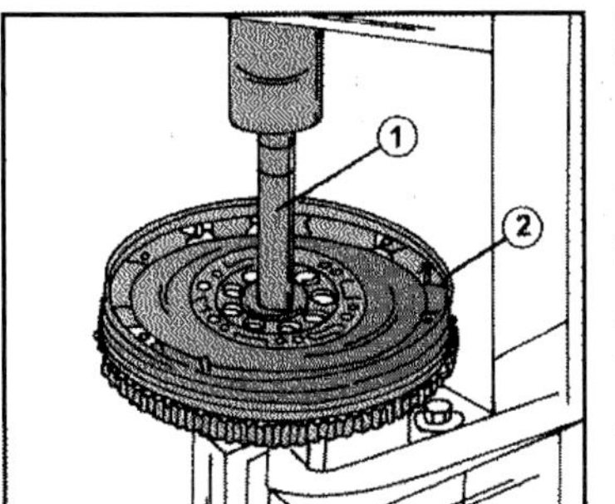

Fig. 1.59 – Pressing the guide bearing inside the flywheel (2). A press mandrel (1) must be used.

1.4.3.5. Crankshaft Pulley and Vibration Damper

The engine is fitted with a once piece crankshaft pulley/vibration damper assembly. The damper is located on the end of the crankshaft by means of a Woodruff key and secured by the central bolt in the end of the shaft. The tightening torque of the bolt is not the same on all models and depends on the marking of the bolt head (see below). Fig. 1.60 shows the attachment of the parts.

Remove the parts as follows, noting that a puller may be necessary to withdraw the hub:

- Remove the sound-proofing panel from underneath the engine compartment and remove the front panel and the radiator as described during the removal of the engine. Release the tension of the drive belt, as described later on.

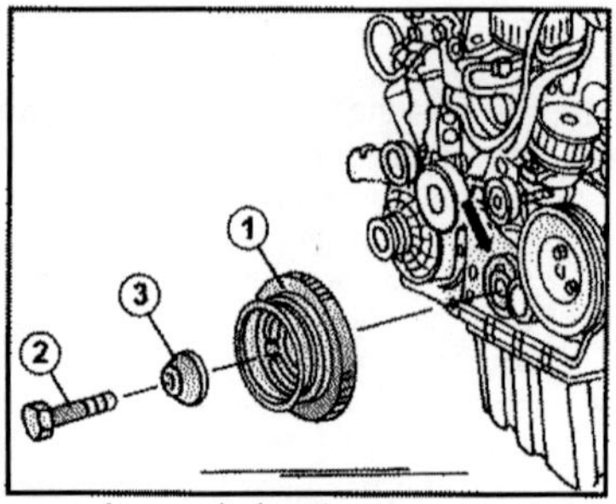

Fig. 1.60 – Details for the removal and installation of the crankshaft pulley/vibration damper assembly. Note the different tightening torque of the bolt (see below).

1 Crankshaft pulley/vibration damper
2 Central bolt
3 Washer

- Engage a gear and apply the handbrake to lock the engine against rotation. In the case of a vehicle with automatic transmission remove the starter motor and lock the starter motor ring gear in suitable manner. The same can be carried out when a manual transmission is fitted and the bolt cannot be removed by engaging a gear.
- Unscrew the crankshaft pulley/vibration damper bolt (2) in Fig. 1.60 and take off the washer (3).
- Withdraw the assembly from the end of the crankshaft. A tight vibration damper can be removed with a suitable puller. You can also try two tyre levers, applied at opposite points of the damper.

The crankshaft pulley has a certain diameter. If replaced, quote the engine type and number.

The installation of the crankshaft pulley and the vibration damper is carried out as follows:

- Rotate the crankshaft until the Woodruff key is visible and slide the crankshaft pulley/vibration damper with the key way over the key and the shaft. Make sure that the Woodruff key has engaged and has not bee dislodged.
- Place the washer (3) over the centre bolt (2), coat the bolt threads with engine oil and fit the bolt. If the bolt head is marked with "8.8" tighten the bolt to 20.0 kgm. If the bolt head is marked with "10.9" tighten the bolt to 32.5 kgm (234 ft.lb.). The crankshaft must still be locked against rotation.
- Fit the drive belt as described later on.
- The remaining operations are carried out in reverse order.

1.4.3.6. Rear Crankshaft Oil Seal and Oil Seal Carrier

The rear crankshaft oil seal is located inside a flange which is bolted to the rear of the crankcase. Flange and oil seal are made together and cannot be replaced separately.

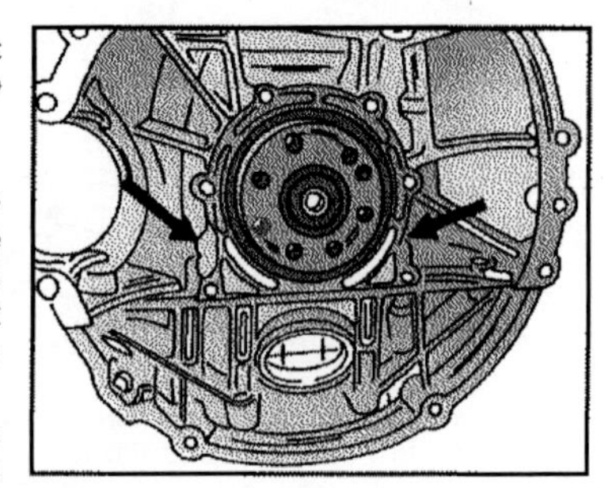

Fig. 1.61 – To remove the oil seal cover insert two screwdrivers carefully underneath two lugs shown by the arrows.

Two dowels locate the flange correctly in relation to the crankshaft centre. The flange is fitted with sealing compound ("Loctite"). Fig. 1.62 shows the attachment of the oil seal flange (cover) and the integral oil seal.

Transmission and flywheel and/or drive plate must be removed to replace the cover/oil seal assembly. Also drain the engine oil.

- Place a jack underneath the engine (wooden block between jack head and engine) and lift the engine approx. 5 cm at the front axle in order to reach he bolts (6) in Fig. 1.62.
- Remove the bolts (5) from the front and the two bolts (6) from below.
- Insert two screwdrivers as shown in Fig. 1.63 and lever off the cover from the crankcase (7) without damaging the flange.

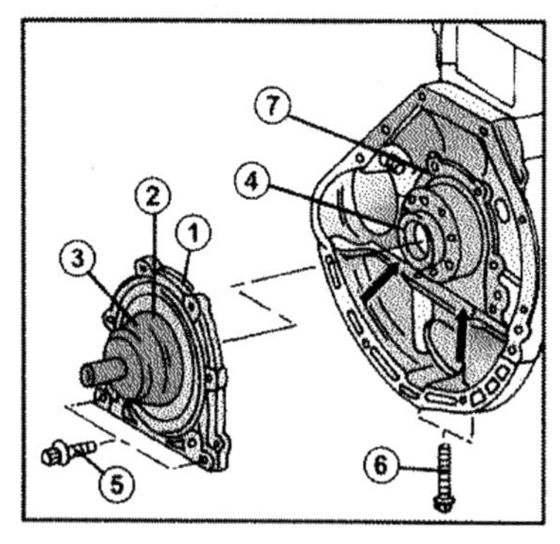

Fig. 1.62 – Removal and installation of the rear oil seal cover together with the oil seal. The faces marked with the arrows must be coated with "Loctite".

1 Cover with oil seal
2 Fitting sleeve (normally in repair kit)
3 Oil seal in cover
4 Crankshaft
5 Screws, inserted from front, 0.9 kgm
6 Screws, inserted from below, 0.9 kgm

- Thoroughly clean the sealing faces and coat the cover face where I rests again the oil sump gasket. Push the cover against the crankcase. You should have a fitting sleeve (2) in Fig. 1.61 which will aid the fitting. Otherwise take care not to damage the seal. After the cover is flush against the crankcase push it upwards until the upper face is against the crankcase and remove the fitting sleeve. The oil seal (3) must have a snug fit.
- Remove the two bolts (6) from below and then the remaining bolts from the front. First tighten the lower bolts and then the remaining bolts. All are tightened to 0.9 kgm (7 ft.lb.).
- All other operations are carried out in reverse order.

1.4.3.7. Front Crankshaft Oil Seal

The front crankshaft oil seal is located in the timing cover. Oil leaks at this position can also be caused by a leaking timing cover gasket. Check before replacing the oil seal.

Fig. 1.63 – Removal of the front crankshaft oil seal.

The crankshaft pulley/vibration damper as already described before the oil seal can be replaced. The seal can be carefully removed with a screwdriver (see Fig. 1.63). Screw a self-tapping screw into the outside of the seal and apply the screwdriver plate under the screw head. It is also possible to unscrew the oil seal cover to replace the seal.

Thoroughly clean the surrounding parts. Burrs on the timing cover bore can be removed with a scraper. Also clean the cylinder block face.

Fill the space between sealing lip and dust protection lip with grease and carefully drive a new oil seal into the timing cover and over the crankshaft until the ring outer face is

flush. Fit the two bolts from below and then the remaining bolts. All bolts are tightened to 7.2 kgm (7.2 ft.lb.).
Refit the vibration damper as described in Section 1.3.3.5. and carry out the remaining operations in reverse order.

1.4.4. TIMING MECHANISM

The parts covered in this section can all be removed with the engine fitted, but the arrangement of the individual component parts are different on the various engines.

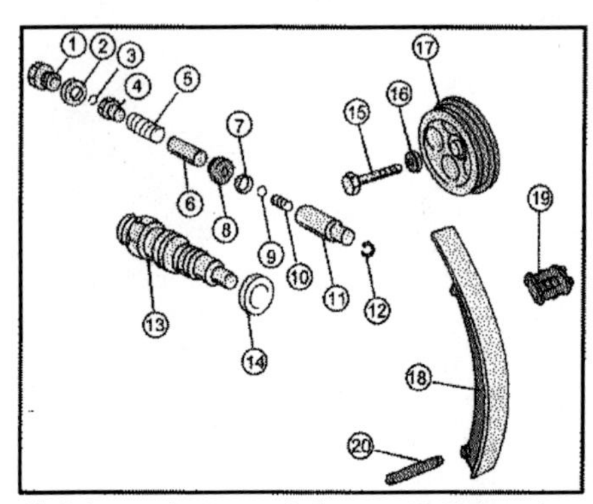

Fig. 1.64 – The component parts on the L.H. side of the timing mechanism (602 engine).

1 Plug
2 Seal
3 Ball
4 Spring seat
5 Thrust spring
6 Plunger
7 "O" sealing ring
8 Relief valve
9 Ball
10 Thrust spring
11 Plunger
12 Circlip
13 Chain tensioner
14 Sealing ring
15 Bolt. sprocket to camshaft
16 Plain washer
17 Camshaft sprocket
18 Glide shoe
19 Chain lock
20 Guide bolt

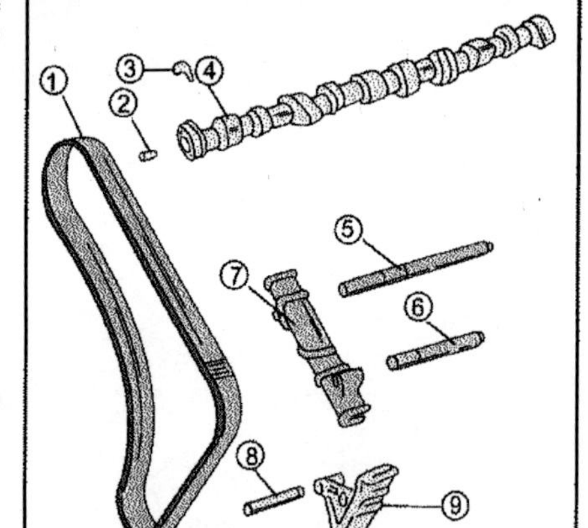

Fig. 1.65 – The component parts on the R.H. side of the timing mechanism (602 engine).

1 Duplex timing chain
2 Dowel pin
3 Camshaft keeper plate
4 Camshaft
5 Guide bolt, slide rail
6 Guide bolt, slide rail
7 Slide rail
8 Dowel pin
9 Chain guide rail

602 Engine

Figs. 1.64 and 1.65 show the timing chain and the associated parts. Fig. 1.66 shows a front view of the fitted timing mechanism.

The endless timing chain is engaged with the camshaft sprocket, the injection pump sprocket and the crankshaft sprocket. The chain is guided by two slide rails. The tension of the chain is insured by means of a hydraulic chain tensioner, which is located in the cylinder head and pushes onto a tensioning rail.

The camshaft sprocket is fitted by means of an M8 bolt and located by a Woodruff key. A second, smaller chain is used to drive the oil pump. The chain is fitted around a second sprocket on the crankshaft and around the oil pump drive sprocket and has its own chain tensioner.

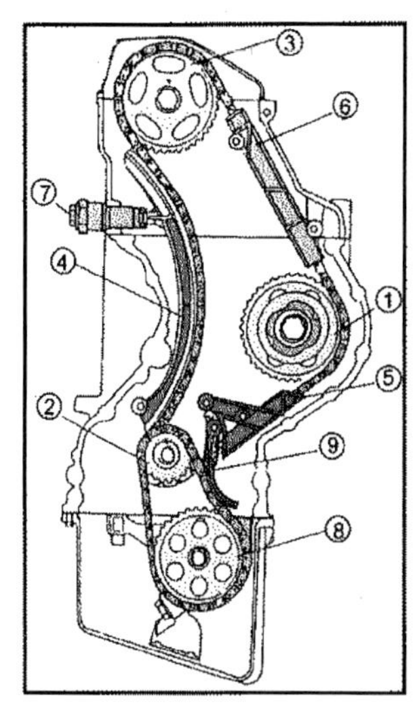

Fig. 1.66 – Timing chain and chain drive in fitted position (602 engine).

1 Injection pump drive gear
2 Crankshaft sprocket
3 Camshaft sprocket
4 Tensioning rail
5 Slide rail
6 Slide rail
7 Chain tensioner
8 Oil pump drive sprocket
9 Tensioning lever for oil pump drive chain

604, 605, 606 Engines

The arrangement is the timing chain is more or less the same as in the case of the 602 engine, with the difference that a second camshaft must be driven. This is carried out by the drive sprocket shown in Fig. 1.67, which is driven by the exhaust camshaft sprocket. The drive of the injection pump and the oil pump is the same as in the case of the 602 engine.

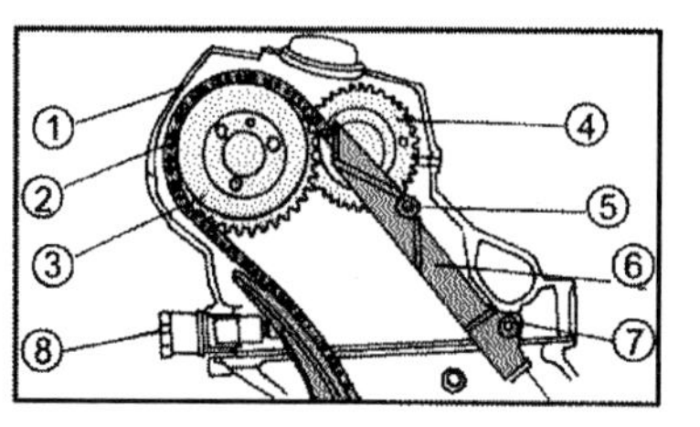

Fig. 1.67 – Camshaft drive of an engine with two camshafts.

1 Cylinder head cover
2 Duplex roller chain
3 Camshaft sprocket (exhaust camshaft)
4 Camshaft sprocket (inlet camshaft)
5 Bearing bolt for slide rail
6 Timing chain slide rail
7 Bearing bolt for slide rail
8 Chain tensioner

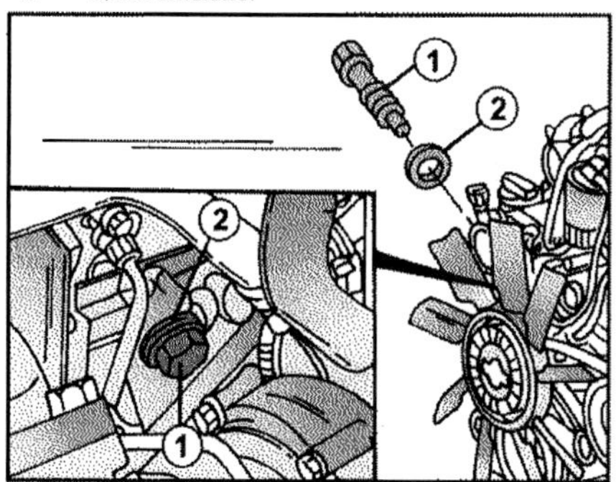

Fig. 1.68 – The chain tensioner (1) is fitted at the position shown and sealed off by means of sealing ring (2).

1.4.4.0. Chain Tensioner – Removal and Installation

The chain tensioner is fitted into the R.H. side of the cylinder head as shown in Fig. 1.68. The tensioning force of the chain tensioner is a combination of the fitted compression spring and the pressure of the engine oil. The oil contained inside the tensioner also absorbs shock loads from the timing chain. A chain tensioner cannot be repaired, i.e. must be replaced if suspect.

The chain tensioner (1) can simply unscrewed from the side of the engine, but a hand press is required to fit it properly. A bench-mounted electric drill is, however, sufficient to pre-load the tensioner. Looking at the chain tensioner you will see a large and a small hexagon. Only apply a socket to the large hexagon. Unscrewing the tensioner by applying the socket to the smaller hexagon will result in the tensioner falling internally apart. Fig. 1.68 shows where the tensioner is located.

The chain tensioner must be filled with oil before installation. This requires the use of a hand press (or the drill mentioned above) and a glass jar, filled with SAE 10 engine oil. Hold the tensioner with the thrust bolt facing downwards into the oil. The oil must be above the flange of the hexagon. Using the hand press, push the thrust bolt about 7 to 10 times into the tensioner. As the tensioner is filled with oil, the pressure required to compress the tensioner will increase. After filling the tensioner with oil, check that some force is required to compress it by hand.

Fit a new gasket seals (2) and screw the chain tensioner in position. Tighten the tensioner to 8.0 kgm (58 ft.lb.). The thrust bolt of the tensioner must engage with the tensioning rail as can be seen in Fig. 1.66.

1.4.4.1. Removal and Installation of Timing Chain

The camshaft of the 602 engine and the exhaust camshaft of the other engines and the injection pump are driven from the crankshaft by means of an endless timing chain. The component parts of the timing gear are shown in Figs. 1.69 for the 602 engine and Fig. 1.70 for the 604, 605 and 605 engines.

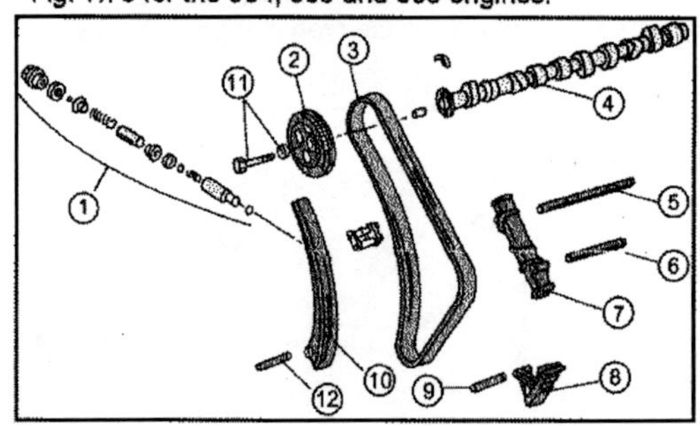

Fig. 1.69 – Details for the removal and installation of the timing chain (602 engine).

1 Chain tensioner
2 Camshaft sprocket
3 Duplex timing chain
4 Camshaft
5 Bearing pin, slide rail
6 Bearing pin, slide rail
7 Slide rail (cylinder head)
8 Slide rail (cylinder block)
9 Bearing pin, slide rail
10 Tensioning rail, chain
11 Bolt and washer
12 Bearing pin, tensioning rail

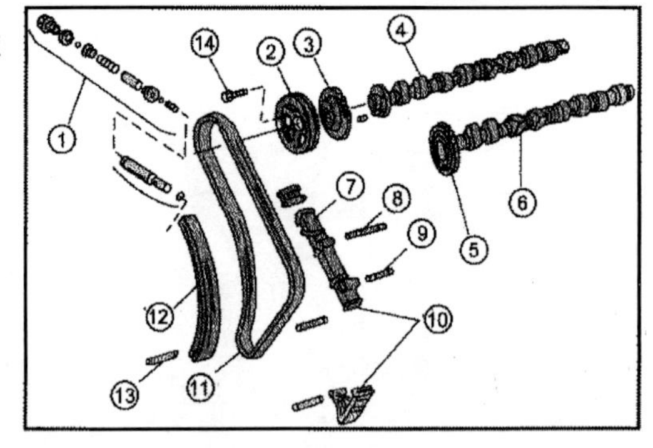

Fig. 1.70 – Details for the removal and installation of the timing chain (604, 605, 606 engines).

1 Chain tensioner
2 Sprocket, exhaust camshaft
3 Drive sprocket, inlet camshaft
4 Exhaust camshaft
5 Sprocket, inlet camshaft
6 Inlet camshaft
7 Upper slide rail
8 Upper bearing pin
9 Lower bearing pin
10 Lower slide rail (block)
11 Duplex chain
12 Timing chain tensioning rail
13 Bearing pin (tensioning rail)
14 Sprocket bolts (1.8 kgm)

The replacement of the timing chain is practically impossible without using the special tools available to Mercedes dealers. As the engine must be removed in any case you will be able to replace the chain in accordance with the instructions given during the removal of the cylinder head, camshaft(s)s, etc. and the instructions given below. The timing chain can then be removed in one piece and refitted. The workshop separates the chain and rivets are used to assemble it.

The following information can be followed if you have some experience with timing chains.

The following operations can be carried out by referring to this illustration. The sprockets for the camshaft and the injection pump have twice as many teeth as the crankshaft sprocket. Two slide rails guide the chain on one side, a long tensioning rail, operating in conjunction with the hydraulic chain tensioner, is fitted to the other side. Note the following points before commencing any operation:

- A hand-held grinding machine must be available to replace the timing chain with the engine fitted. The new timing chain has a chain lock to connect the two chain ends.
- When an engine has been dismantled, always fit an endless timing chain.
- Before a new timing chain is fitted, check all sprockets. Worn sprocket teeth will very soon wear the new chain.
- Timing chains are sometimes changed during the production of the engine. Always quote the engine type, engine number and model year of the vehicle.

Replace the timing chain with the engine fitted as follows:

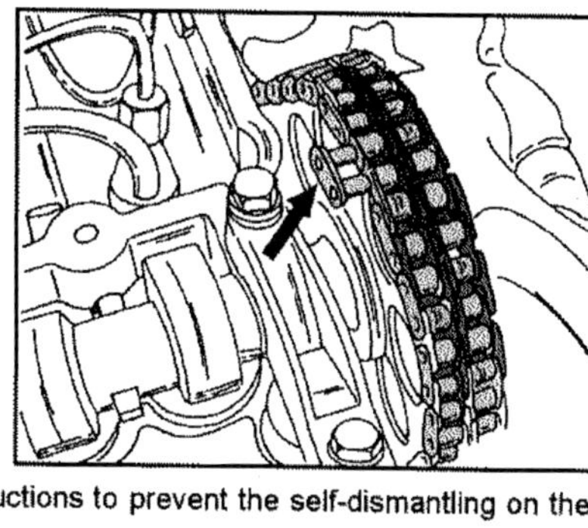

Fig. 1.71 – Fit the connecting link from the inside towards the outside.

602 Engine

- Remove the visco-coupling for the cooling fan as described for the 602 engine.
- Remove the injectors (to facilitate the rotation of the engine) and then remove the cylinder head cover.
- Remove the chain tensioner as described earlier on. Follow the instructions to prevent the self-dismantling on the tensioner.
- Cover the chain chamber with rags to prevent grinding particles from falling inside and cut both chain bolts by grinding down one link of the timing chain. Do not remove the chain at this stage.
- Connect the new timing chain with the connecting link to the old chain, at the same time pushing out the old link.

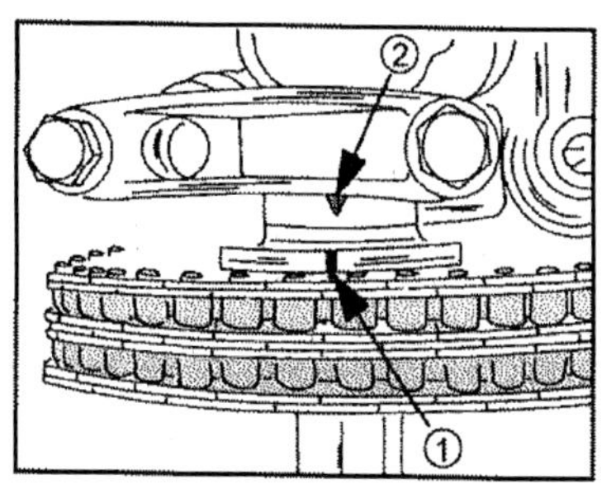

Fig. 1.72 – The notch (1) in the camshaft flange must be aligned with the lug (2) on the first camshaft bearing cap (602 engine).

- Slowly rotate the crankshaft in direction of rotation, using a socket applied to the crankshaft pulley bolt. The timing chain must remain in engagement with the camshaft sprocket whilst the crankshaft is rotated. Do not turn the camshaft by applying a spanner to the sprocket bolt.
- Disconnect the old timing chain from the new chain and push the new connecting link from the inside towards the outside through the two chain ends, as shown in Fig. 1.71. Secure the link with the lock washers from the front.

- Rotate the crankshaft until the piston of No. 1 cylinder is at TDC firing point and check that the timing marks (on the crankshaft pulley or the vibration damper) are aligned, with the camshaft in the position shown in Fig. 1.72. The chain is correctly fitted if this is the case. Otherwise the chain has moved by one tooth and the timing cover must be removed to correct the timing setting.
- Refit the timing chain tensioner and carry out all other operations in reverse order.

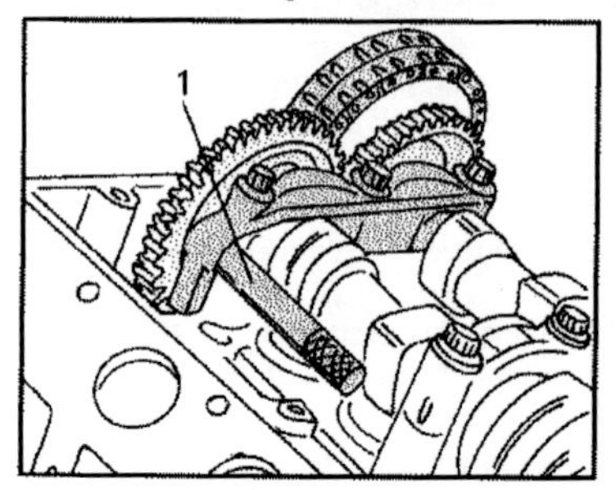

Fig. 1.73 – The arrester pin (1) is inserted from the rear into the camshaft timing gear to lock the camshaft in position.

604, 605. 606 Engines

In the case of an engine with two camshafts it will be necessary to insert an arrester pin from the rear into the camshaft timing gear as shown in Fig. 1.73 to prevent the rotation of the camshafts. This can only be achieved if the camshafts are in the position shown in Fig. 1.74, i.e. the two timing marks must be opposite each other. The pin has a diameter of 6.75 mm (Part No. 111 589 01 25 00). Fig. 1.75 demonstrates the use of the arrester pin.

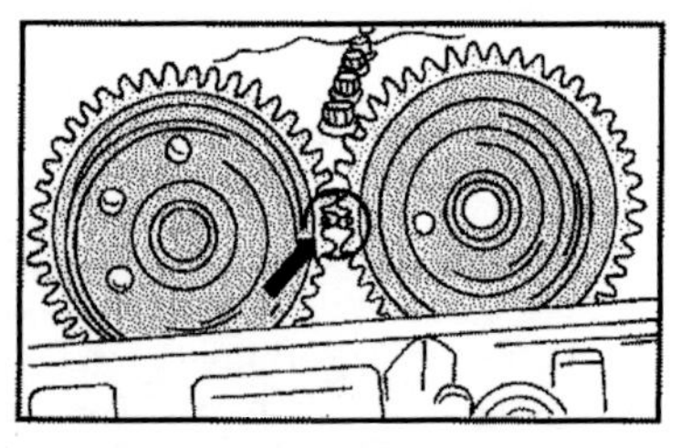

Fig. 1.74 – The two timing marks must be opposite each as shown.

The removal and installation is carried out in a similar manner as described for the 602 engine. The viscous fan coupling must be removed, but note the differences between the 604 engine and the 605/606 engines. Also remove the chain tensioner and the cylinder head cover as already described for these engines.

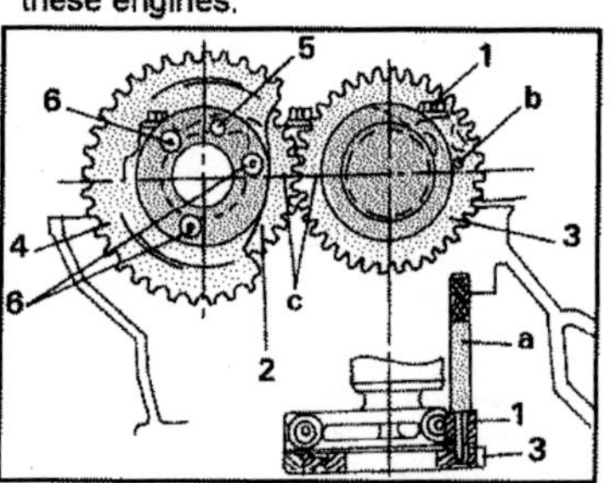

Fig. 1.75 – Engagement of the two camshaft timing sprockets and the locking of the sprockets at the bottom.

1 Camshaft bearing cap
2 Exhaust camshaft sprocket
3 Inlet camshaft sprocket
4 Drive gear for sprocket
5 Dowel pin, 5 mm
6 Torx-head bolt, M7, 1.8 kgm
a Arrester pin
b Hole for pin insertion
c Bore, 1.5 mm, timing adjustment

Removal and Installation of tensioning rail in timing cover

The location of the tensioning rail is shown in Fig. 1.69 and 1.70 for the two engines. The cylinder head and the timing case cover must be removed to replace the tensioning rail. Note that the bolts securing the camshaft sprocket must be replaced in the case of an engine with two camshafts (604/605/606 engines).

- Remove the necessary parts to gain access to the front of the engine and then remove the cylinder head and the timing case cover as already described.
- Mark the crankshaft sprocket and the timing chain opposite each other, using a spot of paint.

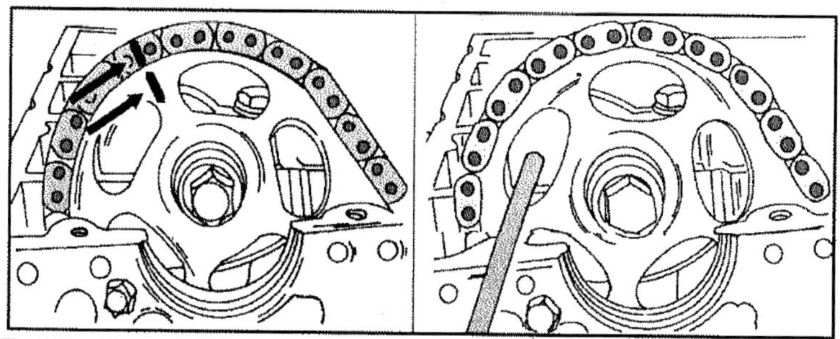

Fig. 1.76 – Removal of the camshaft sprocket of a 602 engine. Mark the chain and sprocket as shown on the left. Prevent the sprocket from rotating as shown on the right.

- Mark the camshaft sprocket(s) and the timing chain opposite each other as shown in Fig. 1.76..
- Remove the bolt or the three bolts (604, 605, 606 engines) and withdraw the camshaft sprocket. The sprocket must be prevented from rotating. This can be carried out as shown in the R.H. view of Fig. 1.76.
- Swing the tensioning rail towards the inside and withdraw it from the bearing pin. Immediately check the condition of the plastic lining. The complete tensioning rail must be replaced.

The installation is a reversal of the removal procedure. Tighten the single bolt of a 602 engine to 6.5 kgm. In the case of the other engines replaced the three bolts and tighten them to 1.8 kgm. In each case prevent the camshaft from rotating as shown in Fig. 1.76. Place the chain over the sprockets (camshaft(s) and crankshaft) so that the marks made before removal (for example Fig. 1.76, left, 602 engine) are opposite each other.

Removal and Installation of slide rails

The position of the slide rails is shown in Fig. 1.69 and 1.70. An impact hammer, together with a M6 threaded bolt of 100 mm in length is required to remove the slide rail bearing bolts. The 6 mm bolt is screwed into the end of the bearing bolt and the impact hammer (slide hammer) attached to the end of the bolt. Provided that these tools can be obtained, the rail can be removed as described below. Proceed as follows to replace the slide rail on the, cylinder head (6 in Fig. 1.66):

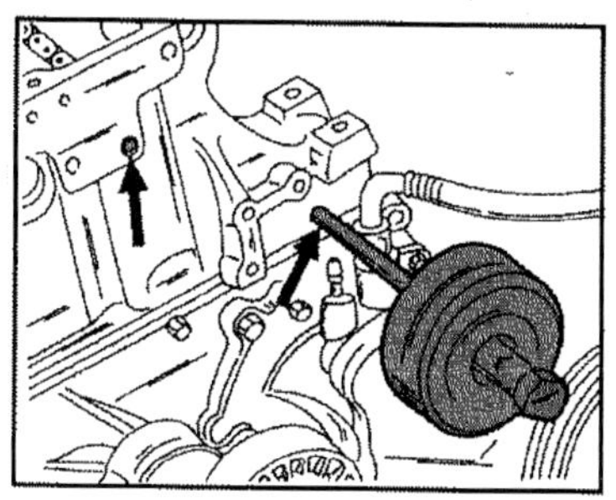

Fig. 1.77 – Remove the two bearing bolts in the manner shown out of the cylinder block.

- Disconnect the battery, the air suction hose, the radiator and the visco-coupling for the cooling fan.
- Remove the tensioning device for the single drive belt. The bearing pin for the tensioning lever of the tensioning device is at the same time the bearing pin for this slide rail.
- Remove the camshaft sprocket as already described.
- Withdraw the two bearing pins with the impact hammer and a threaded insert, as shown in Fig. 1.77 and lift out the slide rail. If no slide hammer is available, try the following: Slide a piece of tube over the bearing pin and place a washer over

the tube. Screw in a 6 mm bolt and tighten it. With the washer pressing against the tube, the bearing pin will be dislodged as soon as the tube is under tension.

Refit the slide rail as follows:

- Coat the two bearing bolts on the flange with sealing compound.
- Fit the slide rail in position and insert the bearing bolts. Fit the 6 mm bolt and the slide hammer to the end of the bearing bolt and knock the bolt in position, this time hitting the weight of the slide hammer towards the front. Fit the second bearing bolt in the same manner. Counterhold the slide rail during the bearing bolt installation with a screwdriver to prevent distortion. A locating nose in the bearing bolt bore of the slide rail will engage in the locating groove of the upper bearing bolt when the bolt is in position.
- The remaining operations are carried out in reverse order to the removal procedure. Pay attention to the paint marks on timing chain and camshaft sprocket when the parts are refitted.

Replace the lower slide rail (5, Fig. 1.66) as follows: The front of the engine must be exposed to gain access.

- Remove the timing case cover.
- Withdraw the tensioning lever together with the spring and the slide rail so that the tensioning lever clears then chain and place it against the crankshaft end as shown in the L.H. view of Fig. 1.78.

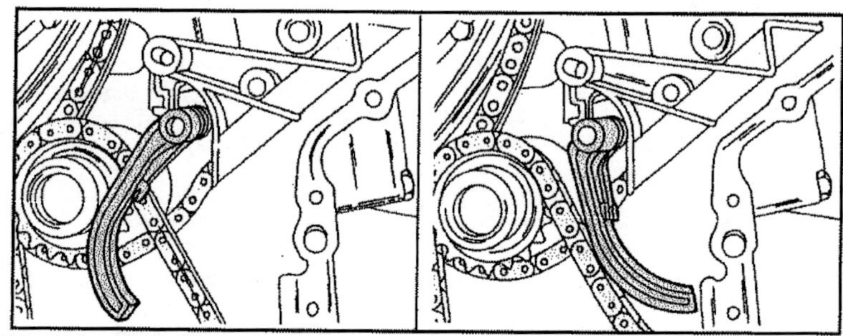

Fig. 1.79 – Guide the tensioning lever passed the timing chain and place it against the crankshaft end before the slide rail is removed. The R.H. view shows how the tensioning lever is placed against the oil pump drive chain.

- Withdraw the tensioning lever from the bearing pin and carefully release the spring tension. Remove the lever together with the spring.
- Withdraw the slide rail off the two bearing pins.
- Fit the new slide rail. Engage the spring into the side rail and the tensioning lever and slide the rail, the spring and the tensioning lever over the two bearing pins, as shown in the R.H. view of Fig. 1.78.
- Refit the timing case cover.

Removal and installation of crankshaft timing sprocket

The removal and installation is similar on all engines.

- Remove all parts to gain access to the front of the engine and remove the timing case cover as described.
- Remove the tensioning lever for the oil pump drive chain as described above and remove the lever (5, Fig. 1.66) – also see Fig. 1.78.

- Remove the bolt securing the oil pump drive sprocket and remove the sprocket from the end of the shaft, at the same time disengaging the chain from the sprocket on the crankshaft.
- Mark the position of the crankshaft sprocket and the oil pump drive sprocket to ensure installation in the same position.
- Mark the camshaft sprocket and the timing chain at opposite points with paint (see Fig. 1.76). Then remove the camshaft sprocket and allow the timing chain to hang down under its own weight.
- Remove the crankshaft sprocket from the end of the crankshaft, using a suitable puller. Immediately check the condition of the Woodruff key in the crankshaft end. If necessary remove the key with a side cutter. The Woodruff key for the oil pump drive sprocket is treated in the same manner.

If a new crankshaft sprocket is to be fitted transfer the markings from the old sprocket to the new sprocket, i.e. the same sprocket tooth in relation to the groove for the Woodruff key must be marked. The sprocket is installed as follows:

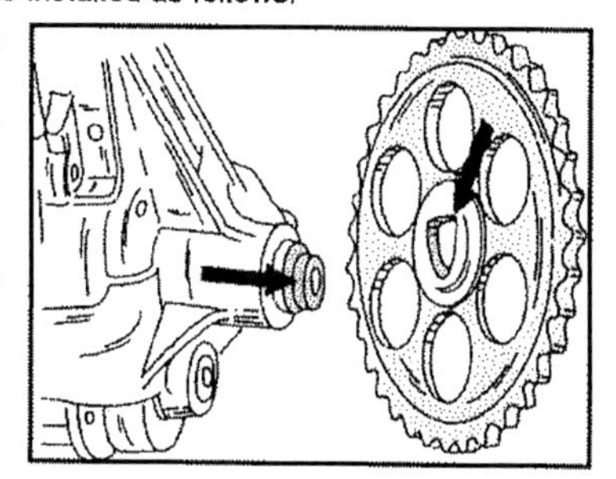

Fig. 1.79 – Align the two flats shown by the arrows when fitting the oil pump drive sprocket.

- Fit both Woodruff keys into the crankshaft end (if removed). The two flats of the keys must be parallel with the crankshaft journal.
- Use a piece of tube and drive the sprocket over the crankshaft end. Make sure the Woodruff key is not dislodged.
- Refit the camshaft sprocket together with the timing chain, noting the marks made before removal.
- Rotate the crankshaft a few times and check that the timing marks on the camshaft(s) are still aligned.
- Fit the oil pump drive sprocket. The inside of the sprocket has a flat which must engage with a similar flat on the pump drive shaft, as shown in Fig. 1.79.
- Refit the tensioning lever for the oil pump drive chain as described above (Fig. 1.78), refit the timing case cover.
- All other operations are carried out in reverse order to the removal procedure.

1.4.4. CAMSHAFT(S) – REMOVAL AND INSTALLATION

602 Engine

The camshaft is not the same for all 602 engines.

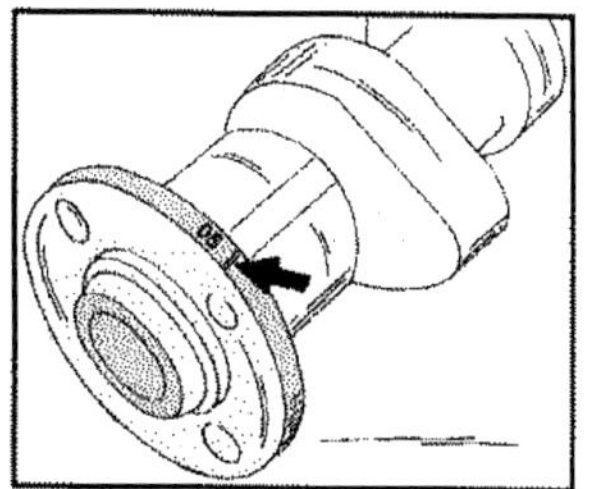

Fig. 1.80 – Camshafts have an identification number at the position shown.

A number stamped into the camshaft flange at the position shown in Fig. 1.80. Repair size camshafts also have a number engraved. If a new camshaft is fitted make sure that it is suitable for the 602 engine in question.

The camshaft is fitted to the cylinder head. The lower part of the camshaft bearings is machined into the cylinder head. The camshaft bearing caps are bolted to the top of the cylinder head.

The camshaft is removed in an upwards direction after the removal of the camshaft bearing caps, with the camshaft sprocket removed. The front of the vehicle must be exposed to gain access to all parts. The battery must be disconnected.

Fig. 1.81 – Rotate the crankshaft in the direction of the arrow until the TDC point has been obtained.

- Remove the bottom engine compartment panel as described earlier on.
- Remove the top charge air pipe at the front.
- Remove the viscous fan clutch as described for the 602 engine.
- Place a protective plate in front of the radiator.
- Remove the top charge air pipe from the top of the cylinder head.
- Remove the engine cover in the centre of the cylinder head cover.
- Remove the cylinder head cover as described during the removal of the cylinder head. Six bolts secure the cover.

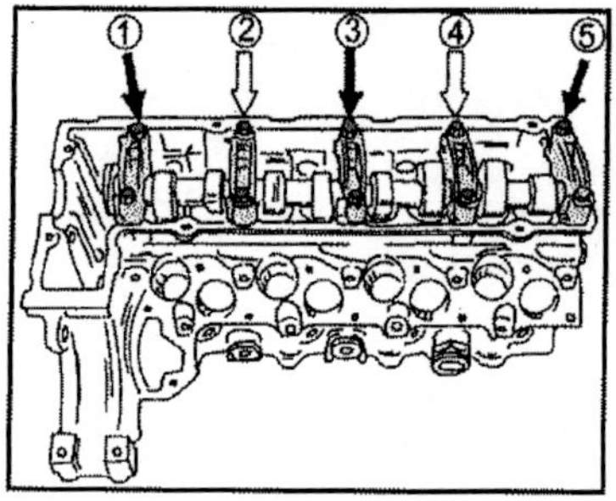

Fig. 1.82 – Slackening the camshaft bearing caps. Remove the caps with the black arrows, slacken the caps with the white arrows.

- Rotate the crankshaft until the piston of No. 1 cylinder is at top dead centre position, i.e. the crankshaft is tuned into the direction shown by the arrow in Fig. 1.81 until the "OT" mark (0) is opposite the pointer on the timing case cover. Apply a socket to the centre bolt of the crankshaft pulley to rotate the shaft – not on the camshaft sprocket bolt. Also note that the crankshaft must be rotated in the direction of rotation (arrow). After the TDC setting has been obtained check the alignment of the camshaft sprocket (Fig. 1.76), i.e. the two timing marks must be in line. Mark the sprocket and the timing chain with a spot of paint in this position.
- Remove the chain tensioner as already described.
- Remove the stretch bolt securing the camshaft sprocket to the camshaft and remove the thrust washer. To prevent rotation of the camshaft insert a drift as shown in Fig. 1.76 (right). It is also possible to remove the cover underneath the flywheel housing (torque converter housing, A/T) to insert a tyre lever into the teeth of the flywheel ring gear.
- Withdraw the camshaft sprocket from the end of the camshaft. The timing chain should be taut to prevent it from disengaging from the crankshaft sprocket teeth. Tie the chain together to keep it tight.
- Slacken the camshaft bearing cap bolts in several stages. Refer to Fig. 1.82 and remove the bearing caps 1, 3 and 5 (bolts with black arrows) followed by the bearing caps 2 and 4 (white arrows) until all tension has been removed. The camshaft can be lifted out after all caps have been removed.

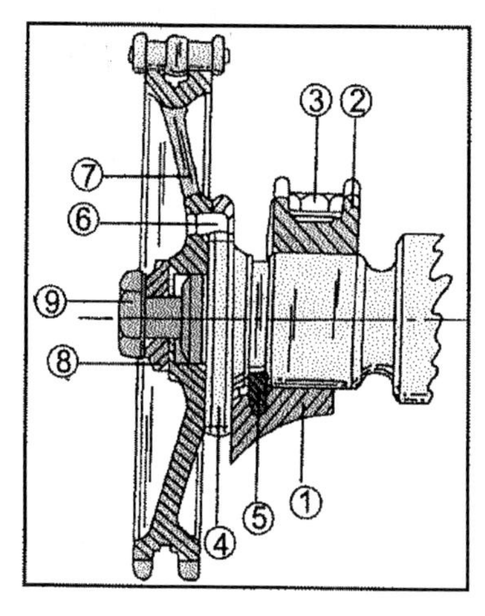

Fig. 1.83 – Sectional view of the camshaft end.

1 Cylinder head
2 Bearing cap
3 Bolt, M8 x 45
4 Camshaft
5 Keeper plate
6 Dowel pin
7 Camshaft sprocket
8 Washer
9 Bolt, M10 x 50

- Remove the camshaft keeper plate from the cylinder head. Fig. 1.83 shows a sectional view of the cylinder head with the location of the plate. Fig. 1.84 shows where you will find the plate.
- Valve tappets can be removed from their bores with a suction tool. Mark the installation position of each tappet to ensure installation in the same tappet bores.

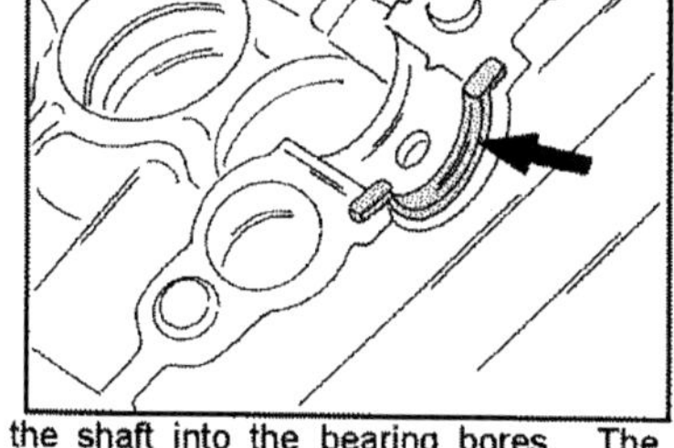

Fig. 1.84 – The camshaft keeper plate is located at the position shown in the cylinder head.

- If a new camshaft is fitted, refer to the identification number (shown in Fig. 1.80) or consult your dealer. A new shaft should be inserted into the bearing bores (well oiled) to carry out the following checks:
- Insert the keeper plate into the cylinder head. The collar must have sharp edges, otherwise replace it.
- Coat the camshaft journals and place the shaft into the bearing bores. The tappets should not be fitted.
- Fit the camshaft bearing caps in accordance with their numbers and tighten the bolts from the centre towards the outside with 2.5 kgm (18 ft.lb.).
- Screw a bolt, M10 x 30, into the end of the camshaft and tighten it until the shaft begins to turn. Rotate the shaft to notice any "bearing crush". If you feel that the camshaft shows "hard spots", slacken the bearing caps one at a time. When the troublesome bearing cap has been reached, check the bearing clearance with "Plastigage" as described for the crankshaft. The clearance must be between 0.050 and 0.081 mm.
- Remove the camshaft.
- Lubricate the tappets with engine oil and insert them into their original bores.
- Insert the camshaft into the bearing bores, with the groove entering the locking collar.
- Fit the bearing caps and tighten caps No. 2 and No. 4 in Fig. 1.82 to 2.5 kgm. Then fit the remaining caps and tighten the bolts to the same torque. The tightening must be carried out carefully to prevent distortion of the shaft through the pressure of the tappets.
- Fit the camshaft sprocket together with the timing chain over the camshaft (hold the chain tight during installation). The dowel pin in the shaft must engage with the hole in the sprocket. If a stretch bolt is used (M11 thread), measure the length of the bolt between the underside of the bolt head and the end of the bolt. If the length exceeds

53.6 mm, use a new bolt. Fit the bolt and tighten it to 2.5 kgm (18 ft.lb.) and from the final position by a further quarter of a turn. The camshaft must be prevented from rotating (for example as shown in Fig. 1.76 (right).

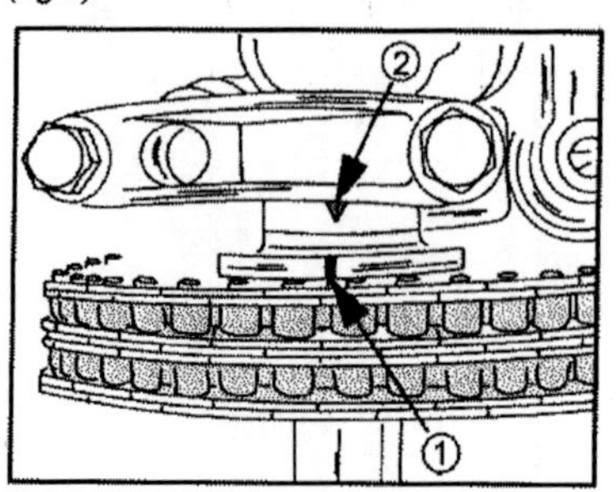

Fig. 1.85 – The notch in the camshaft and the mark in the bearing cover must be aligned as shown.

- Refit the chain tensioner as described. Tighten it to 8.0 kgm (58 ft.lb.).
- Check the timing marks on the camshaft. The marks must align when the piston of No. 1 cylinder is at top dead centre. The camshaft has a notch which must be, when seen from above, in line with a mark in the bearing cap as shown in Fig. 1.85.
- Refit the cylinder head cover.
- All other operations are carried out in reverse order to the removal procedure. After the engine has been assembled start it up and check the areas which were separated for oil leaks.

604, 605 and 606 Engines

The camshafts are fitted to a camshaft housing and can be removed with the housing or without the housing. The removal of the camshaft housing is described below. Remove the camshafts as follows:

- Remove the cylinder head cover as described during the removal of the cylinder head.
- Set the piston of No. 1 cylinder to the top dead centre position as described earlier on. The "OT" (TDC) mark on the crankshaft pulley must be opposite the pointer on the timing case cover, the two timing marks in the camshaft sprockets must be opposite each other as shown in Fig. 1.74. Rotate the crankshaft not the camshaft.

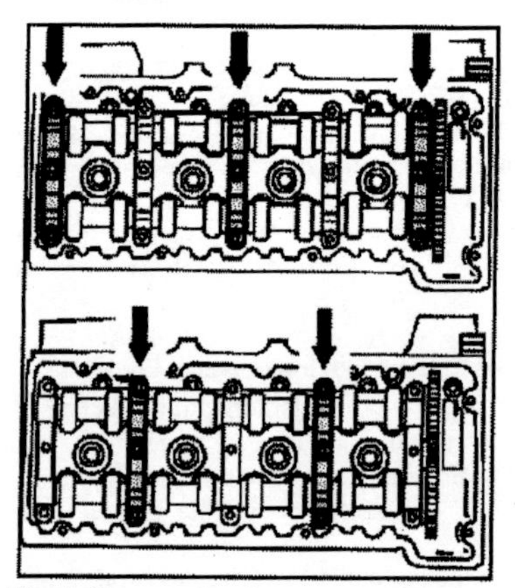

Fig. 1.86 – Removing the camshaft bearing caps of a four-cylinder engine (604). First the caps in the upper view, followed by the caps in the lower view.

- Insert the locking pin (arrester pin) as shown in Fig. 1.73 through the first camshaft bearing cap and into the hole of the inlet camshaft sprocket.
- Mark the timing chain and the camshaft sprocket at opposite points with paint to ensure installation in the same position.
- Remove the chain tensioner as described earlier on.
- Remove the three bolts securing the camshaft sprocket from the exhaust camshaft and withdraw the sprocket from the shaft with the chain engaged.

The bolts securing the camshaft bearing caps must be slackened in a certain order. Each bearing cap is marked with a number, which shows where each cap is located. Follow the sequence below to slacken the camshaft bearing caps.

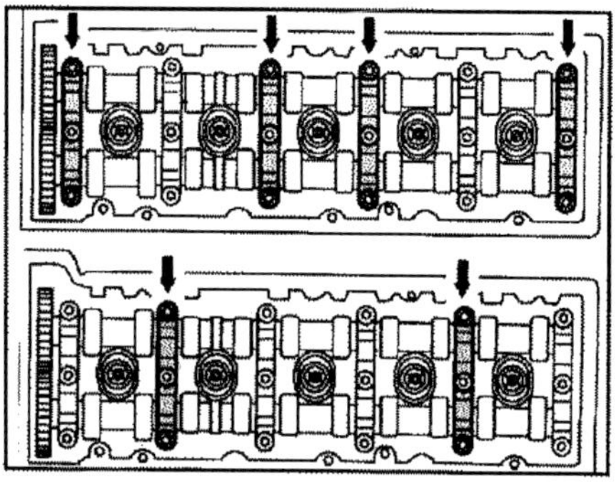

Fig. 1.87 – Removing the camshaft bearing caps of a five-cylinder engine (605). First the caps in the upper view, followed by the caps in the lower view.

- ***In the case of a four-cylinder engine (604)*** refer to Fig. 1.86 remove the bearing cap bolts shown by the arrows, but do not slacken the No. 2 and 4 bearing caps. Remove the caps.
- Refer to the lower view of Fig. 1.86 and slacken the bearing cap bolts of No. 2 and 4 (arrows) bearing caps a turn at a time until all pressure has been released. The camshafts must not be rotated during the removal of the caps. Remove the caps after they are free.

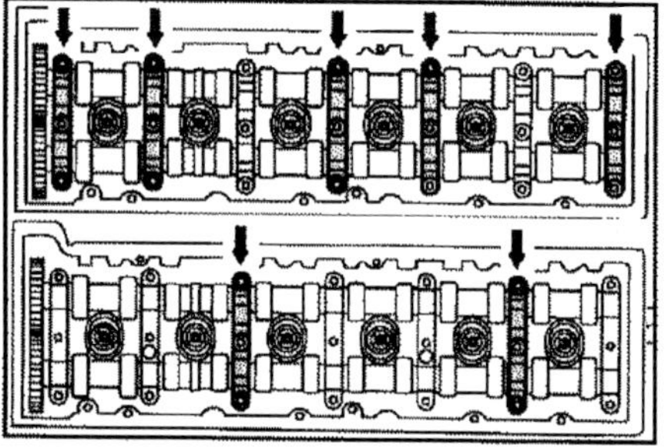

Fig. 1.88 – Removing the camshaft bearing caps of a six-cylinder engine (606). First the caps in the upper view, followed by the caps in the lower view.

- ***In the case of a five-cylinder engine (605)*** refer to Fig. 1.87 and remove the bearing cap bolts shown by the arrows, but do not slacken the No. 2 and 5 caps. Remove the caps.
- Refer to the lower view of Fig. 1.87 and slacken the bearing cap bolts of No. 2 and 5 (arrows) bearing caps turn at a time until all pressure has been released. The camshafts must not be rotated during the removal of the caps. Remove the caps after they are free.
- ***In the case of a five-cylinder engine (606)*** refer to Fig. 1.88 and remove the bearing cap bolts shown by the arrows, but do not slacken the No. 3 and 6 caps. Remove the caps.
- Refer to the lower view of Fig. 1.88 and slacken the bearing cap bolts of No. 3 and 6 (arrows) bearing caps turn at a time until all pressure has been released. The camshafts must not be rotated during the removal of the caps. Remove the caps after they are free.
- Lift out the two camshafts and the two adjusting washers fro the camshaft end float from the centre bearing.
- Unscrew the camshaft housing and lift it off. Bolts are also located in the centre of the removed camshaft bearing caps. Remove the sealing rings underneath the housing. It is not necessary to remove the tappets when removing the camshaft housing as a tab prevents the flat tappets from falling out.

The installation of the camshafts is carried out as follows:

- Place the camshaft housing over the cylinder head with the four "O" sealing rings between housing and cylinder head. Tighten the bolts in the centre to 1.5 kgm (11 ft.lb.).
- Insert the washers for the end float control and place the camshafts into the well oiled bearing bores. Rotate the shafts a few times to settle the bearing journals in

the bearing bores. After fitting both shafts rotate them until the two timing marks shown in Fig. 1.74 are opposite each other (the illustration shows the sprockets fitted to the shaft ends). The markings on the camshaft and camshaft bearing cap must also be in line.

Fig. 1.89 – Sectional view of the front end of the two camshafts (604, 605, 606 engines).

1 Exhaust camshaft
2 Inlet camshaft
3 Camshaft bearing cap
4 Bolt, 1.5 kgm
5 Exhaust camshaft sprocket
6 Drive gear for inlet camshaft sprocket
7 Inlet camshaft sprocket
8 Dowel pin
9 Torx-head bolt, 1.8 kgm
10 Duplex roller timing chain

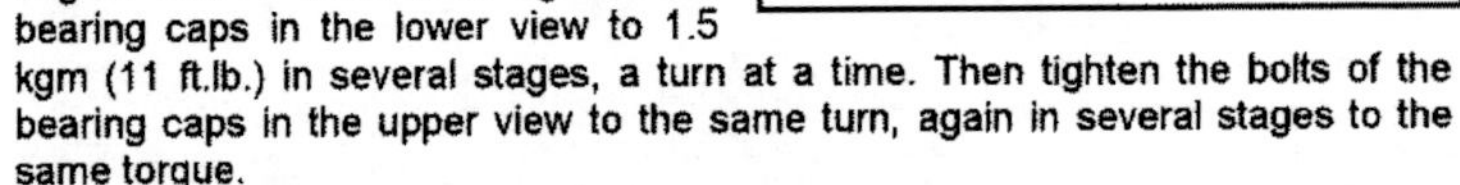

- Fit the camshaft bearing caps in accordance with the numbers. Refer to Figs. 1.86 to 1.88, depending on the engine and in each case tighten the bearing caps in the lower view to 1.5 kgm (11 ft.lb.) in several stages, a turn at a time. Then tighten the bolts of the bearing caps in the upper view to the same turn, again in several stages to the same torque.

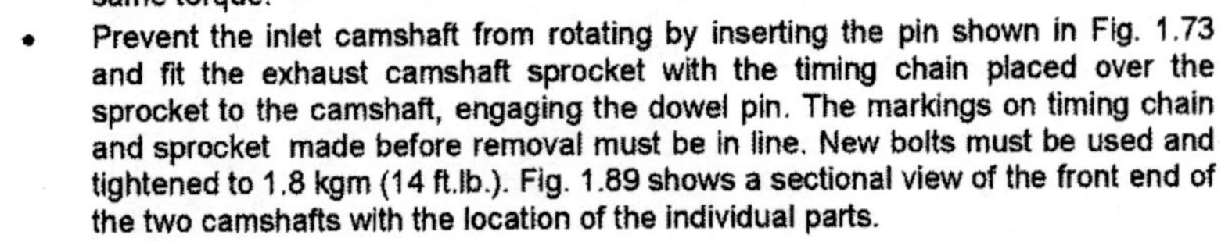

- Prevent the inlet camshaft from rotating by inserting the pin shown in Fig. 1.73 and fit the exhaust camshaft sprocket with the timing chain placed over the sprocket to the camshaft, engaging the dowel pin. The markings on timing chain and sprocket made before removal must be in line. New bolts must be used and tightened to 1.8 kgm (14 ft.lb.). Fig. 1.89 shows a sectional view of the front end of the two camshafts with the location of the individual parts.

1.4.5. Valve Timing

As the valve timing cannot be adjusted, ft is sufficient to obtain the timing mark alignment in to ensure that valves, cams, etc. are in the correct position. Only re-start the engine after you have carried out the check.

1.4.6. CYLINDER BLOCK

The cylinder block consists of the crankcase and the actual block with the cylinder bores. Special attention should be given to the cylinder block at each major overhaul of the engine, irrespective of whether the bores have been re-machined or not.

Thoroughly clean all cavities and passages and remove all traces of foreign matter from the joint faces. If any machining of the bores has taken place, it is essential that all swarf is removed before assembly of the engine takes place.

Measurement of the cylinder bores should be left to an engine shop, as they have the proper equipment to measure cylinder bores. It is, however, feasible to check the cylinder block face for distortion, in a similar manner as described for the cylinder head. The max. permissible distortion is 0.05 mm.

1.4.6.1. Replacing Welsh Plugs in the Crankcase

Welsh plugs are fitted into the side of the cylinder block. These plugs will be "pushed" out if the coolant has been allowed to freeze and can be replaced with the engine fitted, provided the special tool 102 589 07 15 00 can be obtained.

Welsh plugs can be replaced as follows:

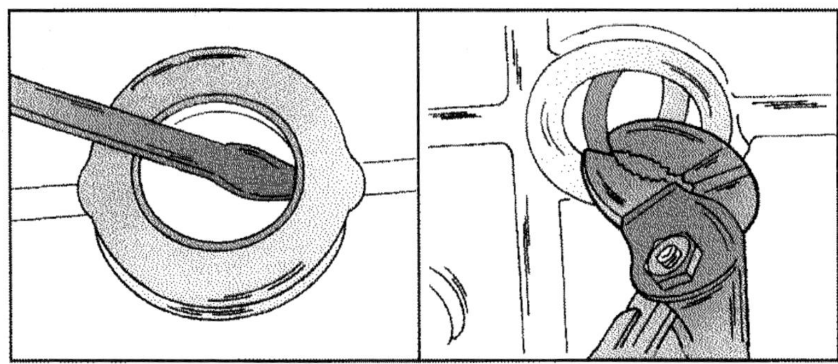

Fig. 1.90 – Removal of a welsh plug. Operation 1 on the left, operation 2 on the right.

- Drain the cooling system and remove all parts obstructing the welsh plug in question, i.e. transmission, intermediate flange, injection pump, etc.
- The removal of a welsh plug is carried out in two stages. First place a small chisel or strong screwdriver blade below the lip of the welsh plug, as shown in Fig. 1.90 on the L.H. side and push the screwdriver in the direction of the arrow until it has swivelled by 90° Then grip the plug with a pair of pliers, as shown in Fig. 1.90 on the right and remove it.

Thoroughly clean the opening in the cylinder head from grease and coat the locating bore with "Loctite" (obtain from a dealer if possible). Fit the large new welsh plug with the special tool mentioned above or use a drift of suitable size and drive it in position until flush with the cylinder block face.

Refit all removed parts and allow the vehicle to stand for at least 45 minutes before the cooling system is filled and the engine is started. Then start the engine and check for coolant leaks.

1.5. Tightening Torque Values

Note: Most tightening torque values are given during the removal and installation of the various engine parts. The following is a list of the most important values:

Cylinder Head Bolts (see Figs. 1.13 and 1.14):
- 1st stage: .... 1.5 kgm (10 ft.lb.)
- 2nd stage: .... 3.5 kgm (25 ft.lb.)
- 3rd stage: .... Angle-tighten by 90°
- 4th stage: .... Wait 10 minutes
- 5th stage: .... Angle-tighten by 90°

Cylinder head bolts (in timing case): .... 2.5 kgm (18 ft.lb.)

Connecting Rod Bearing Caps:
- 1st stage: .... 4.0 kgm (29 ft.lb.)
- 2nd stage .... Angle-tighten 90 to 100°

Main Bearing Cap Bolts:
- 1st stage .... 5.5 kgm (40 ft.lb.)
- 2nd stage .... Angle-tighten 90° (quarter of a turn)

Camshaft sprocket bolts:
- 602 engine: .... 2.5 kgm (18 ft.lb.) + 90°
- 604, 605, 606 engine (new bolts): .... 1.8 kgm (13 ft.lb.)

Camshaft bearing caps:
- 602 engine: .... 2.5 kgm (18 ft.lb.) + 90°
- 604, 605, 606 engine: .... 1.5 kgm (11 ft.lb.)

Camshaft bearing housing:
- 604, 605, 606 engines: .... 1.5 kgm (11 ft.lb.)

Chain tensioner.. 8.0 kgm (58 ft.lb.)
Cylinder head cover bolts:.. 1.0 kgm (7.2 ft.lb.)
Injectors into cylinder head:
- 602 engine:.. 8.0 kgm (58 ft.lb.)
- 604, 605, 606 engines:..7.0 + 1.0 kgm (58 + 7 ft.lb.)
Threaded ring for pre-combustion chamber in cylinder head:.. 13.0 kgm (94 ft.lb.)
Vacuum pump to cylinder head:.. 1.0 kgm (7.2 ft.lb.)
Centre bolt of crankshaft pulley (bolt head marked "8,8"):
- 1st stage.. 20 kgm (145 ft.lb.)
- 2nd stage..Angle-tighten 90°
Centre bolt of crankshaft pulley (bolt head marked "10.9"):
- 1st stage.. 32.5 kgm (234 ft.lb.)
- 2nd stage..Angle-tighten 90°
Crankshaft pulley to hub:.. 2.3 kgm (16.5 ft.lb.)
Flywheel or driven plate – 602 engine:
- 1st stage:.. 3.5 kgm (25 ft.lb.)
- 2nd stage..Angle-tighten by 90 to 100°
Flywheel or driven plate – 604, 605, 606 engines, Torx-head bolts:
- 1st stage:.. 4.5 kgm (32.5 ft.lb.)
- 2nd stage..Angle-tighten by 90 to 100°
Oil sump drain plug:
- M12 thread:.. 3.0 kgm (22 ft.lb.)
- M14 thread:.. 2.5 kgm (18 ft.lb.)
Oil sump to cylinder block:
- M6 thread:.. 1.0 kgm (7.2 ft.lb.)
- M8 thread:.. 2.5 kgm (18 ft.lb.)
Rear oil seal flange:.. 0.9 kgm (7 ft.lb.)
Centre bolt of fan to water pump (L.H. thread):.. 4.5 kgm (32.5 ft.lb.)
Rear oil seal flange:.. 0.9 kgm (7 ft.lb.)
Front oil seal cover:.. 1.0 kgm (7.2 ft.lb.)
Alternator mounting bracket to cylinder block:
- Bolt inserted from front:.. 2.5 kgm (18 ft.lb.)
- Bolt inserted from the side:.. 4.5 kgm (32.5 ft.lb.)
Bolt for T.D.C. transmitter (sensor):.. 1.0 kgm (7.2 ft.lb.)
Oil drain plug in sump:.. 3.0 kgm (21 ft.lb.)
Exhaust manifold to cylinder head:.. 2.5 kgm (18 ft.lb.)
Centre bolt, fan to water pump:.. 4.5 kgm (32.5 ft.lb.)
Engine mounting bracket to engine mounting, front.. 8.3 kgm (60 ft.lb.)
Engine mounting, side, to gearbox.. 4.5 kgm (32 ft.lb.)
Engine mounting, rear, to mounting bracket.. 4.5 kgm (32 ft.lb.)
Exhaust manifold to exhaust pipe flange:.. 3.0 kgm (22 ft.lb.)
Exhaust pipe to transmission bracket:.. 2.0 kgm (14,5 ft.lb.)
Oil pipes to automatic transmission:.. 2.5 kgm (18 ft.lb.)
Earth strap to chassis:.. 2.3 kgm (17.0 ft.lb.)
Coolant drain plug (block):.. 3.0 kgm (21.5 ft.lb.)
Water pump to cylinder block:.. 1.0 kgm (7.2 ft.lb.)
Pulley to water pump:.. 1.0 kgm (7.2 ft.lb.)
Thermostat housing cover:.. 1.0 kgm (7.2 ft.lb.)
Oil filter cover to oil filter:.. 2.5 kgm (25 ft.lb.)
Injection pump bolts:.. 2.0 kgm (14 ft.lb.)
Injection pump drive sprocket:.. 2.5 kgm (25 ft.lb.)
Fuel filter housing bolts:.. 2.5 kgm (25 ft.lb.)
Timing case cover:
- M6 thread:.. 1.0 kgm (7.2 ft.lb.)
- M8 thread:.. 2.5 kgm (18 ft.lb.)
TDC sensor bolt:.. 1.9 kgm (7.2 ft.lb.)
Oil pump drive sprocket bolt:.. 2.5 kgm (25 ft.lb.)

1.6. Lubrication System

The lubrication system used in all engines is a pressure-feed system. A gear-type oil pump is driven via the crankshaft by means of a separate chain. Fig. 1.91 shows where you will find the oil pump and drive chain with the front parts of the engine exposed. Fig. 1.92 shows the component parts of the oil pump, but note that some of the parts are not fitted to all engines.

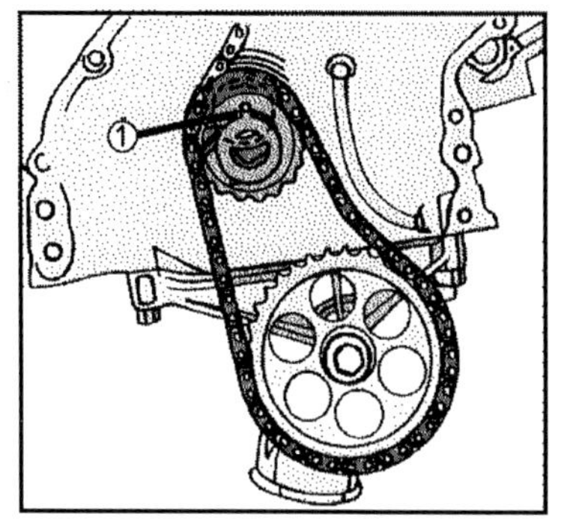

Fig. 1.91 – Vie of the fitted oil pump. The Woodruff key (1) guides the pump drive sprocket.

A warning light in the instrument panel will light up, when the oil level is approaching the lower limit of the oil dipstick. At least 1 litre of oil should be filled in as soon as possible.

Oil pressure indication is electrically by means of a contact switch, fitted to the lower part of the oil filter. Increased oil pressure increases the resistance in the switch and changes the reading in the instrument accordingly.

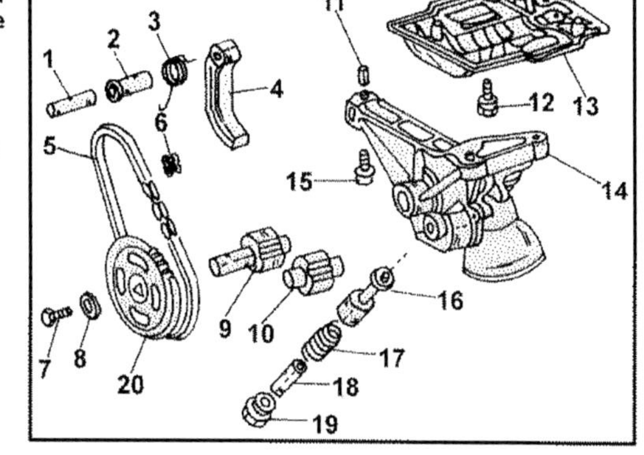

Fig. 1.92 – Component parts of the oil pump. Not all parts are fitted to all engines.

1 Dowel pin, chain tensioner
2 Guide bush
3 Tensioning spring
4 Chain tensioner, pump chain
5 Pump drive chain
6 Chain lock
7 Bolt, pump sprocket
8 Washer
9 Drive shaft with gearwheel
10 Driven gearwheel
11 Dowel sleeve
12 Bolt, oil baffle plate
13 Oil baffle plate
14 Oil pump
15 Bolt, oil pump
16 Plunger, relief valve
17 Pressure spring 18 Pressure pin 19 Closing plug 20 Pump drive sprocket

The oil filter is fitted in upright position to the cylinder block. The oil flows through the filter element from the outside to the inside. A return shut-off valve prevents oil from flowing back through the oil pump into the oil sump when the engine is switched off. A by-pass valve opens when the pressure differential between the dirty and the clean end the filter exceeds a certain value. The oil is then directed to the oil gallery without being cleaned. It should be noted that the oil filter element must be changed after 600 - 1000 miles if the engine has been overhauled.

The oil pump can be removed with the engine fitted to the vehicle. Note that the removal of the sump, however, is not the same on all engines. We would also like to point that the removal of the oil sump and the oil pump is a fairly complicated operation and we advise you to read the instructions in full before commencing.

The oil pump cannot be repaired and must be replaced in case of wear or damage.

1.6.0. TECHNICAL DATA

Oil Capacities:.. See Page 10
Oil pressure at idle speed: ....7 psi.
Oil pressure at 3000 rpm:...43 psi.

1.6.1. OIL SUMP - REMOVAL AND INSTALLATION

The workshop uses special equipment to lift the power unit put of its mountings when the oil sump is removed. Under DIY conditions you will either need to lift up the engine with a jack or use a suitable lifting hoist (hand crane with beam) to release the tension on the mountings when the latter are removed. As already mentioned different operations are necessary to remove the sump, depending on the engine.

602 Engine

We must point out that both front springs must be removed on the model with the 602 engine in order to remove the oil sump. Therefore only remove the oil sump if absolutely necessary.

- Remove the bottom engine compartment trim panel, disconnect the battery earth cable.
- Drain the engine oil.
- Remove the top charge air pipe at the front.
- Remove the fan shroud (section "Cooling System"),
- Remove the engine trim panel at the top of the cylinder head and disconnect the upper charge air pipe.
- At the front left of the frame floor on the outside disconnect an electrical cable.
- Remove the front suspension springs as described in the relevant section.
- Unbolt the front engine mounts on both sides of the engine. Tighten the bolts to 3.5 kgm (25 ft.lb.) during installation. The engine must be lifted out of its mountings in order to carry out the operation or the front suspension support brackets must be lifted up with a suitable appliance.
- Unscrew the clamp bolt from the steering shaft universal joint and separate the joint. The bolt is tightened to 2.0 kgm (14 ft.lb.) during installation.
- Unbolt the front suspension support bracket and lower the engine.

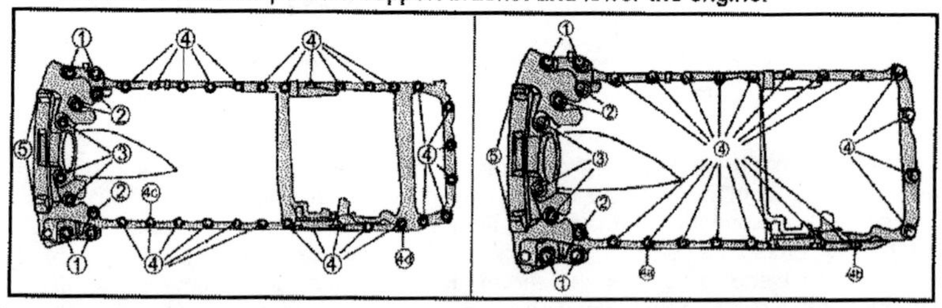

Fig. 1.93 – The location of the various oil sump bolts. On the L.H. side for the 602 and 605 engine, on the R.H. side for the 604 engine.

1 M8 x 40 mm bolt and washer
2 M6 x 35 mm bolt and washer
3 M6 x 85 mm bolt and washer
4 M6 x 20 mm bolt and washer
4a M6 x 20 mm hexagon socket bolt (A/T)
4b M6 x 75 mm hexagon socket bolt (A/T)
4c M6 x 20 mm hexagon socket bolt (A/T)
5 M10 x 40 mm bolt and washer

- Disconnect the cable plug from the oil level sensor at the side of the oil sump.

- Remove the bolts securing the oil sump. Bolts have a different length and a different diameter. Make a note where they are inserted. Fig. 1.93 shows on the L.H. side the diameter and length of the bolts (just in case), also valid for the 605 engine.
- Remove the oil sump and take off the gasket. The sump is taken out towards the front in the direction of travel. You may have to rotate the crankshaft slightly in order to clear all crankshaft webs. The sump gasket must always be replaced.

The installation is a reversal of the removal procedure. Clean the sealing face of the sump and the cylinder block before the sump is refitted. Coat the sump face with sealing compound. Before inserting the bolts (refer to Fig. 1.93) and tightening of the bolts push the sump towards the rear until it contacts the transmission. Tighten the M6 bolts to 1.0 kgm (7 ft.lb.) and M8 bolts to 2.5 kgm (18 ft.lb.) in several stages around the sump face. Finally fill the engine with the correct quantity of engine oil.

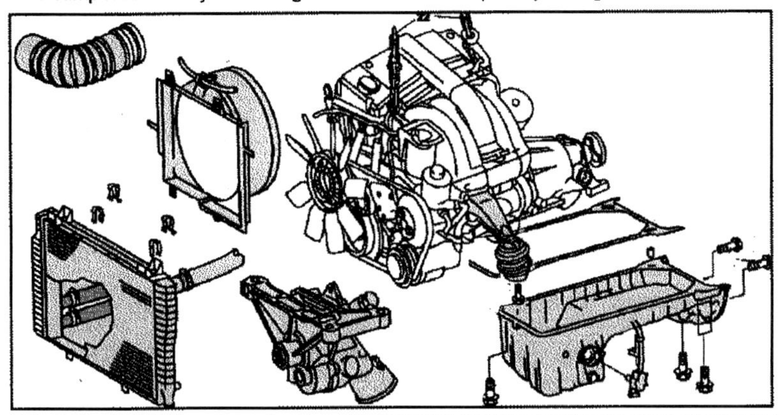

Fig. 1.94 – Parts to be removed as described below to remove the oil sump (and oil pump) of a 604 engine.

604 Engine

Fig. 1.94 shows the parts to be removed on this engine.

- Disconnect the battery earth cable.
- Detach the fan shroud from its fixtures and place it over the fan.
- Unbolt the power steering fluid pipe from the front side member.
- Disconnect the electrical cable at the front left glow plug relay.
- Detach the radiator from its mountings, as it will be in the way when the engine is raised.
- The engine must be lifted out of its mountings in order to unscrew the front engine mountings from the front axle carrier on both sides of the engine. The bolts are tightened to 3.5 kgm (25 ft.lb.) during installation.
- Disconnect the cable plug (Fig. 1.95from the oil level sensor at the side of the oil sump.
- Remove the bolts securing the oil sump. Bolts have a different length and a different diameter. Make a note where they are inserted. Fig. 1.93 shows on the RH. side the diameter and length of the bolts (just in case).
- Lower the oil sump. The oil pump must be removed in order to take out the sump. The oil sump is now removed towards the front in the direction of travel. Take off the sump gasket. The sump gasket must always be replaced.

The installation is a reversal of the removal procedure. Clean the sealing face of the sump and the cylinder block before the sump is refitted. Coat the sump face with grease to stick it to the crankcase. Before inserting the bolts (refer to Fig. 1.93, R.H. side) and tightening of the bolts push the sump towards the rear until it contacts the transmission. Tighten the M6 bolts to 1.0 kgm (7 ft.lb.) and M8 bolts to 2.5 kgm (18 ft.lb.) in several stages around the sump face. Finally fill the engine with the correct quantity of engine oil.

605, 606 Engines

We must point out that both front springs must be removed on the model with the 605 and 606 engines in order to remove the oil sump. Therefore only remove the oil sump if absolutely necessary. The battery earth cable must be disconnected.

- Remove the bottom engine compartment trim panel, disconnect the battery earth cable.
- Drain the engine oil.
- Unscrew the R.H. charge air pipe from the front axle carrier and in the case of a 605 engine remove the top charge air pipe.
- Remove the top charge air pipe at the front.
- Remove the fan shroud (section "Cooling System") and place it over the fan. The fan shroud must be completely removed in the case of a 606 engine (300 TD).
- Extract the fluid from the power steering pump reservoir and disconnect the two hoses from the reservoir (necessary to lower the front axle carrier).
- At the front left unscrew the electrical cable from the inlet manifold.
- Remove the front suspension springs as described in the relevant section.

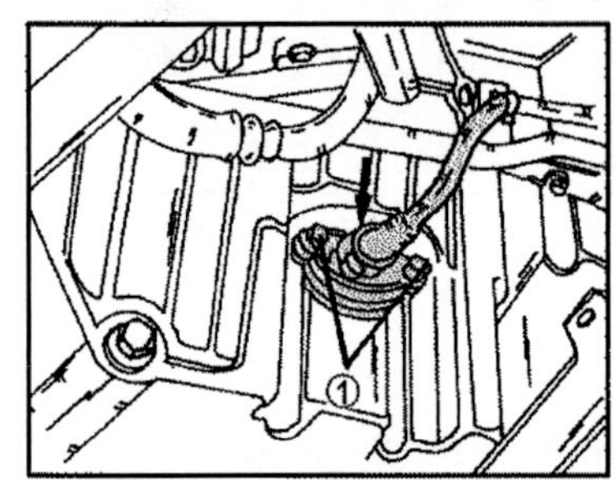

Fig. 1.95 – The oil level sensor is attached to the side of the engine with the two screws (1) – All engines.

- Unbolt the front engine mounts on both sides of the engine. Tighten the bolts to 3.5 kgm (25 ft.lb.) during installation. The engine must be lifted out of its mountings in order to carry out the operation. Place a mobile jack underneath the front axle carrier and lift it up as it must be unbolted.
- Unscrew the clamp bolt from the steering shaft universal joint and separate the joint. The bolt is tightened to 2.0 kgm (14 ft.lb.) during installation.
- Unbolt the front axle carrier from the front of the engine.
- Disconnect the cable plug from the oil level sensor at the side of the oil sump (Fig. 1.95).
- In the case of the 606 engine with automatic transmission drain the fluid and disconnect the fluid line from the transmission.
- Remove the bolts securing the oil sump. Bolts have a different length and a different diameter. Make a note where they are inserted. Fig. 1.96 shows on the L.H. side the diameter and length of the bolts (just in case). Pay attention to the bracket for the oil level sensor electrical cable and the earth cable fastening at the transmission bell housing. On a vehicle with automatic transmission pay attention to the bracket for the automatic transmission and the wiring harness for the transmission.
- Remove the oil sump, attached as shown in Fig. 1.94. The oil sump is now removed towards the front in the direction of travel. The crankshaft may have to

be rotated to clear the crankshaft webs. In the case of an automatic transmission push the fluid lines to one side Take off the sump gasket. The sump gasket must always be replaced.

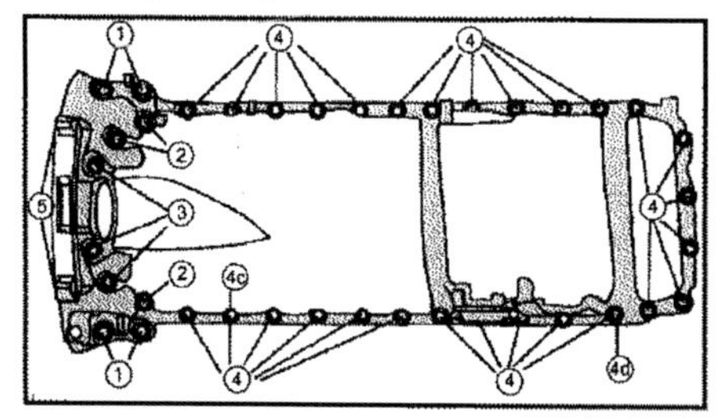

Fig. 1.96 – The length of the oil sump bolts of a 606 engine. Numbers correspond to Fig. 1.93.

The installation is a reversal of the removal procedure. Clean the sealing face of the sump and the cylinder block before the sump is refitted. Coat the sump face with grease to stick it to the crankcase. Before inserting the bolts (refer to Fig. 1.93 or Fig. 1.96, depending on the engine) and tightening of the bolts push the sump towards the rear until it contacts the transmission. Tighten the M6 bolts to 1.0 kgm (7 ft.lb.) and M8 bolts to 2.5 kgm (18 ft.lb.) in several stages around the sump face. The bolts securing the transmission to the bell housing are tightened to 4.0 kgm (29 ft.lb.). Finally fill the engine with the correct quantity of engine oil. Also fill the steering fluid reservoir after connecting the fluid lines and bleed the system.

1.6.2. OIL PUMP - REMOVAL AND INSTALLATION

The removal of the oil pump is the same on all engines, but the pump has been modified during production of the engines. The difference is the oil pump drive sprocket. On the earlier version the sprocket is removed during the removal of the pump, on the second version the pump is removed together with the sprocket. Remove the pump as follows:

- Remove the oil sump as described for the engine in question,
- On the earlier version (bolt in the centre of the sprocket) remove the bolt and remove the pump together with the oil pump drive chain as one assembly. The pump is secured with three bolts, one on one side and two on the other side. The sprocket securing bolt is tightened to 3.2 kgm (23 ft.lb.). The pump bolts to 2.5 kgm (18 ft.lb.). Fig. 1.97 shows the later pump type.

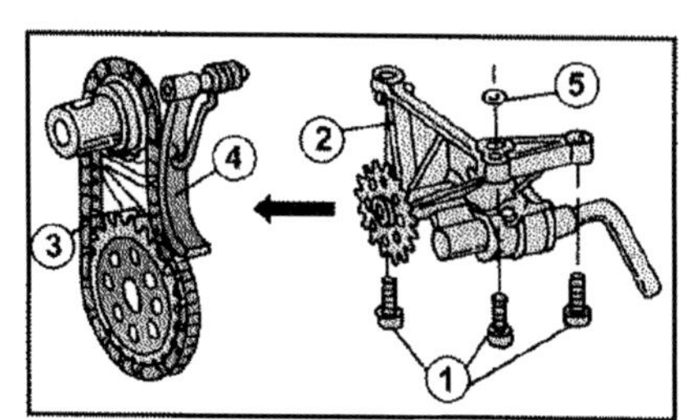

Fig. 1.97 – Removal and installation of the oil pump (all engines).

1 Pump securing bolts, 2.5 kgm
2 Oil pump
3 Oil pump drive chain
4 Tensioning lever
5 Pump oil seal

- On the later version unscrew the three pump securing bolts and remove the pump downwards.

The installation is a reversal of the removal procedure, noting the tightening torques already given above and the following:

- On all types clean the oil strainer and fill the pump before engine oil before installation.
- On the earlier version fit the pump drive sprocket with the flat on the pump drive shaft and the flat inside the sprocket aligned in accordance with Fig. 1.79. On this type you will find two dowel sleeves on the two front bolts which must engage

when the pump is placed against the crankcase. Hexagon bolts with washers are used to secure the pump.

- On the later version an "O" sealing ring (5, Fig. 1.97) is fitted between oil pump and crankcase which must be inserted into the groove (inside the pump). Torx-head bolts (size T45) are used to secure the pump.
- Finally refit the oil sump as described and fill the engine with oil.

1.6.3. OIL PUMP - REPAIRS

Oil pumps should not be dismantled and/or repaired. If a new pump is fitted check the drive chain and the chain sprocket to prevent fitting a new pump together with new drive components. Also make sure that the correct pump type is fitted (earlier tap or later type).

1.6.4. OIL FILTER

Fig. 1.98 shows where the oil filter together with the oil-water heat exchanger is located in the case of a 604, 605 and 606 engine, in the latter case for models without turbo charger. A fuel pre-heater is fitted to the 602 engine, which is not shown in the illustration. The filter of the 606.962 (300 TD) is covered under separate heading.

The following text describes the removal and installation of the filter as applicable to the various engines.

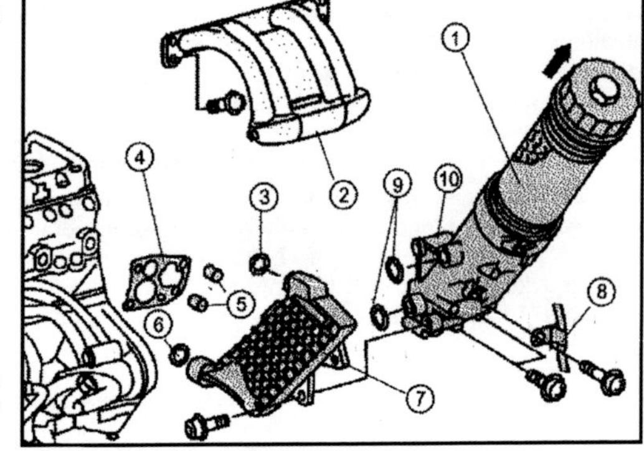

Fig. 1.98 – Oil filter installation on models with heat exchanger.
1 Oil filter housing
2 Intake manifold
3 Sealing ring
4 Gasket
5 Dowel sleeves
6 Sealing ring
7 Oil-water heat exchanger
8 Bracket for electric cable
9 Sealing rings
10 Oil filter attachment bracket

605, 605, 606 Engines (except 606 engine for 300 TD)

- Disconnect the battery earth cable.
- Remove the intake manifold in the case of a 604 engine and the resonance intake manifold on the 606.912 engine (E300D). On the 605 engine remove the charge air distribution pipe. The operations are described during the removal of the cylinder head.
- Drain the engine oil by unscrewing the oil filter cap and removing the filter element. The oil will be allowed to return to the oil sump.
- Drain the cooling system on an engine with oil-water head exchanger on the filter.
- On the 605 engine there is a bracket for the intake manifold support which must be removed. Remember to refit it after the manifold has been installed.
- Remove the starter motor as it will not be possible to reach the lower bolt of the filter housing with the starter motor fitted.
- Unbolt the oil filter housing from the engine and lift it out together with the oil-water head exchanger. If necessary unscrew the heat exchanger from the filter

housing. The "O" sealing rings must be replaced during installation. The heat exchanger must be fitted over the dowel sleeves, the gasket must always be replaced.

The installation is a reversal of the removal procedure. Finally fill the engine with oil.

606 Engine (300 TD)

Fig. 1.99 shows the parts to be removed on this engine, i.e. an engine with oil cooler. The removal and installation is similar as described above, with the following differences. Some operations may be repeated for easier understanding. There is no need to remove the intake manifold or the resonance intake manifold.

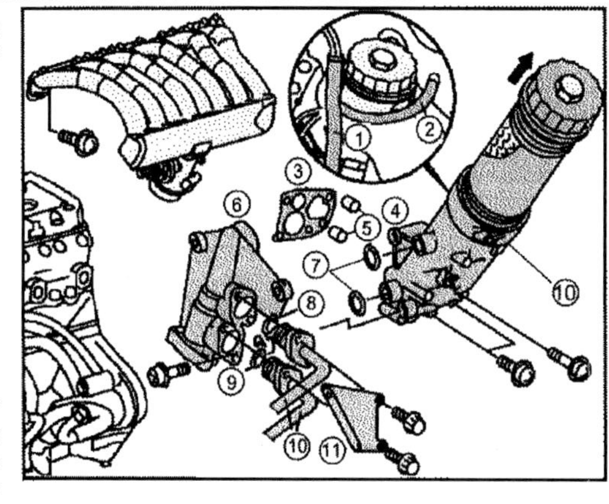

Fig. 1.99 – Oil filter installation of a 606 engine with turbo charger.
1 Cable duct
2 Coolant pipe
3 Gasket, always replace
4 Oil filter housing attachment
5 Dowel pins
6 Connecting piece for engine oil cooler lines
7 Sealing rings
8 Sealing ring
9 Sealing ring
10 Oil filter housing
11 Bracket, EGR pipe

- Remove the charge air distribution pipe as described during the removal of the cylinder head.
- After draining the filter as described above disconnect the oil cooler lines (10) from the connecting piece (6). Some oil will flow out and must be collected. The oil sealing "O" rings (8) and (9) must be replaced during installation.
- Remove the bracket (11) for the EGR r-circulation pipe.
- Unbolt the cable duct of the engine wiring harness (1) and the coolant pipe (2) at the oil filter housing.
- Unplug the cable connector from the crankshaft position sensor.
- Remove the connecting piece (6) for the engine oil cooler lines.
- Remove the starter motor and remove the oil filter housing (10) as described above. No oil-water heat exchanger is fitted.

Installation is a reversal of the removal procedure. Finally fill the engine with oil.

602 Engine

The removal and installation is carried out in a similar manner as described for the 604, 605 and 606 engine, but on the side of the oil-water heat exchanger you will find the fuel pre-heater which must be removed together with the connected fuel lines. Installation is a reversal of the removal procedure. Tighten the filter housing to 2.5 kgm (18 ft.lb.) and the pre-filter to 1.0 kgm (7.2 ft.lb.).

Oil Filter Element – Replacement

The replacement of the oil filter element is straight forward. The workshop uses a special wrench to apply to the hexagon at the top of the filter. Otherwise use s suitable socket and extension. A torque wrench is necessary for the installation. After unscrewing the cap remove it together with the filter element. The seal underneath the cap must always be replaced. Insert the new filter element into the filter housing, screw on the cap and tighten it to 2.5 kgm (18 ft.lb.).

1.6.4. ENGINE OIL COOLER – E300TD

The oil cooler is located at the front L.H. side of the vehicle, hidden behind the bumper. The removal is rather complicated but can be carried out as follows if required.

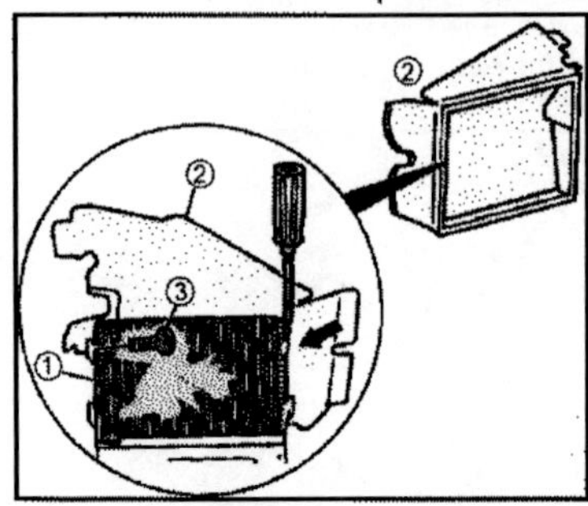

Fig. 1.100 – Removal of the oil cooler. The surround gasket (2) is attached to the oil cooler (1) with the screw (3). The locking detent must be disengaged as shown (arrow).

The L.H. front wheel, the bottom section of the noise encapsulation at the front and the inner wheel housing liner on the left-hand side must be removed to gain access to the oil cooler. After disconnecting the pipes connected to the cooler (oil will run out), unscrew the bolt securing the front bumper bracket inside the L.H., front wing and turn the bumper downwards.

Disconnect the fan motor connector. Unscrew the nuts securing the pipe bracket and remove the oil cooler downwards. The oil cooler is surrounded by a gasket which must be removed. To do this, unscrew a screw on one side and press the locking detent with a screwdriver to one side as shown in Fig. 1.100.

The installation is a reversal of the removal procedure.

1.6.5. ENGINE OIL CHANGE

The engine oil should be changed in accordance with the instructions in your Owners Manual. Remember that there are a few litres of engine oil to handle and the necessary container to catch the oil must be large enough to receive the oil. Dispose of the old oil in accordance with the local laws. You may be able to bring it to a petrol station. **Never discharge the engine oil into a drain.** Drain the oil as follows, when the engine is fairly warm:

- Jack up the front end of the vehicle and place the container underneath the oil sump. Unscrew the oil drain plug (ring spanner or socket). Take care, as the oil will "shoot" out immediately. Remove the oil filler cap to speed-up the draining.
- Check the plug sealing ring and replace if necessary. Clean the plug and fit and tighten to 3.0 kgm (22 ft.lb.).
- Fill the engine with the necessary amount of oil. Make sure that the oil is suitable for diesel engines.
- Refit the oil filler cap and drive the vehicle until the engine operating temperature is reached. Jack up the vehicle once more and check the drain plug area for oil leaks.

1.6.6. ENGINE OIL PRESSURE

The oil pressure can only be checked with an oil pressure gauge, which is fitted with a suitable adapter in place of the oil pressure switch. We recommend to leave the oil pressure check to a workshop. Low oil pressure can also be caused through a low oil level in the sump.

1.7. Cooling System

The cooling system operates with an expansion tank, at the R.H. side of the engine compartment. A coolant level indicator is fitted into the expansion tank. If the level drops below the "Min" mark for any reason, the switch contacts will close and light up a warning light in the instrument panel. A check of the coolant level is therefore redundant.

The water pump is fitted to the front at the bottom of the cylinder block, is sealed off by means of an "O" sealing ring and secured by means of four bolts. The pump cannot be repaired. The thermostat is mentioned later on.

1.7.0. TECHNICAL DATA

Type: .... Water pump-assisted thermo-siphon system with impeller-type water pump
Filling Capacity: .... See Page 10
Anti-freeze amount: .... See Section 1.7.1.2
Thermostat:
- Opens at .... 87° C
- Fully opens at .... 102° C

1.7.1 COOLANT - DRAINING AND REFILLING

- If the engine is hot open the expansion tank cap or the radiator cap to the first notch and allow the pressure to escape. The coolant must have a temperature of less than 90° C. but we recommend to wait until the temperature has dropped to 50° C. In all cases use a thick rag to cover the expansion tank cap to protect your hand.
- Remove the noise dampening panel from underneath the vehicle.
- Unscrew the coolant drain plug at the bottom of the radiator. You can push a hose of 12 mm inner diameter over the nipple (1) in Fig. 1.101 and guide the hose into a container to drain and collect the coolant. If the anti-freeze solution is in good condition, your can re-use it. Poke a piece of wire into the bore of the drain hole to dislodge sludge, if the coolant flow is restricted. There is also a plug in the side of the cylinder block, which can be drained as described above (14 mm hose). After draining the coolant tighten the radiator plug to 1.5 kgm, the cylinder block plug to 3.0 kgm.

Fig. 1.101 – The arrow shows the small connection where the hose can be connected before the drain plug (1) is opened.

To ensure that the cooling system is filled without air lock, proceed as follows when filling in the coolant. Refer to Section 1.7.1.0 for the correct anti-freeze amount to be added. Anti-freeze marketed by Mercedes-Benz should be used, as this has been specially developed for the engine.

- Set both heater switches to the max. heating capacity, by moving the controls. If an automatic climate control system is fitted, press the "DEF" button.
- Fill the pre-mixed anti-freeze solution into the expansion tank filler neck or the radiator until it reaches the "Cold" mark. Do not fit the expansion tank or radiator cap at this stage.

- Start the engine and run it until the operating temperature has been reached, i.e. the thermostat must have opened. Fit the cap when the coolant has a temperature between 60 to 70° C. A thermometer can be inserted into the expansion tank filler opening to check the temperature.
- Check the coolant level after the engine has cooled down and correct if necessary.
- Refit the noise dampening panel underneath the vehicle.

1.7.1.0. Anti-freeze Solution

The cooling system is filled with anti-freeze when the vehicle leaves the factory and the solution should be left in the system throughout the year. When preparing the anti-freeze mixture, note the following ratio between water and anti-freeze solution. We recommend to use the anti-freeze supplied by Mercedes-Benz. It may cost you a little more, but your engine will thank you for it. Refer to page 10 for the total capacity of the engine in question. The following ratios should then be observed:

To –30° C:

- 50 % anti-freeze, 50 % of water

1.7.2. RADIATOR AND COOLING FAN

1.7.2.0. Checking Radiator Cap and Radiator

The cooling system operates under pressure. The expansion tank cap is fitted with a spring, which is selected to open the cap gasket when the pressure has risen to value engraved in the cap. If the cap is replaced, always fit one with the same marking, suitable for the models covered.

To check the radiator cap for correct opening, a radiator test pump is required. Fit the pump to the cap and operate the pump until the valve opens, which should take place near the given pressure (1.4 kg/sq.cm). If this is not the case, replace the cap. Fig. 1.79 shows the working principle of such a test pump.

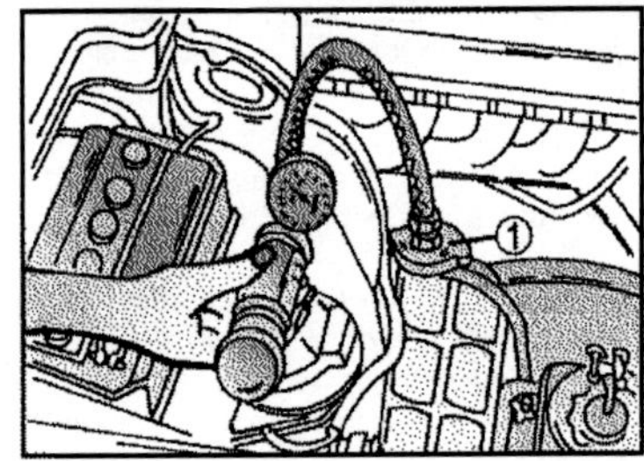

Fig. 1.101 – A radiator test pump (1) is used to check the cooling system for leaks and the expansion tank cap for correct opening. The pump is connected to the expansion tank.

The same pump can also be used to check the cooling system for leaks. Fit the pump to the expansion tank filler neck and operate the plunger until a pressure of 1.0 kg/sq.cm. is indicated. Allow the pressure in the system for at least 5 minutes. If the pressure drops, there is a leak in the system.

1. 7.2.1. Radiator – Removal and Installation

Fig. 1.102 shows details of the parts to be removed to take out the radiator, fitted to all models. When the radiator is replaced make sure you obtain the correct one for your vehicle.

- Disconnect the battery negative cable.
- Remove the bottom part of the noise encapsulation.
- Drain the cooling system as described in Section 1.1.1. On a model with automatic transmission, use clamps and clamp off the two hoses for the oil cooler

(14) and (17). Disconnect the hoses from the R.H. side of the radiator. Some fluid will drip out. The connection (8) must be slackened to separate the hoses. Immediately plug the hose ends and connections in suitable manner.

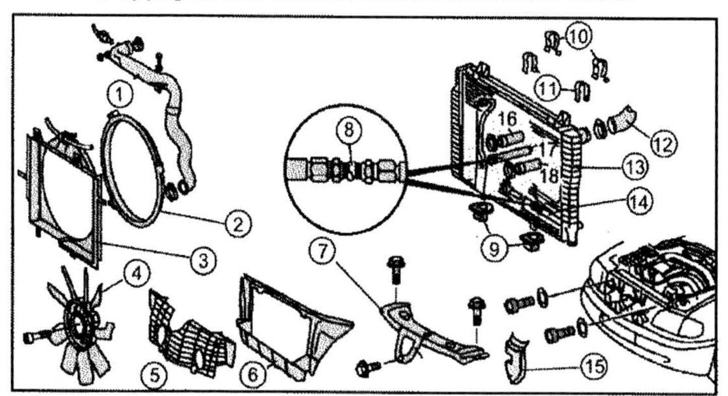

Fig. 1.102 – Details for the removal and installation of the radiator. The numbers are referred to in the text.

- Remove the charge air pipe (1) if a turbo engine is dealt with, as described during the removal of the cylinder head.
- Pull off the flat shaped springs (10) and (11),
- Remove the viscous fan (4) as described below for the different engines.
- Remove the fan shroud (2) and (3).
- Disconnect the coolant hoses (18), (16) and (12) from the radiator elbows. Inspect hoses and hose clamps before re-using them.
- In the case of a fitted A/C system remove the fan grille (5) and the air scoop (15).
- Unscrew the intercooler from the radiator (13) in the case of a turbo diesel engine from the radiator (13), in the case of a vehicle with A/C system separate the condenser from the radiator. The air guide (6) and the bracket (7) must also be removed. The bracket is fitted to models with A/C system.
- Remove the radiator by listing it upwards. To protect the radiator during removal we recommend to make up a protection plate from sheet metal, 1 mm thick and with dimensions of 400 mm x 600 mm and insert it in front of the radiator. During installation insert the studs at the bottom of the radiator into the rubber grommets (9) in the lower crossmember.

The installation is a reversal of the removal procedure. The transmission fluid pipes are tightened to 2.0 kgm (14.5 ft.lb.). Finally refill the cooling system as described in Section 1.7.1. Start the engine and check all cooling hose and drain points for leaks.

1.7.3. WATER PUMP

The water pump is fastened to a light-alloy housing which is bolted to the lower front of the crankcase. To remove the water pump proceed as follows:

- Drain the cooling system as described earlier on.
- In the case of a turbo diesel engine (602/E290 TD) and 605/E250 TD) remove the charge air pipes at the upper and lower ends.
- Remove the viscous fan as described for the different engines (see below).
- Slacken the bolts securing the water pump pulley to the pump hub. The pulley must be prevented from rotating. A screwdriver can be inserted between the bolt

heads as shown in Fig. 1.103 to lock the pulley. Withdraw the pulley from the pump after all bolts have been removed.

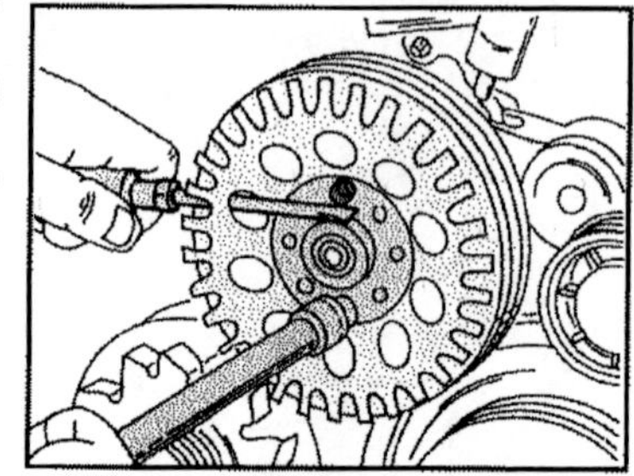

Fig. 1.103 – Counterholding the water pump pulley.

- Remove the poly V-belt as described further on in this section.
- Disconnect the coolant hose from the pump.
- Rotate the crankshaft in the direction of rotation until No. 1 pistons is at top dead centre, indicated when the "OT" (TDC) mark on the crankshaft pulley is opposite the pointer on the timing case cover.
- If a TDC sensor is fitted mark the position of the TDC sensor bracket with a scriber and remove it from the timing case cover. During installation make sure that the marks are in line once more.
- Remove the bolts securing the pump to the cylinder block and take off the pump. Immediately clean the pump and the block faces. The gasket underneath the pump must be replaced.

The installation is a reversal of the removal procedure. The pump is located by a dowel sleeve and must engage. Tighten the pump bolts to 1.0 kgm (7.2 ft.lb.). Different torques apply to the pulley bolts. 1.5 kgm (11 ft.lb.) in the case of a 602 engine or 1.0 kgm (7.2 ft.lb.) in the case of the other engines. Finally fill the cooling system, start the engine and check for leaks.

The water pump cannot be overhauled and must be replaced in case of damage or wear.

1.7.4. VISCOUS FAN CLUTCH

The attachment of the viscous fan clutch is different on the 604 engine fitted to the E200 and E220 and the remaining engines.

604 Engine

Two different types are used. On one type the clutch is fitted with a centre bolt, as shown in Fig. 1.104, on the other type a union nut is used to secure the fan clutch. Remove the fan clutch as follows:

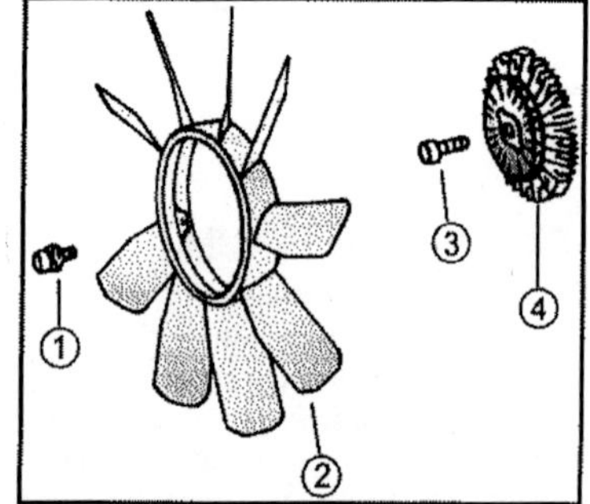

Fig. 1.104 – Cooling fan and viscous clutch.

1 Fan securing bolt, 0.9 kgm
2 Fan blade
3 Fluid clutch securing bolt, 4.5 kgm
4 Fluid clutch

- On the type shown in Fig. 1.104 counterhold the water pump pulley as shown in Fig. 1.103 and remove the fan clutch together with the fan. The bolt has a left-hand thread, must therefore be slackened as you would tighten a bolt.

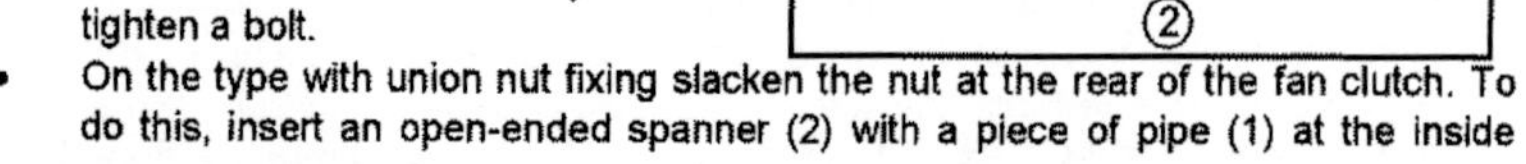

- On the type with union nut fixing slacken the nut at the rear of the fan clutch. To do this, insert an open-ended spanner (2) with a piece of pipe (1) at the inside between the fan unit and the pulley, as shown in Fig. 1.105. The gap is very small and the spanner must have the necessary width (approx. 9 mm). A 36 mm A/F spanner must be used. Again the union nut has a left-hand thread.

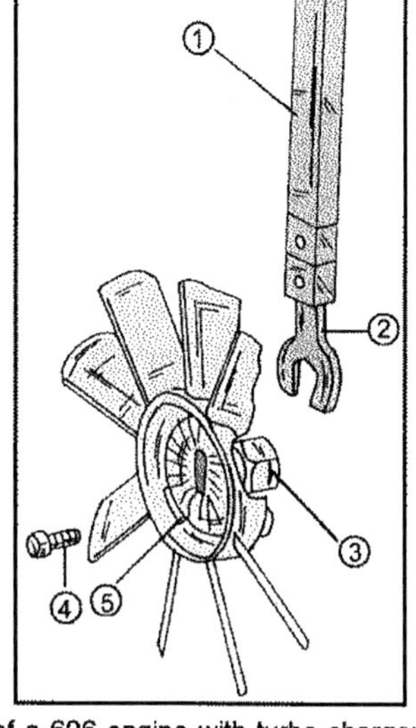

Fig. 1.105 – Removal details for the viscous fan clutch when a union nut is used. The numbers are referred to below.

- Place the spanner over the nut (3) as shown and slacken the fan clutch. Again the clutch must be prevented from rotating.
- The fan (5) can be removed from the fan clutch (4) if required.

The installation is a reversal of the removal procedure. The installation position of the fan is marked by a lug on the fan ring and is marked "Vorne Front" in the direction of travel on the fan ring. The inner nut is tightened to 4.5 kgm, but the torque will have to be estimated as it will be difficult to apply a torque wrench. The fan bolts are tightened to 1.0 kgm (7.2 ft.lb.). The bolt (3) in Fig. 1.104 is tightened to 4.5 kgm.

602, 605, 606 Engines

The fan clutch is secured by a centre bolt as shown in Fig. 1.104. The charge air pipe must be removed to gain access to the fan clutch. Also remove the fan shroud and place it over the fan or in the case of a 606 engine with turbo charger lift it out. The remaining operations are the same as described for the 604 engine. During installation tighten the fan securing bolt to 4.0 kgm (30 ft.lb.).

1.7.5. THERMOSTAT

The thermostat is fitted inside a housing fitted to the water pump housing, as shown in Fig. 1.106. The thermostat is removed as follows:

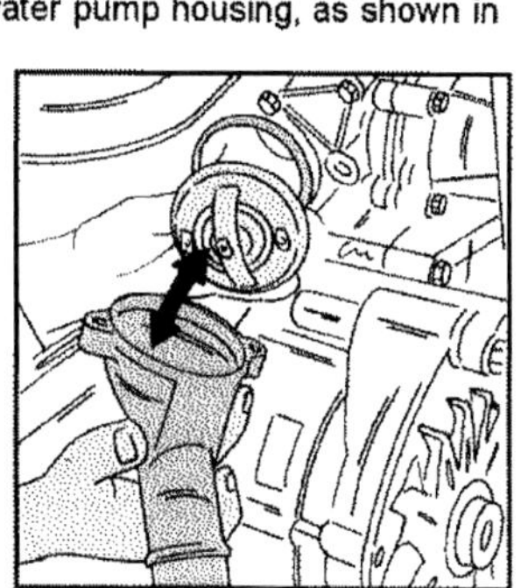

Fig. 1.106 – Thermostat position.

- Drain the cooling system. Always wait until the temperature is below 90° C. Even then remove the expansion tank/radiator cap very slowly. Use a thick rag to protect your fingers and hands. Only turn the cap to its first stop to allow all vapour to escape. It is enough to drain the cooling system to the level of the thermostat.
- Remove two bolts.
- Disconnect the coolant hose after slackening of the hose clamp and remove the thermostat housing from the water pump housing.
- Remove the thermostat from the opening, making a note of the fitted position.

The installation is a reversal of the removal procedure. The sealing ring between thermostat housing and water pump housing must always be replaced.

Checking the Anti-freeze Strength

The frost protection of the anti-freeze should be checked before the beginning of the winter period – know never know or cant remember if you have topped up with plain water during the summer. The anti-freeze tester shown in Fig. 1.107 is one of many available to check the mixture strength. Whichever tester is used it will show you the strength of the mixture in the cooling system and you can correct it as necessary. We strongly recommend the use of anti-freeze marketed by Mercedes-Benz. It may be more expensive, but has been made especially for your engine.

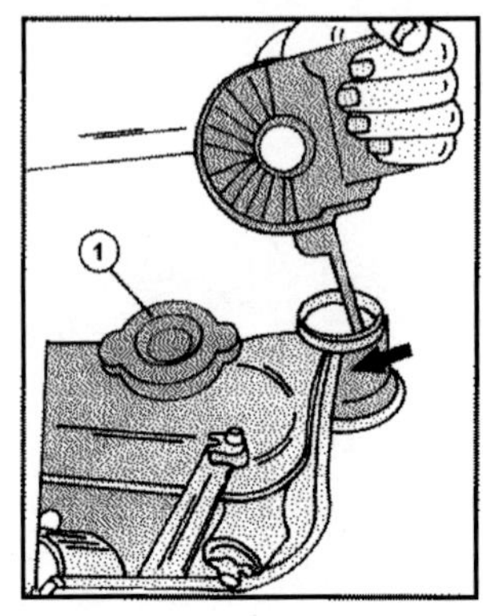

Fig. 1.107 – Checking the anti-freeze strength.

A thermostat cannot be repaired, but can be tested by immersing it in a container of cool water and gradually raising the temperature to check the opening temperature.

Suspend the thermostat on a piece of wire so that it does not touch the sides or the bottom of the container. Suspend a thermometer in a similar manner.

Gradually heat the water and observe the thermometer. The thermostat should begin to open at 87° C and should be fully open at 105° C.

Otherwise replace the thermostat. The thermostat pin must emerge at least 7 mm from the thermostat. Allow the thermostat to cool down and check if it closes properly.

1.7.6. POLY V-BELT – REMOVAL AND INSTALLATION

A single drive belt, also known as poly V-belt, is fitted to the front of the engine, but the layout of the belt is not the same on all engines and depends on a fitted of an air conditioning system or a vehicle without A/C system.

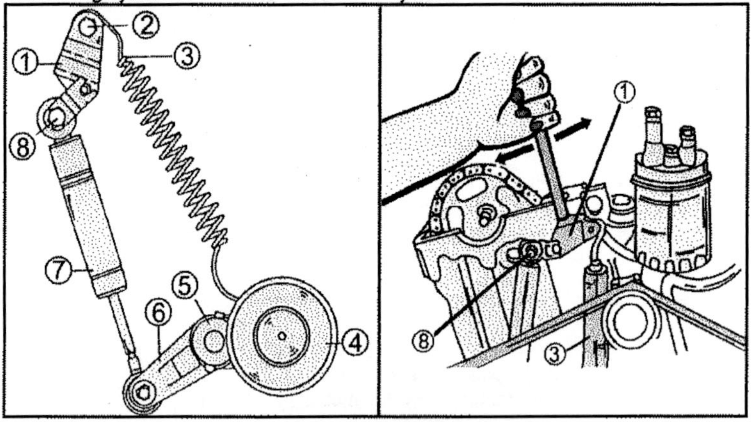

Fig. 1.108 – Removal of the poly V-belt.

1 Tensioning lever
2 Securing nut
3 Tensioning spring
4 Tensioning roller
5 Closing cover
6 Tensioning roller lever
7 Shock absorber
8 Upper shock absorber attachment

If a compressor is fitted it will also be driven by the same belt. The same belt drive system is, however, fitted to all engines and is held in its correct tension by means of a tensioning device.

Different is also the length of the belt, i.e. without A/C system and with A/C system. If a new belt is fitted make sure your obtain the correct one.

On all engines

- In the case of an engine with turbo charger remove the top charge air pipe.
- Remove the viscous fan clutch as already described.
- Slacken the nut (2) securing the spring tensioning lever (1) in the L.H. view of Fig. 1.108 and insert a cylindrical drift of 13 mm diameter into the tensioning lever as shown in the R.H. view. Move the lever towards the left until the bolt can be pushed back and then to the right to slacken the tensioning spring.
- Push the tensioning roller downwards until the belt can be removed.
- Check the removed V-belt for damage and traces of wear before installation. Replace the belt if necessary, fitting the correct belt.

Installation is a reversal of the removal procedure. Place the belt first over the tensioning roller and then over the remaining pulleys in the numbered order shown in the various belt layouts. The belt is finally placed over the water pump pulley. The lever shown in Fig. 1.108 is again used to move the spring tensioning lever.

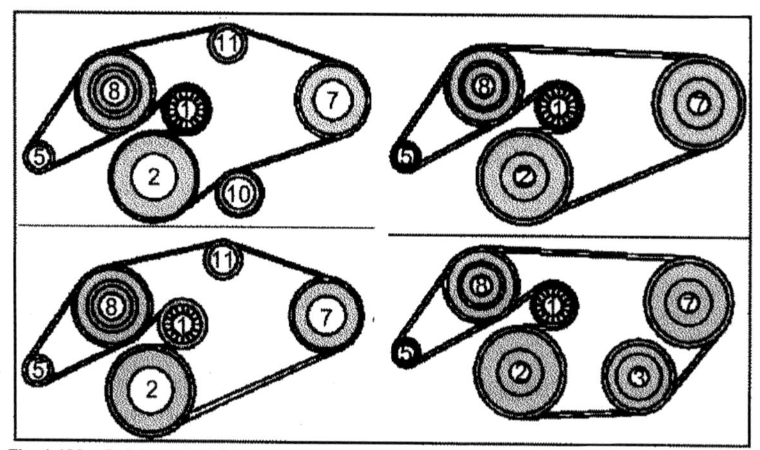

Fig. 1.109 – Belt layout of the various engines. Top left: and bottom left: Engine 602 without A/C, top right: 604, 605, 606 engines without A/C, bottom right: all engines with A/C.

1 Tensioning pulley	5 Alternator	9 A/C compressor
2 Crankshaft	7 Power steering pump	10 Bottom guide pulley
3 A/C compressor	8 Coolant pump	11 Upper guide pulley

1.7.6.1. V-Belt Tensioning Device – Removal and Installation

The component parts of the tensioning device are shown in Fig. 1.110. The assembly can be removed as follows:

- Remove the charge air pipe in the case of a turbo diesel engine (602 and 605 engines.
- Remove the viscous fan clutch as described for the different engines and then remove the poly V-belt as described above.
- Remove the spring tensioning lever and the tensioning spring. The spring has a colour mark which must be at the upper end during installation.
- Unscrew the shock absorber from the cylinder head (Fig. 1.111).

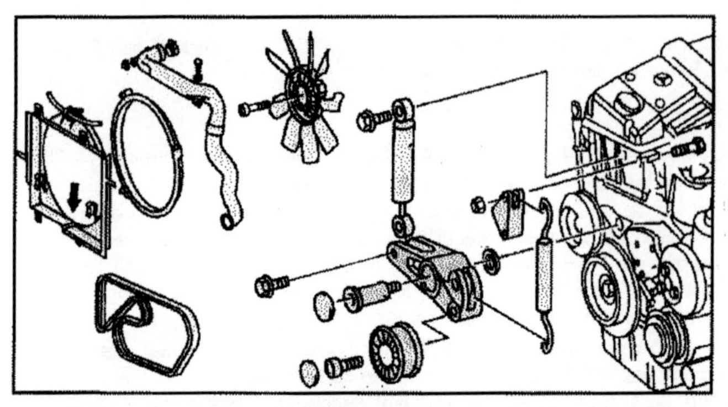

Fig. 1.110 – The component parts of the belt tensioning device.

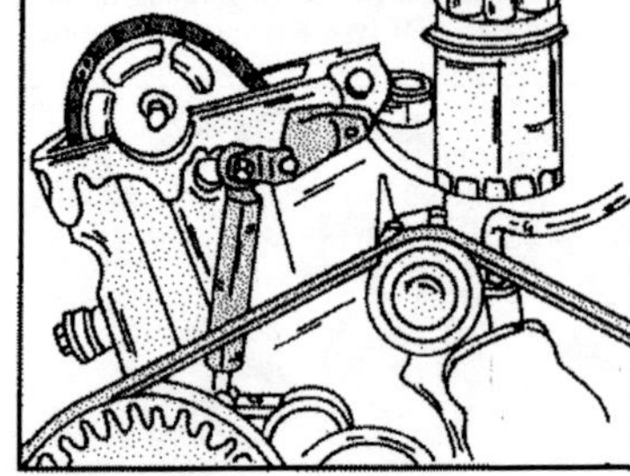

Fig. 1.111 – The shock absorber is attached at the upper end to the cylinder head.

- Remove the end covers from the tensioning pulley and the bearing pin and remove the tensioning pulley.
- Unscrew the bearing pin from the crankcase. Before installation clean the thread as it must be free of grease. Coat the thread with sealing compound. The bolt as specified below.
- Remove the tensioning lever together with the shock absorber from the tensioning roller. If a new shock absorber is fitted, move the piston rod several times up and down to bleed the shock absorber. The piston rod of the shock absorber is fitted at the lower end. Note the shim behind the tensioning lever.

The installation is a reversal of the removal procedure. Due to the different tightening torques the values will be listed below:

604, 605, 606 Engines

Tensioning pulley securing bolt:... 3.0 kgm (22 ft.lb.)
Bearing bolt for tensioning lever: .... 10.0 kgm (72 ft.lb.)
Shock absorber to cylinder head: .... 2.5 kgm (18 ft.lb.)
Shock absorber to tensioning lever: .... 2.0 kgm (14 ft.lb.)

602 Engine

Shock absorber to cylinder head: .... 2.5 kgm (18 ft.lb.)
Shock absorber to tensioning lever: .... 2.0 kgm (14 ft.lb.)
Tensioning pulley securing bolt:... 3.0 kgm (22 ft.lb.)
Bearing bolt for tensioning lever: .... 10.0 kgm (72 ft.lb.)

1.8. Diesel Fuel Injection System

1.8.0. INTRODUCTION

Absolute cleanliness is essential during any repairs or work on the diesel fuel injection system, irrespective of the nature of the work in question. Thoroughly clean union nuts before unscrewing any of the injection pipes.

Injection pump and injectors are the main components of the fuel injection system. The injection system is basically the same for all engines.

A fuel lift pump sucks the fuel out of the tank and delivers it via a fuel filter to the suction chamber in the injection pump. The injection pump delivers the fuel through high-pressure pipes to the injectors and injector nozzles. Diesel injection systems operate on the principle of direct injection or indirect injection. The Mercedes diesel engines covered in the manual operate with indirect injection, i.e. pre-combustion chambers are used in the cylinder head, with the exception of the E290 TD, which is classified as "Turbodiesel direct". The fuel in injected into these pre-combustion chambers which are connected to the main combustion chambers. The combustion of the fuel commences in the pre-combustion chamber. Through the resulting pressure increase, the burning fuel particles are pushed into the main combustion chamber and complete the combustion.

Fig. 1.112 shows the arrangement of the major parts, in this case for the four-cylinder engine.

The fuel injection pump is a piston pump and is fitted to the L.H. side of the cylinder block and is driven from the timing chain. A pump element, consisting of cylinder and piston, serves each of the engine cylinders.

A governor is fitted to the rear of the injection pump. The governor consists of a system of levers and springs and a centrifugal mechanism, fitted to the rear of the injection pump camshaft. A vacuum unit for the switching off of the engine, a vacuum unit for the idle speed increase (if applicable) and a stop lever on the side of the governor all act on the governor. The injection pump cannot be repaired or overhauled and an exchange pump or a new pump must be fitted in case of malfunction or damage.

The adjustment of the injection timing and also the removal and installation of the injection pump requires certain special tools and these operations should not be undertaken if these are not available. The following text describes these operations, in case that the listed special tools can be obtained or hired.

1.9.0.0. PRECAUTIONS WHEN WORKING ON DIESEL INJECTION SYSTEMS

Whenever repairs are carried out on a diesel fuel injection system, whatever the extent, observe the greatest cleanliness, apart from the following points:

- Only carry out work on diesel injection systems under the cleanest of conditions. Work in the open air should only be carried out when there is no wind, to prevent dust entering open connections.
- Before removal of any union nut clean all around it with a clean cloth.
- Removed parts must only be deposited on a clean bench or table and must be covered with a sheet of plastic or paper. Never use fluffy shop rags to clean parts.
- All open or partially dismantled parts of the injection system must be fully covered or kept in a cardboard box, if the repair is not carried out immediately.
- Check the parts for cleanliness before installation.
- Never use an air line to clean the exterior of the engine when connections of the injection system are open. With the availability of air compressors which can be plugged into a cigar lighter socket, you may be tempted to use air for cleaning.
- Take care not to allow diesel fuel in contact with rubber hoses or other rubber parts. immediately clean such a hose if it should happen accidentally.

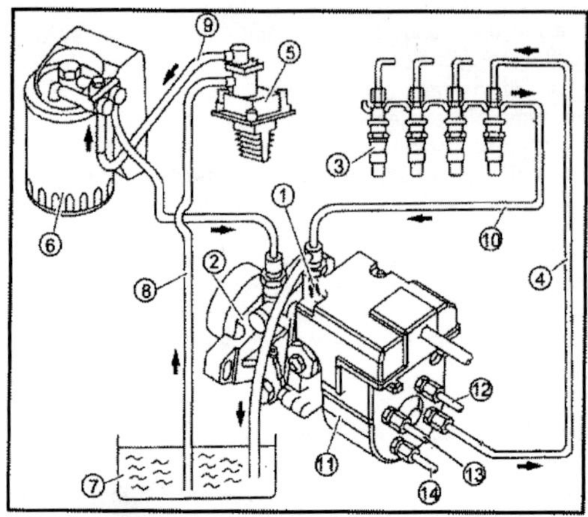

Fig. 1.112 – Connections of fuel pipes and other component parts of a 604 engine.
1 Position of throttle for pump ventilation
2 Position of fuel pump
3 Fuel injectors
4 Fuel pipe to cylinder 1
5 Pre-heating device for fuel
6 Full-flow filter
7 Fuel tank
8 Fuel inlet, cold fuel
9 Fuel return, warm fuel
10 Leak-off pipe
11 Fuel injection pump

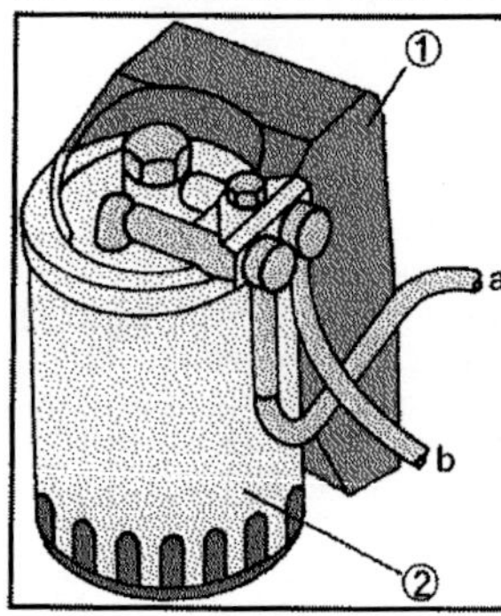

Fig. 1.113 – The full-flow fuel filter.
1 Fuel filter bracket
2 Fuel filter
a From fuel pre-heater
b To injection pump

1.9.1. FUEL FILTER

The fuel filter element should be replaced in accordance with the intervals specified in your maintenance manual. Some vehicles are built with a full-flow filter with water separator, which can be recognised on the drain plug at the bottom of the filter. Fig. 1.113 shows the fitted fuel filter.

The fuel filter can be removed towards the bottom after unscrewing the bolt in the centre.

Thoroughly clean the bottom bowl, if the filter housing remains on the engine, and fit a new filter insert. Always replace the sealing ring. Make sure that the seal is correctly located underneath the head of the centre bolt. Tighten the bolt to 1.6 kgm (11.5 ft.lb.).

1.9.2. INJECTION PUMP - REMOVAL AND INSTALLATION

A special tool is required to fit the injection pump. This tool (refer to the description) must be obtained before the pump is removed. Also available must be a 27 mm socket (to rotate the crankshaft), a 14 mm ring spanner, with a slot cut into the ring to undo the injection pipe union nuts and a serrated wrench to rotate the injection pump shaft. Provided these tools can be obtained, proceed as follows. Note that the instructions do not refer in detail to all engines. The following instructions are baed on the 605 and 606 engines:

- Disconnect the battery earth cable, remove intake manifold, the charge air distribution pipe (turbo diesel) and the resonance intake manifold (E300) as applicable.
- Disconnect the injection pipes and the fuel pipes from the injection pump and carefully bend them to one side. Seal the open connections and fuel hoses at the injection pump in suitable manner, by pushing push-on caps over the connections. Make absolutely sure that no dirt can enter the open connections.

- Rotate the engine (apply the socket mentioned above to the crankshaft pulley) until the piston of No. 1 cylinder is at top dead centre and then slightly further until the 15° mark in the crankshaft pulley/vibration damper is in line with the pointer.

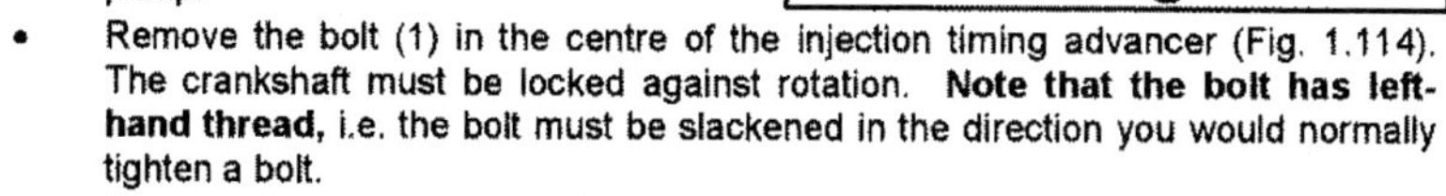

Fig. 1.114 – The bolt (1) <u>with left-hand thread</u> secures the injection timing advancer to the front of the injection pump.

- Remove the single drive belt and the tensioning device as described earlier on.
- Remove the timing chain tensioner.
- Disconnect the hoses connected to the vacuum pump and remove the pump.
- Disconnect the regulating rod for the accelerator control from the injection pump.
- Remove the bolt (1) in the centre of the injection timing advancer (Fig. 1.114). The crankshaft must be locked against rotation. **Note that the bolt has left-hand thread,** i.e. the bolt must be slackened in the direction you would normally tighten a bolt.
- Remove the pump securing bolts and the bolt securing the support bracket and withdraw the pump towards the rear.

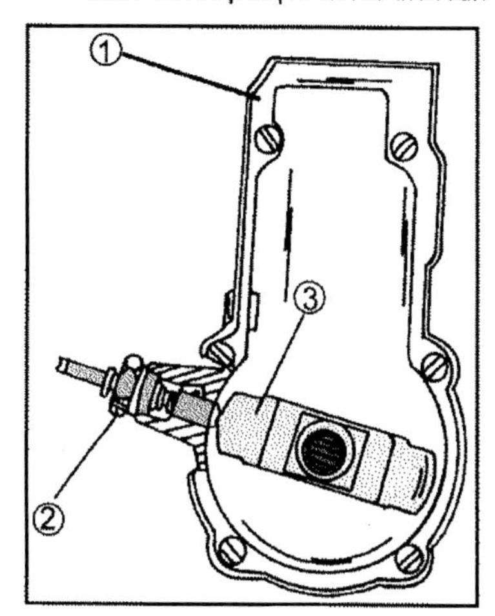

Fig. 1.115 – Locking the governor inside the injection pump, using the special locking pin.
1 Injection pump
2 Locking bolt
3 Governor

Refit the injection pump in the following order. If the special tool mentioned cannot be obtained, you will be able to have the commencement of injection adjustment is carried out in a workshop.

- Check that the engine is still in the position described above. Otherwise make the necessary corrections, by carefully turning the crankshaft with the 27 mm socket. Remember that the timing chain is engaged with the wheel.
- Remove the plug from the side of the injection pump at the position shown in Fig. 1.115. The serrated wrench 601 589 00 08 00 must now be used to rotate the pump until the lug (3) of the governor rotor is visible. In this position insert the locking tool 601 589 05 21 00, as shown in the illustration. Hand-tighten the nut of the locking bolt. If the tools can be obtained, carry out the operation by studying Fig. 1.115. **Read the note below.**
- The injection pump can now be refitted. Tighten the pump flange bolts to 2.0 - 2.5 kgm (14.5 - 18 ft.lb.). Immediately remove the locking pin from the injection pump and refit the plug. Tighten the plug to 3.0 - 3.5 kgm (22 - 25 ft.lb.).

NOTE: Immediately after fitting the pump in position remove the locking pin from the side of the pump. The pump can be seriously damaged if this is forgotten.

- Re-connect the injection pipes one after the other. Make sure to engage the threads properly before the union nuts to 1.0 – 2.0 kgm (7.2 – 14.5 ft.lb.).
- All other operations are carried out in reverse order.

1.9.2.0. Checking the Start of Injection Delivery

The start of the injection delivery can be checked in various ways, but every method requires the use of special tools. For this reason we recommend to have the work carried out in a workshop. The following description is only intended to give you some idea what the workshop will do.

- The first method requires a pump, which is used to determine the exact injection point by means of pressure.
- The second method uses a checking gauge, which is inserted into the injection pump at the position shown in Fig. 1.96. By using an Instrument with two test lamps, the exact injection point is shown on the degree scale on the crankshaft pulley/vibration damper.
- The third method requires a test appliance with digital indicator.

NOTE: Whenever possible have the idle speed adjusted at a Dealer who has the necessary equipment to do it. Remember that a correct idle speed can save you fuel.

1.9.3. INJECTORS

1.9.3.0. Removal and Installation

The removal and installation of the injectors is different on the 602 engine and the remaining engines.

602 Engine (E290 TD)

Fig. 1.116 shows the attachment of the five injectors. An injector can be removed as follows:

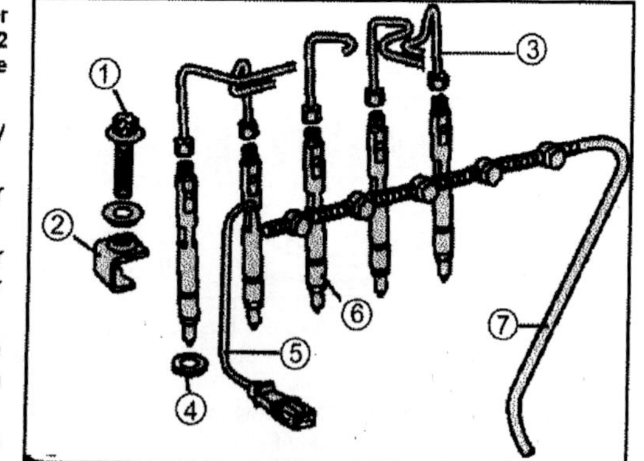

Fig. 1.116 – Injector installation as fitted to the 602 engine. The numbers are referred to below.

- Disconnect the battery earth cable.
- Remove the charge air pipe.
- Remove the dust cover above the cylinder head cover.
- Disconnect or unclip the fuel leak hoses (7) on one end.
- Unscrew the injection pipes (3) from the injector. The injector of No. 2 cylinder has a needle lift sensor (5). Disconnect the cable plug connector when this injector is removed.
- Remove the injector securing bolt (1) and take off the washer and the injector clamp bracket (2). The injector (6) can now be withdrawn.

The injector must be inserted with special grease during installation. The clamps (2), the bolts (1) and the sealing washer (4) must always be replaced. The fuel injection pipe union nuts are tightened to 2.7 kgm (19.5 ft.lb.). The bolts (1) are tightened to only 0.7 kgm (5 ft.lb.) and from the final position a further one half of a turn (180°).

604, 605 and 606 Engines with and without Turbo Charger

Fig. 1.117 shows the attachment of the injectors on these engines. An injector can be removed as follows, noting that a special injector wrench (No. 606 689 00 09 00) is used in the workshop to remove the injectors:

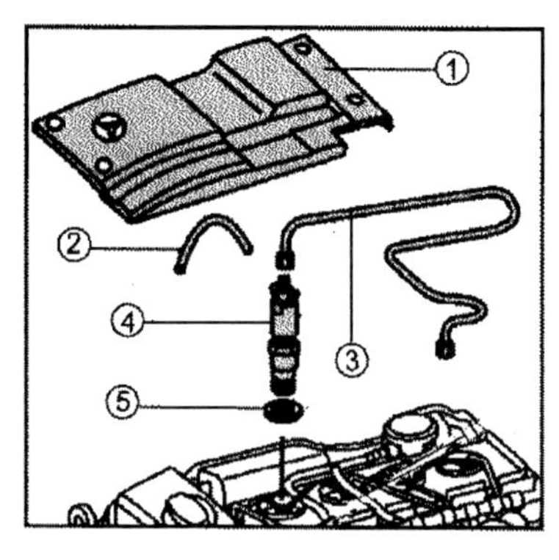

Fig. 1.117 – Injector installation in the case of the 604, 605 and 606 engines. The numbers are referred to below.

- On a 604 and 605 engine remove the air cleaner cross pipe or the charge air pipe, depending on the engine (none turbo and turbo respectively).
- On the non-turbo E300 (engine 606.912) remove the resonance intake pipe as described during the removal of the cylinder head.
- Remove the trim panel (1) from the cylinder head cover. During installation take care to route the leak fuel hose from the cylinder 1 to the fuel return hose.
- Detach the fuel leak hoses (2) from the injectors.
- Unscrew the injector pipes. A suitable ring spanner can be cut for this purpose. Slide the open gap over the injection pipe and slacken the union nuts. If an ordinary open-ended spanner is used, take care not to damage the union nuts.

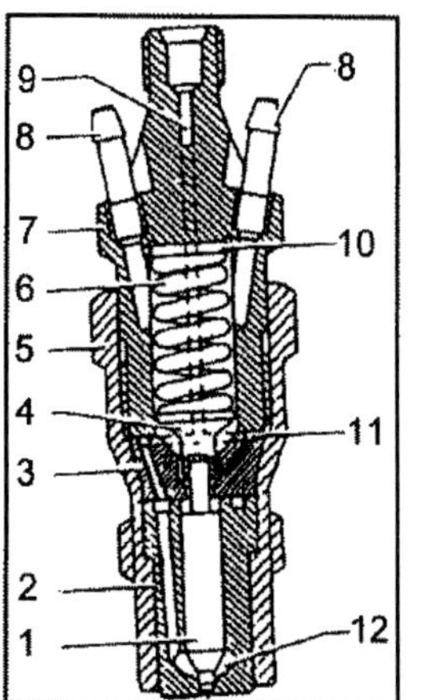

Fig. 1.118 – Sectional view of an injector as fitted to a four-cylinder engine.

1 Injector nozzle
2 Nozzle body
3 Injector holder insert
4 Thrust pin
5 Union nut
6 Pressure spring
7 Injector holder
8 Leak-off connection
9 Fuel feed connection
10 Steel washer
11 Ring groove and feed bore
12 Pressure chamber and injector nozzle

- Unscrew the injectors with the special tool available for this purpose or use a long 27 mm socket. Take out the injector sealing gaskets and the nozzle plates. Injector nozzle plates must be replaced once an injector has been removed.

The installation of an injector is a reversal of the removal procedure. Tighten them to 4.0 kgm (29 ft.lb.). Attach the injector pipes and tighten the union nuts to 2.3 kgm (16.5 lt.1b.). Make sure that none of the injector pipes is tightened under tension or strain.

1.9.3.1. Injector Repairs

A special test pump is required to check the injectors. If a faulty injector is suspected, take the set to a specialist and have them checked and their injection pressure adjusted. Different injectors are fitted to different engines. It is of advantage to check with your parts supplier against the engine number to make sure that the correct injectors are fitted. Figs. 1.118 shows a sectional view of the injectors fitted to engines covered in this manual for reference to give you some idea of the component parts.

1.9.4. GLOW PLUGS

As shown in the wiring diagram in Fig. 1.119, the glow plug system consists of the glow plugs, the glow time relay and the warning lamp in the instrument panel.

When the ignition is switched on, the glow plug relay will receive current via terminal "15" ' The relay "a" directs the current from the plus terminal "30" via a fuse (80 amps) to the glow plugs. The glow plugs receive a voltage of at least 11.5 volts and a current of 30 amps, which is however, reduced to 8 to 15 amps by means of a regulator, thereby preventing burning out of the plugs. The glow plugs heat up to 90° C within 10 seconds and can reach a temperature of 1180° C after 30 seconds.

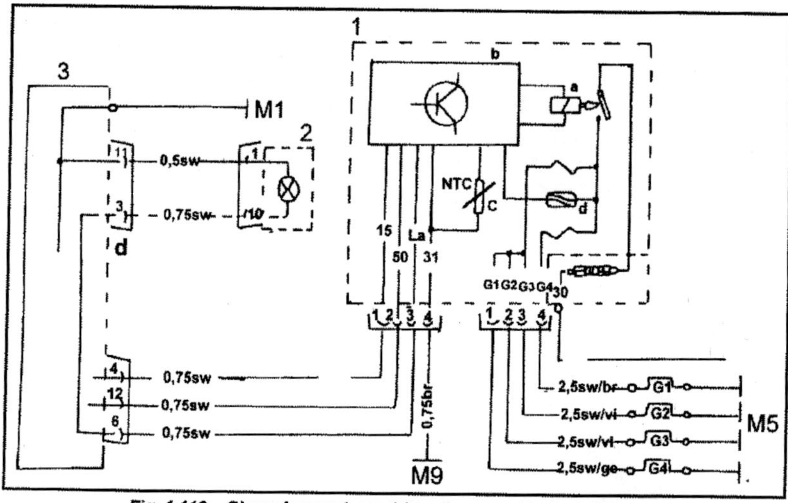

Fig. 1.119 – Glow plug system wiring diagram – Typical diagram.

1 Glow plug relay
2 Warning light
3 Central electrics
a Supply relay
b Electronic control unit
d Temperature sensor
d Reed relay
G Glow plug
M Earthing points

The glow plug time relay determines the operating time of the glow plugs. This relay senses the outside temperature, i.e. at very low temperatures, for example – 30°) C, the plugs can glow as long as 25 seconds. During the summer months, however, glowing time may be as little as 2 seconds. If the engine is not started immediately after the warning light has gone "off", the current feed will be interrupted through a safety circuit. Subsequent starting of the engine will switch in the glow plug circuit via starter motor terminal "50".

The connectors and the 80 amp fuse are accessible after removal of the cover. The warning lamp circuit, the relay "a" in Fig. 1.100 and the safety circuit are controlled by the relay. If the vehicle has been built as model year 1989 or later, the 80 amp fuse is replaced by an electronic short-circuit fuse. The current supply will be interrupted in case of a short circuit. Again this Arrangement is not fitted to all vehicles.

Glow plugs are not the same for all engines and we advise you to check with your parts supplier.

The glow plugs are fairly hidden below the tubes of the inlet manifold and a socket and extension are required to reach them. A ratchet is of advantage.

Unscrew all nuts from the glow plugs and take off the connecting line. The nuts cannot be fully removed. Unscrew the glow plug with the special wrench, available for this purpose, or use an ordinary long socket of the correct size.
Before installation of a glow plug, clean out the plug channels and the holes in the pre-combustion chambers with a reamer. Again a special reamer is used by Mercedes workshops. Pack the flutes of the reamer with grease when reaming out the plug channels. The installation of the glow plugs is a reversal of the removal procedure. Plugs are tightened to 2.0 kgm (14.5 ft.lb.). Do not over-tighten the cable securing nuts.

1.9.4.0. Faults in the Glow Plug System

Difficult starting of the engine can in many cases be traced back to the glow plug system. Failure for the warning light to light up will obviously indicate a fault in the system which may be traced with a few simple operations. If a 12 volt test lamp can be made available and you have some experience with electrical systems. First remove the cover from the glow plug relay.
If the warning light does not come on, but the engine starts, check as follows:

- Withdraw the small connector plug from the glow plug relay and switch on the ignition.
- Connect the test lamp to the two terminals inside the plug. The warning lamp should come on. Otherwise check the cables or replace the warning light bulb. To do this remove the combination instrument: Remove the upper foot well covering on the driver's side, the air hose to the side air jet and disconnect the speedometer cable from the rear. Push out the combination instrument from the rear. It is held in position by spring clips.

If the engine does not start and the warning light does not come on:

- Check the fuse strip has burnt out.
- Switch on the ignition and check that there is current at terminal "15". Check with the test lamp on the small terminal with the red/black cable.
- Withdraw the large connector plug with the cables from the glow plug relay.
- Connect the test lamp to earth and hold the other lead of the lamp to the other relay terminals in turn. Each terminal must show current with the ignition switched on. Remember to have the ignition switched on and off every time to re-start the glowing process.
- The relay must be replaced if it fails any of the tests.

A defective glow plug can be suspected if the engine starts, but is not firing on all cylinders immediately. If the "missing" cylinder comes in after a while, replace the glow plug.

1.9.4. AIR CLEANER

1.9.4.1. Removal and Installation of Air Cleaner

It is obvious that not all engines have the same air cleaner installation. Figs. 1.119 and 1.120 show two versions, in this case for a 605 engine without turbo charger and with turbo charger.

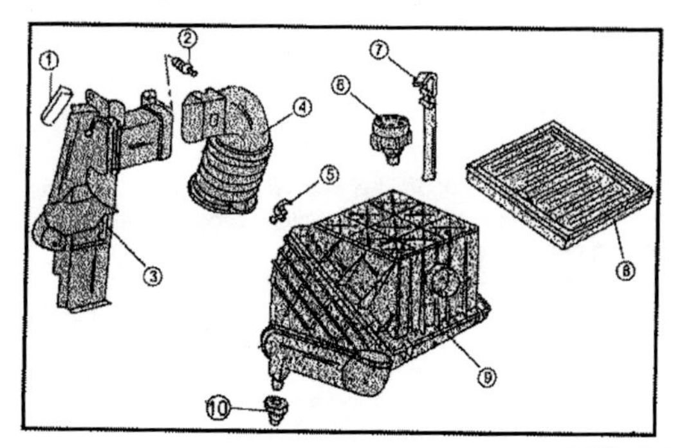

Fig. 1.119 – Air cleaner installation of a 605 engine without turbo charger (typical installation).

1 Connecting hose
2 Rivet
3 Fresh air scoop
4 Air intake hose
5 Air mass meter in filter
6 Maintenance indicator
7 Filter housing clamp
8 Filter insert
9 Filter housing
10 Rubber grommet

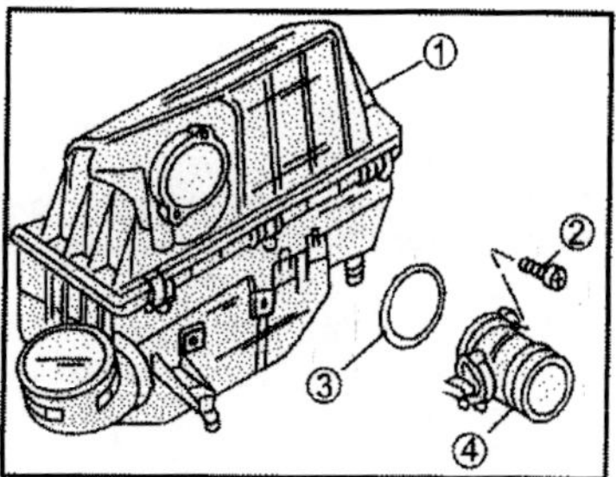

Fig. 1.120 – Typical air cleaner installation as fitted to a 605 engine with turbo charger.

1 Air cleaner housing
2 Securing bolt
3 Sealing ring
4 Noise dampener

Removal and installation is fairly easy by following the diagrams. The air cleaner housing is lifted out of the rubber grommets. The installation is a reversal of the removal procedure. The guide studs at the bottom of the air cleaner housing must be inserted into the three rubber grommets.

1.9.4.2. Replacing the Air Cleaner Element – All Engines

The element must be replaced in accordance with the time interval given in the maintenance booklet. The replacement of the filter element is carried out as described above. In all cases it will be necessary to unclip the upper air cleaner housing from the lower housing to take out the filter element.

Clean the inside of the air cleaner housing with a damp rag before fitting the new element. Make sure that the element is correctly seated at the sealing surfaces.

Fit the filter housing cover, making sure that the three plates on the cover engage into the recesses of the bottom housing. Snap the clamps in position but ensure that they are securely fastened.

1.9.5. CHARGE AIR COOLER – ALL ENGINES

Fig. 1.121 shows the parts to be removed to take out the charge air cooler with the 602, 605 and 606 engine, fitted E290 TD, E250 TD and E300 TD respectively. The front bumper must be removed to gain access to the charge air cooler. This is a rather complicated operation. With the bumper removed, however, proceed as follows:

Fig. 1.121 – Details for the removal and installation of the charge air cooler. The numbers are referred to below.

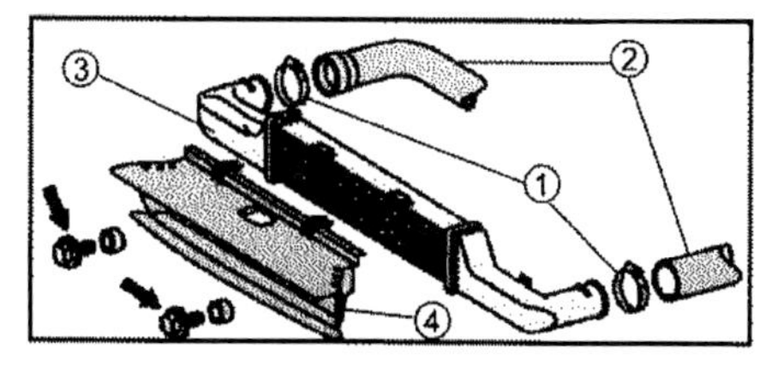

- Remove the two bolts indicated by the arrows and remove the air scoop (4).
- Disconnect the charge air hoses (1) after slackening the hose clamps (2) from the charge air cooler (3). Slight quantities of oil in the charge air hoses can be ignored as they come from the crankcase ventilation. The charge air hoses are connected on each side of the charge air cooler. During installation make sure that the charge air cooler is fitted in its holder on the radiator.

The installation is a reversal of the removal procedure. Hose clamp screws are only tightened to 0.4 kgm.

1.9.6. TURBO CHARGER – ALL ENGINES

Fig. 1.122 shows the turbo charger installation in the case of the 606 engine, as fitted to the E300 TD. Similar parts have to be removed if the turbo charger of a 602 engine (E290 TD) is dealt with. The description is combined for both models.

Fig. 1.122 – Details for the removal and installation of the turbo charger, shown in the case of the 606 engine. The numbers are referred to in the description.

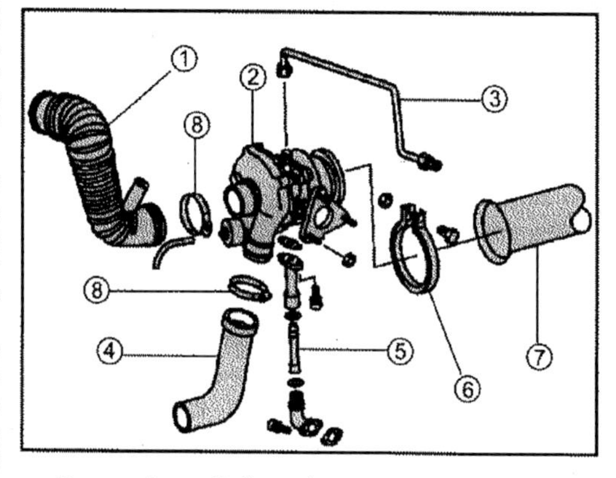

- Separate the intake hose (1) and the charge air hose (4) from the turbo charger after slackening the hose clamps (8).
- Unscrew the union nuts securing the oil supply pipe (3) to the turbo charger (2) and the cylinder block and remove the pipe. The connections underneath the union nuts must be prevented from rotating by counterholding them with an open-ended spanner.
- Remove the exhaust pipe clamp (6) and disconnect the exhaust pipe. Tighten the clamp to 2.5 kgm (18 ft.lb.) during installation.
- Remove the two bolts and withdraw the oil drain pipe (4) from the bottom of the turbo charger.
- Remove the turbo charger from the exhaust pipe. A support bracket is used in the case of the 602 engine. The nuts are tightened to 2.0 kgm (14.5 ft.lb.) during installation. The support bracket is tightened to 2.0 kgm and from the final position a further 90°.

The installation is a reversal of the removal procedure, noting the point already given above.

1.9.7. INJECTION PIPES – ALL ENGINES

Fig: 1.123 shows the injection pipes as fitted to a 606 engine. Four-cylinder and five-cylinder engines have, of course, less injection pipes. The union nuts are tightened in all cases to 2.3 kgm (16 ft.lb.). During installation it is important to install the retaining clips (a) the correct way round, i.e. make a note how they are fitted before removal.

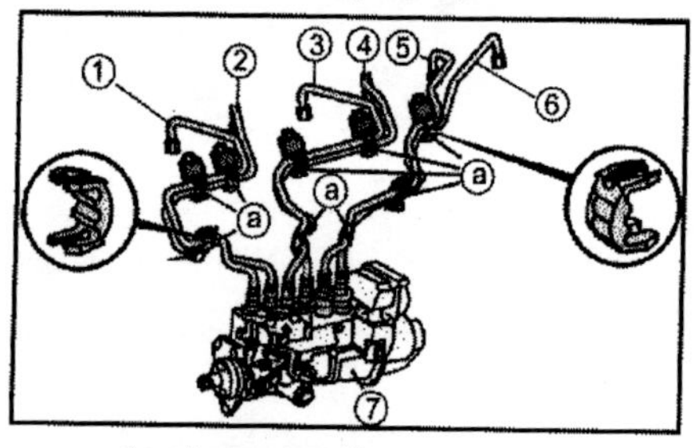

Fig. 1.23 – The injection pipes of a six-cylinder (606) engine. The numbers at the top show the cylinder connections on the injection pump (7).

Note the following when an injection pipe or all pipes are to be removed:

- On all engines remove the trim panel of the cylinder head cover. During installation of the panel take care to route the leak-off fuel pipe from the cylinder No. 1 to the fuel return hose correctly.
- Remove the intake manifold in the case of the 604 engine and the 605 engine without turbo charger if all injection pipes are removed.
- Also if all pipes have to be removed remove the charge air distribution pipe on a turbo charged engine and the resonance intake manifold in the case of a 300D (606 engine). Refer to the removal of the cylinder head for details.
- Remove the retaining clips (a) if an individual pipe is removed or unscrew the bolts securing the injection pipe bracket if all pipes are removed. Tighten the bolts to 1.0 kgm (7.2 ft.lb.) during installation-
- Unscrew the union nuts at the injector and injection pump ends. Protect the open connections to prevent entry of foreign matter.
- Take of the pipe(s). The retaining clip of the pipe in question must always be replaced.

The installation is a reversal of the removal procedure, noting the tightening torques given above.

2 CLUTCH

2.0. Technical Data

Type: .... Single dry plate diaphragm clutch
Operation: .... Hydraulic system
Clutch diameter: .... Refer to parts lists

2.1. Removal and Installation

To remove the clutch unit, it will be necessary to separate the transmission from the engine, with the assembly fitted to the vehicle.

- Remove the transmission (Section 3.1).
- Mark the clutch in its fitted position on the flywheel if there is a possibility that the clutch unit is re-used. To remove the clutch, unscrew the six bolts securing the pressure plate to the flywheel plate and lift off the flywheel and then the driven plate, now tree. Before removing the driven plate, note the position of the longer part of the driven plate hub, as the driven plate must be refitted in the same way.

Fig. 2.1 – The component parts of the clutch.
1 Clutch driven disc
2 Clutch pressure plate
3 Clutch release bearing
4 Clutch release lever
5 Allen-head bolt

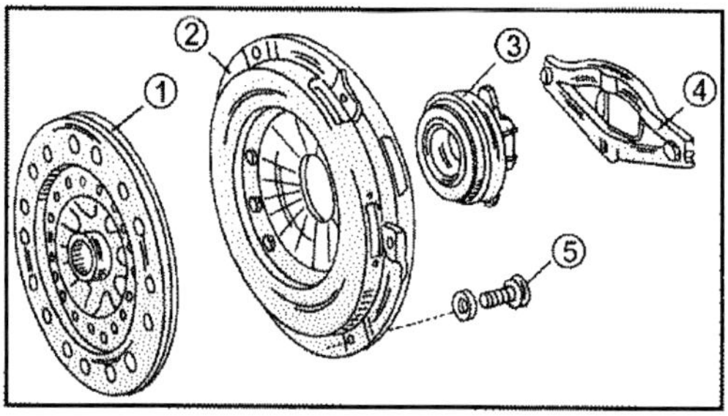

Install in the reverse sequence to removal, noting the following points:

- If the old clutch unit is fitted, align the marks made before removal. A new clutch can be fitted in any position.
- A centering man-drel is required to centre the clutch driven plate inside the flywheel. Tool hire companies normally have sets of mandrels for this purpose. An old transmission (clutch) shaft, which you may be able to obtain from a Mercedes workshop, can also be used. Experienced D.I.Y. mechanics will also be able to align the clutch plate without the help of a mandrel.
- Fit and tighten the six clutch to flywheel bolts to a torque reading of 2.5 kgm (18 ft.lb.). The flywheel must be locked against rotation when the clutch bolts are tightened.

2.2. Servicing

The cover assembly - pressure plate and diaphragm spring must not be dismantled. Replace, if necessary with a complete assembly from your dealer or distributor.

Fig. 2.2 – To check the driven plate for run-out, clamp it between the centres of a lathe and check with a dial gauge.

Inspect the driven plate and the linings, replacing the complete plate if the linings are worn down close to the rivets. A driven plate with the linings contaminated with grease or oil cannot be cleaned successfully and should also be replaced. All rivets should be tight and the torsion springs should be sound and unbroken. Check the condition of the driven plate splines. Clamp the driven plate between the centres of a lathe and apply a dial gauge to the outside of the plate as shown in Fig. 2.2, at a diameter of approx. 175.0 mm (6.4 in.). The max. run/out of the driven plate should be no more than 0.5 mm (0.02 in.).

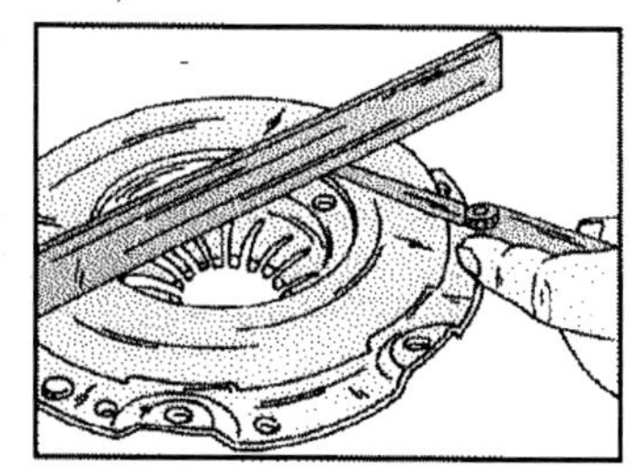

Fig. 2.3 – Checking the clutch pressure plate for distortion. The gap should not be more than given below.

Check the rivet fastening of the clutch pressure plate and replace the plate, if loose rivets can be detected.

Place a straight edge (steel ruler) over the friction face of the pressure plate and insert feeler gauges between the ruler and the surface. If the gap at the innermost spot of the friction face is no more than 0.03 mm (0.012 in.), the plate can be re-used. Fig. 2.3 shows this check.

2.2. Clutch Release Mechanism

Engagement and disengagement of the clutch is by means of the slave cylinder push rod, acting on the clutch release lever and sliding the ball bearing-type release bearing along a guide tube on the clutch shaft of the transmission. The release system is free of play, as the wear of the clutch linings is compensated automatically.

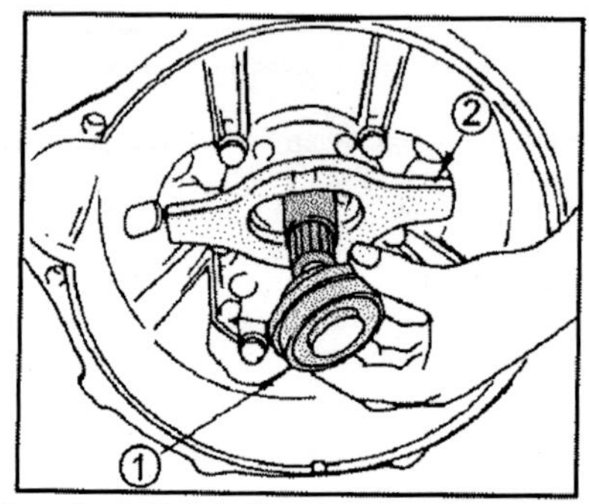

Fig. 2.4 – Removal of the clutch release bearing (1) from the release lever (2).

2.2.0. REMOVAL AND INSTALLATION

The transmission must be removed to replace the release bearing. Remove the bearing from the bearing sleeve on the front transmission housing cover, as shown in Fig. 2.4. To remove the release fork, refer to Fig. 2.5 and move it in direction of arrow (a) and then pull it from the ball pin in the clutch housing in direction of arrow (b).

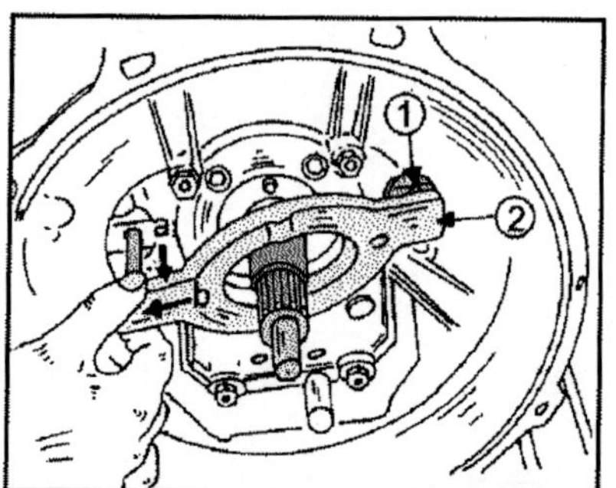

Fig. 2.5 – Removal and installation of the clutch release lever (see text).
1 Ball pin 2 Release lever

Thoroughly grease the guide sleeve on the front transmission cover, the ball pin and all of the parts of the release mechanism in contact with the release bearing with long term grease. Push the release lever in reverse direction of arrow (b) over the ball pin until the spring clip of the release lever engages with the ball pin. Check for secure fitting. Then move the lever in reverse direction of arrow (a) until the slave cylinder push rod is engaged with the ball-shaped cut-out in the release lever.

Grease the release bearing on the inside and on both sides at the rear, where it rests against the release lever and slip the bearing over the guide sleeve. Rotate the bearing until it snaps in position into the release lever. Check that the bearing is properly fitted and refit the transmission.

2.3. Clutch Master Cylinder

As the removal and installation of the clutch master cylinder is a complicated operation and the cylinder will have a long service lift, we suggest to have the cylinder replaced in a workshop.

2.4. Clutch Slave Cylinder – Removal and Installation

The clutch slave cylinder is fitted to the top of the transmission and is attached as shown in Fig. 2.6. The cylinder can be removed as follows:

- Unscrew the fluid pipe (3) from the slave cylinder (1), using an open-ended spanner. Close the end of the pipe in suitable manner to prevent fluid leakage (rubber cap for bleeder screw).
- Remove the two cylinder securing screws (4) and take off the cylinder. Observe the fitted plastic shim (2).
- When fitting, insert the shim with the grooved side against the clutch housing and hold in position. Fit the slave cylinder, engaging the push rod into the ball-shaped cut-out of the clutch release lever, and insert the two screws (4). Tighten the screws. Finally bleed the clutch system as described in the next section.

Fig. 2.6 – Attachment of the clutch slave cylinder on the transmission.
1 Clutch fluid pipe
2 Bleed valve
3 Securing screws
4 Slave cylinder
5 Plastic washer

2.5. Models with Central Clutch Operator

Some E-Class models are fitted with a central clutch release bearing, fitted to the inside of the transmission (transmission type 716.6), replacing the standard clutch release bearing (not applicable to type 717). The transmission must be removed to replace the clutch release bearing. To remove the bearing, remove the four securing bolts and withdraw the bearing. The bolts are tightened to 0.8 kgm during installation.

2.6. Bleeding the Clutch System

A pressure bleeder is used by Mercedes workshops. The following description involves the brake system and is therefore to be treated with caution. Make absolutely sure that the brakes have correct operating pressure after the clutch has been bled. A transparent hose of approx. 1 metre (3 ft.) in length is required. Proceed as follows:

- Fill the brake/clutch fluid reservoir.
- Remove the dust cap of the bleeder screw on the R.H. front brake caliper, push the hose over the bleeder screw and open the bleeder screw.
- Ask a second person to operate the brake pedal until the hose is completely filled with brake fluid and no more air bubbles can be seen. Place a finger over the hose end to prevent fluid from running out.
- Push the free end of the hose over the bleeder screw on the slave cylinder and open the bleeder screw. Remove the dust cap first.
- The following operations must now be carried out in exactly the given order: Depress the brake pedal, close the bleeder screw on the wheel brake cylinder, allow the brake pedal to return and open the bleeder screw on the wheel brake cylinder. Repeat this operation until no more air bubbles can be seen in the fluid reservoir. During the pumping, keep an eye on the reservoir to make sure it has enough fluid.

- Close the bleed screws on caliper and slave cylinder and remove the hose. Refit the dust caps (easily forgotten).
- Check the fluid level in the reservoir and, If necessary, top it up to the "Max." mark. Start the engine, depress the clutch pedal and engage reverse. No grating noises should be heard.

3 Manual Transmission

The four-, five- and 6 cylinder models covered in this manual are fitted with a five-speed transmission, with the technical designation type "717", with different end numbers, depending on the fitted engine (see Section 3.0) or an automatic transmission with five speeds. Again a different transmission is fitted, allocated to the different engines. This section covers the manual transmission. The type is important when it is intended to fit a second-hand transmission during the life-span of your vehicle. The transmission is also identified by different letters/numbers which can be identified by means of the description below:

The overhaul of the transmission is not described in this manual. The description in the overhaul section is limited to some minor repair operations, not involving the gear train or the gear shafts. If the transmission appears to be damaged or faulty, try to obtain an exchange unit. Transmission overhaul is now limited to specialised workshops which are equipped with the necessary special tools.

3.0. Technical Data

Fitted transmission:
- E200 D (model 210.003): .... 717.416 or 717.418
- E220 D (model 210.004: .... 717.416 or 717.418
- E250 D (models 210.010): 717.417
- E250 TD (model 210.015): .... 717.460 or 417.466
- E290 TD (model 210.017 and 210.617): .... 717.460 or 417.466
- E300 D (model 210.020): 717.446

Transmission Ratios (typical ratios, not all listed)

Transmission Ratios:	**E200/E220/E250D**	**E250 TD**
- First gear:	3.910 : 1	3.860 : 1
- Second speed	2.170 : 1	2.180 : 1
- Third speed	1.370 : 1	1.380 : 1
- Fourth speed	1.000 : 1	1.000 : 1
- Fifth speed	0.810 : 1	0.750 : 1
- Reverse speed	4.270 : 1	4.220 : 1
- Axle drive ratio:		3.460 : 1
- E200D:	3.810 : 1	
- E220D:	3.710 or 3.640 : 1	
- T model:	3.910 : 1	

Transmission Ratios:	**E300D**	**E290 TD**
- First gear:	3.386 : 1	3.86 : 1 or 4.100 : 1
- Second speed	2.180 : 1	2.180 : 1
- Third speed	1.380 : 1	1.380 : 1
- Fourth speed	1.000 : 1	1.000 : 1
- Fifth speed	0.800 : 1	0.800 : 1
- Reverse speed	4.220 : 1	4.220 : 1
- Axle drive ratio:	3.460 : 1 (T 3.570 : 1)	3.070 : 1

Oil capacity: .... 1.5 litres
Lubrication oil: .... As in automatic transmissions

3.1. Removal and Installation

The following text describes the general removal and installation operations. The transmission is heavy and the necessary precautions must be taken when it is lifted out.

- Disconnect the battery earth cable from the battery.
- To prevent damage to the insulation on the engine compartment bulkhead during the later lowering of the transmission insert a thick piece of cardboard between the engine and the bulkhead. As this is, however, not necessary in the case of all models you will have to decide after checking the area. Also if the vehicle is fitted with an additional heating system make sure that the heater hose cannot be damaged when the transmission is lowered.

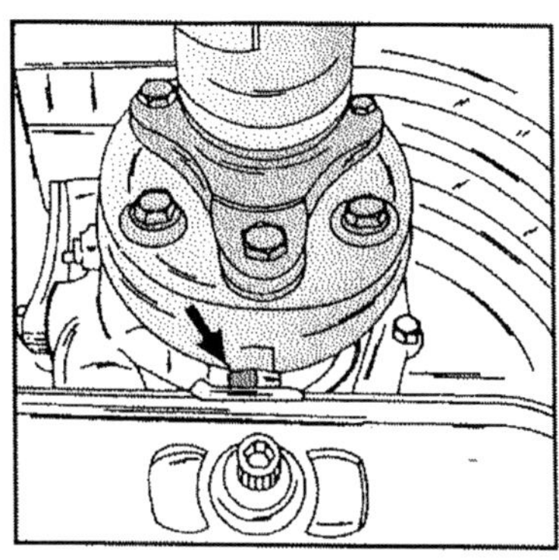

Fig. 3.1 – The arrow points to the location of the nut and bolt where the gearbox is secured to the rear gearbox cover.

Place the front end of the vehicle on chassis stands.

- Remove the engine compartment undertray (noise insulation) after unscrewing the six screws.
- Place a mobile jack underneath the gearbox, with a block of wood between the jack head and the gearbox and carefully lift the gearbox, until the latter is just under tension.
- Remove the mounting at the position shown in Fig. 3.1.
- Remove the two bolts shown in Fig. 3.2 and release the mounting crossmember from the gearbox and the vehicle floor. Make absolutely sure that the gearbox is well supported from underneath on the jack/wooden block combination.

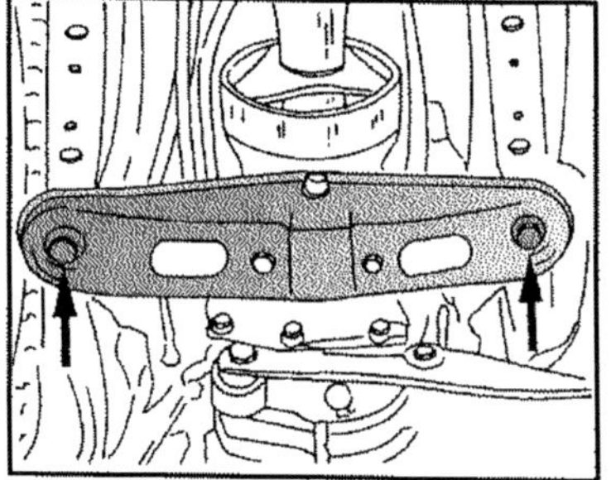

Fig. 3.2 – The two arrows show the bolts to be removed to release the mounting cross-member to the vehicle floor.

- Release the exhaust attachment from the gearbox. Note the fitted washers.
- Remove the nuts securing the exhaust pipe clamp and release the exhaust mounting.
- Remove the covering plate above the propeller shaft intermediate bearing and slacken the propeller shaft clamp nut.
- Slacken the bolts securing the propeller shaft intermediate bearing without removing them.
- Unscrew the propeller shaft from the gearbox so that the drive flange remains on the shaft. Before the drive flange is separated from the gearbox flange you will have to remove the dowel sleeves out of the joint flange. This is carried out with a drift of 10 mm in diameter and 150 mm length. Drive the sleeves from the rear towards to the front out of their locations. As this operation is not necessary in the case of all models, refer to the section dealing with the propeller shaft for further details.
- Push the propeller shaft as far as possible towards the rear as this is enabled by the loose intermediate bearing and clamp piece.

- Slacken the knurled nut securing the speedometer drive shaft to the rear gearbox cover and withdraw the shaft. Remove the retaining clamp from the drive shaft.
- Remove the bolt securing the hydraulic pipe from the clutch housing.
- Unscrew the clutch slave cylinder and pull it together with the connected pipe towards the rear, until the cylinder operating plunger is out of the clutch housing.
- Remove the retaining clips from the gearchange levers on the side of the transmission and disconnect all gearchange linkages.
- Disconnect the electrical leads from the starter motor, remove the starter motor mounting bolts and withdraw the starter motor from the engine and transmission.
- Remove the transmission-to-intermediate flange bolts. The two upper bolts are removed last. Under one of the bolts an earth cable is connected. Remember which one.
- Rotate the transmission towards the left and remove it horizontally towards the rear and away from the clutch housing before it is lowered. Make absolutely sure that the clutch shaft has disengaged from the clutch driven plate before the gearbox is lowered. Failing to observe this can lead to distortion of the clutch shaft or damage to the driven plate.

The installation of the gearbox is carried out as follows:

- Coat the dowel pin and the clutch shaft splines with long-term grease and lift the gearbox until it can be pushed in horizontal position against the engine. The clutch slave cylinder with the pipe should be placed over the gearbox before the box is joined.
- Engage a gear, slightly rotate the gearbox towards the left and rotate the drive flange at the end of the transmission to and fro until the splines of the clutch shaft have engaged with the splines of the clutch driven plate. Now push the gearbox fully against the engine until the gap is closed.
- Bolt the gearbox to the gearbox, remembering the earth cable under one of the bolt heads.
- Fit the starter motor and re-connect the cable connections.
- Push the clutch slave cylinder in position and fit it to the gearbox. Make sure that the plastic insert is in position. Fit the fluid pipe with the clamp and the bolt.
- Fit the gearchange linkages to the gearchange levers and secure them with the retaining clips.
- Re-connect the speedometer cable and attach the cable.
- Extend the propeller shaft as far as possible on the slide piece and re-connect the shaft to the gearbox. The gearbox should be lifted or lowered with the jack to align the flange connection.
- Refit the rear mounting crossmember to the gearbox and then on the outside to the vehicle floor.
- Refit the exhaust system.
- Tighten the propeller shaft intermediate bearing.
- Tighten the propeller shaft clamp nut to 3.0 – 4.0 kgm (22 – 29 ft.lb.).
- Refit the cover plate above the propeller shaft intermediate bearing.
- The remaining operations are carried out in reverse order.

3.2 Transmission Repairs

Many repairs can be carried out without dismantling the transmission completely. The following sections described some of the jobs which do not need special tools and some of them are described below, which can be carried out after the gearbox has been removed. Complete dismantling of the transmission requires special tools. A damaged transmission should be replaced by an exchange unit.

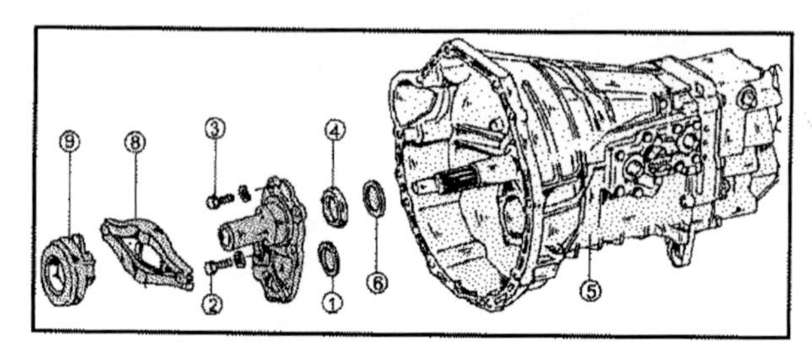

Fig. 3.5 – The component parts of the front gearbox cover.

1 Gearbox cover
2 Cover bolt
3 Cover screw
4 Oil seal
5 Gearbox case
6 Adjusting shim
7 Adjusting shim
8 Release lever
9 Clutch release bearing

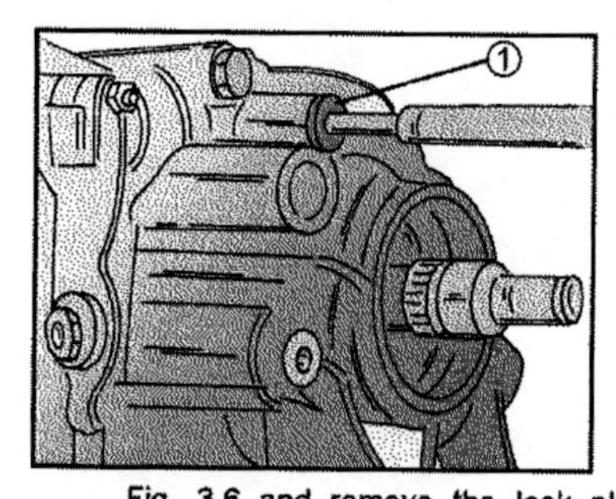

Fig. 3.6 – After removal of the plug (1) the parts shown in Fig. 3.7 can be removed. An Allen key is required to remove the plug.

Removal:

- Unlock, slacken and remove the nut securing the propeller shaft flange. A metal rod can be inserted into one of the flange holes to prevent the flange from rotating. Then remove the drive flange. A two-arm puller can be used to remove a tight flange.
- Unscrew the plug for the 5th speed and reverse gear lock at the position shown in Fig. 3.6 and remove the lock plunger and the spring. Fig. 3.7 shows this operation.

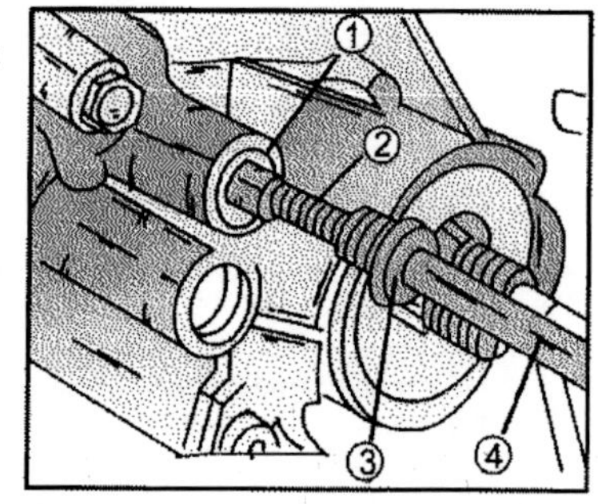

Fig. 3.7 – Removal of the lock for the 5th and reverse gear.

1 Locking pin
2 Spring
3 Bolt with internal hexagon
4 Allen key

- The selector lever for the 5th speed and the reverse gear must be removed. Remove the circlip and the washer from the selector shaft as shown in Fig. 3.8 and push the shaft towards the inside. Also remove the bolt shown by the arrow in the R.H. view of Fig. 3.8 securing the reverse gear shaft.
- The cover must be removed with a two-arm puller. The claws of the puller must be placed under the lugs of the cover, the centre spindle must press against the cover. A few tabs with a rubber or plastic mallet will facilitate the removal.
- Remove the oil seal carefully with a screwdriver.
- If further dismantling is required remove the speedometer drive gear (3), the support washer (2) and the bearing race (1) in Fig. 3.9.
- Push the selector shaft for the 5th gear towards the inside and turn the reverse gear shaft so that the bore is in line with the bore in the gearbox (Fig. 3.8). To facilitate the operation you can fit the cover provisionally and look through the bore.

All parts must be thoroughly cleaned before the cover is refitted.

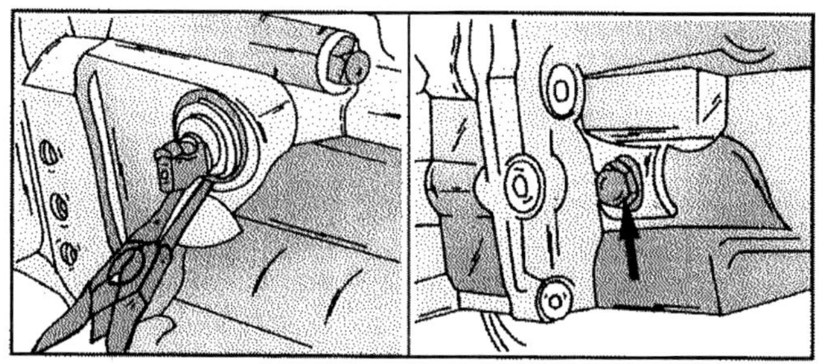

Fig. 3.8 – Removal of the retaining circlip in the L.H. view. Push the shaft towards the inside after removal. The R.H. view show the plug, underneath which you will find the reverse gear shaft.

Fig. 3.9 – Removal of the speedometer drive gear. The numbers are referred to in the text.

- Coat the cover gasket with sealing compound and fit the cover over the gearbox shaft and against the gearbox.
- Clean the threads of the cover bolts and coat them with "Loctite". Fit the bolts and tighten them evenly to 2.0 kgm (14 ft.lb.). Note that not all bolts have the same length.
- Coat the threads of the bolt for the reverse gear shaft. Insert the bolt and tighten it to 4.5 kgm (32.5 ft.lb.).
- Withdraw the selector shaft once more and fit the washer and the retaining ring as shown in Fig. 3.8. The selector lever and the lock for the 5th/reverse gear is fitted in accordance with Figs. 3.7 an d 3.6. The plug is tightened to 4.0 kgm (29 ft.lb.).
- Heat the bearing race (1) in Fig. 3.9 to 80° C (boiling water) and drive it onto the shaft. Place the washer (2) over the shaft end against the bearing race.

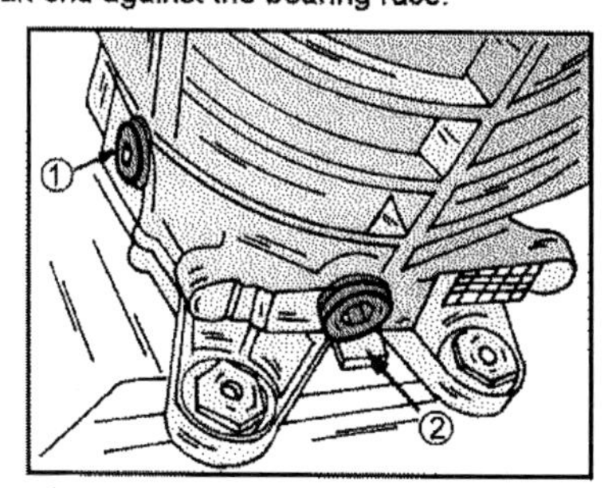

Fig. 3.10 – Location of the gearbox oil filler plug (1) and drain plug (2).

- Slide the speedometer drive gear (3) over the shaft and engage it with the drive pinion. The cut-out in the drive gear must face towards the bottom.
- Coat the sealing lip and the outside of a new sealing ring grease and drive the oil seal from the outside into the cover until the outside of the seal is flush with the cover face.
- The remaining operations are carried out in reverse order.

3.2.2. GEARBOX OIL LEVEL

The gearbox oil is only changed after the first 6000 miles and is then filled for life. To check the oil level remove the plug (1) in Fig. 3.10 (Allen key) and insert the tip of the forefinger to reach for the oil. If necessary top-up with the recommended oil. You may have to clean an oil can of old oil to use for the topping-up operation.

3.3 Gearbox - Tightening Torques

Front gearbox cover:	2.8 kgm (20 ft.lb.)
Rear gearbox cover:	2.0 kgm (14 ft.lb.)
Drive flange nut:	16.0 kgm (115 ft.lb.)
Bolt for gearchange lever:	2.5 kgm (18 ft.lb.)
Clamp nut for propeller shaft:	3.5 kgm (25 ft.lb.)
Oil drain and filler plug:	6.0 kgm (43 ft.lb.)
Propeller shaft intermediate bearing to vehicle floor:	2.5 kgm (18 ft.lb.)

4 Propeller Shaft

All vehicles are fitted with a two-part propeller shaft, with a vibration damper fitted to the front shaft. The diameter of the front and rear shaft is 60 mm in the case of most models. Only the E200D has a shaft diameter of 50 mm. The bolt circle of the flange connection is also different. The shaft is balanced for vibration-free running. The front and rear shaft are marked to each other. The shaft is correctly assembled when the yoke with the two marks are in line with the single mark on the rear shaft. The alignment is shown in Fig. 4.1.

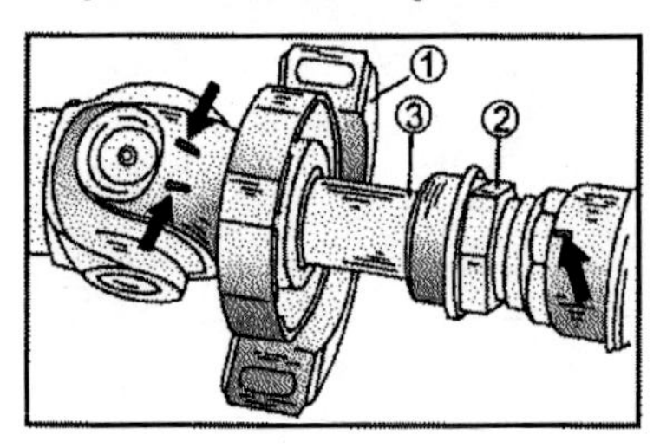

Fig. 4.1 – Marking of the propeller shaft. The markings of the front shaft (arrows) must be aligned with the mark on the rear shaft (R.H. arrow). Always assemble the shaft as shown.
1 Intermediate bearing
2 Locknut
3 Rubber sleeve

Different rubber couplings are used on the various models. Remember this when a shaft is replaced.

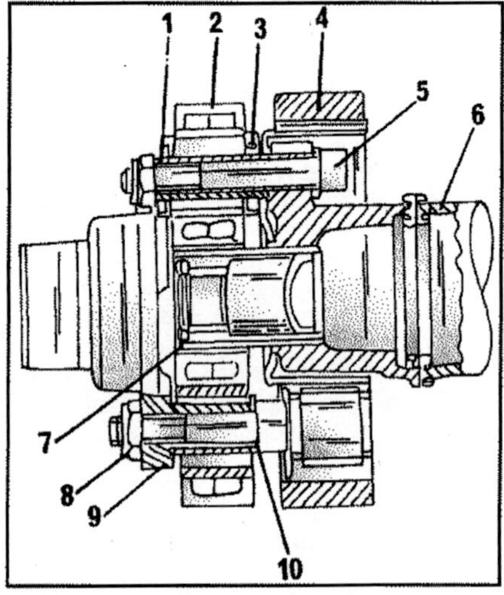

Fig. 4.2 – Sectional view of the propeller shaft with vibration damper in front of the three-arm flange.
1 Washer
2 Flexible coupling
3 Centering sleeve
4 Vibration damper
5 Socket-head bolt
6 Front propeller shaft
7 Centering sleeve
8 Hexagon nut, self-locking
9 Flexible joint coupling flange
10 Washer

The vibration damper is fitted to the front shaft, but again depending on the model version is constructed in a different manner. On vehicles with manual transmission roller sleeves are fitted, on models with automatic transmission no roller sleeves are fitted. Further differences will be found in the attachment of the vibration damper. In the case of the E200 the damper is fitted in front of the three-arm

flange, on other models, , the damper is fitted behind the flange. Figs. 4.2 and 4.3 show the differences between the two types.
Note the following points in the case of a four-cylinder model:

- In the case of some vehicles the vibration damper must be released and removed to remove the propeller shaft.
- On other models the damper must also be released to remove the front propeller shaft towards the front and the rear propeller shaft towards the rear.

4.1. Propeller Shaft – Removal and Installation

- Remove the noise dampening panel underneath the front end of the vehicle.
- Unscrew the covering panel above the exhaust silencer (four bolts in the corners) and take it out.
- At the rear of the shaft remove the transverse bracket. Tighten the nuts to 4.0 kgm (29 ft.lb.) during installation.

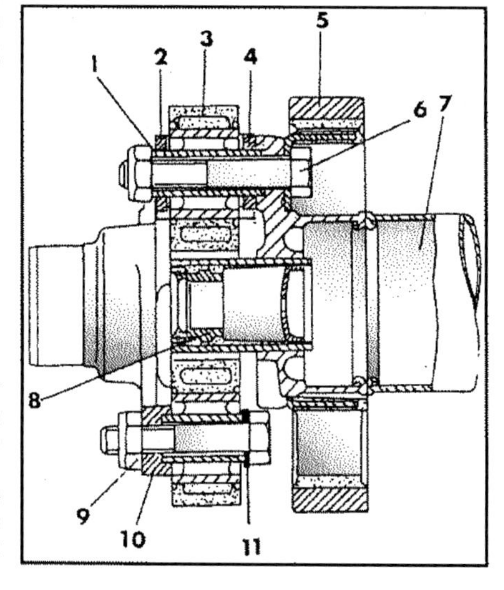

Fig. 4.3 – Sectional view of the propeller shaft with the vibration damper behind the three-arm flange, for example fitted to six-cylinder models.
1 Washer
2 Stop washers
3 Flange coupling
4 Stop washers
5 Vibration damper
6 Hexagon head bolt
7 Front propeller shaft
8 Centering sleeve
9 Self-locking nut
10 Flexible flange
11 Washer

- At the front of the shaft unscrew the reinforcement strut. Tighten the bolts to 2.5 kgm (18 ft.lb.) during installation.
- Place a jack with a suitable wooden block underneath the transmission and lift the unit slightly without forcing it.
- Remove the hexagon bolt securing the rear engine crossmember from the vehicle floor (left and right outside). Fully unscrew the bolts.
- Remove the nut in the centre of the crossmember from the mounting rubber and lift out the complete crossmember.
- Separate the propeller shaft flange from the transmission. This requires a strong Allen key and a socket and ratchet as shown in Fig. 4.4.
- Slacken the propeller shaft clamp nut by approx. 1 turn without pushing back the rubber gaiter (the gaiter will move by itself).
- Before the propeller shaft is pushed to the rear you will have to knock the dowel sleeves in the shaft flange from the rear towards the front, using a mandrel of 150 in length and 10 mm in diameter (only shaft flange with dowel sleeves).
- Disconnect the propeller shaft flange from the rear axle drive flange. The operation is carried out as shown in Fig. 4.5.

Fig. 4.4 – Removal of the flexible flange at the front of the propeller shaft. An Allen-key is required.

- Remove the bolts securing the propeller shaft intermediate bearing from the floor panel. The bearing must be tightened after the shaft has been re-connected to the transmission and gearbox flanges.

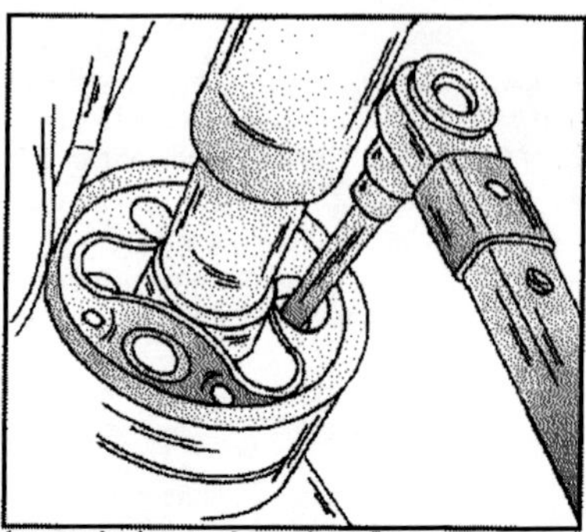

Fig. 4.5 – Removal of the propeller shaft from the rear axle flange. A socket and an open-ended spanner are required.

- Force the propeller shaft from the centering pin in the rear axle flange. Take care not to separate the propeller shaft if it is to be removed for other reasons than repair. Otherwise you will have to re-align it as shown in Fig. 4.1.
- Check all parts for damage or other faults before the shaft is refitted. If the coupling disc is damaged, replace it. This is most important, as it acts as damper during acceleration and braking. If during removal of the coupling disc the vibration damper is separated from the propeller shaft, mark the vibration damper and the three-arm flange to position them into their original position. Fig. 4.6 shows the two types of vibration dampers which can be fitted.

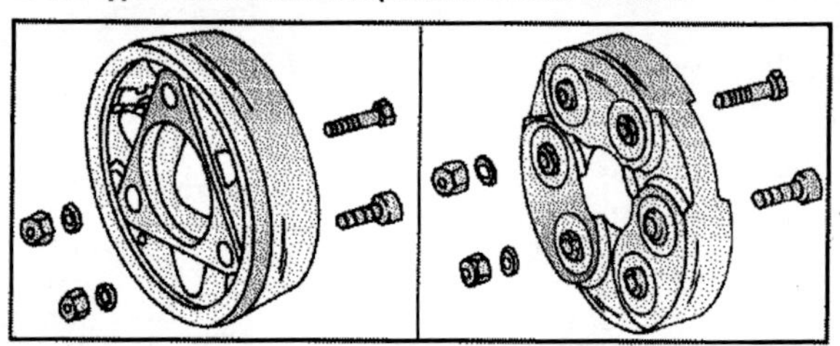

Fig. 4.6 – View of the different vibration dampers which can be fitted to your vehicle.

The installation of the propeller shaft is carried out as follows:

- Fill the cavities of the two centering sleeves with grease (approx. 2 grams per sleeve). Fit the propeller shaft with the companion plates over the centering pin of the transmission and rear axle.
- Refit the propeller shaft intermediate bearing to the vehicle floor, without tightening the bolts fully.
- Refit the vibration damper to the three-arm flange as applicable.
- Connect the front and rear ends of the propeller shaft to the drive flanges. Counterhold the bolts and tighten them to 4.0 kgm (29 ft.lb.) in the case of M10 threads or 6.0 kgm (43.5 ft.lb.) in the case of M12 threads. Measure the diameter if not sure (10 or 12 mm).
- Pay attention to the correct seat of the rollers in the vibration damper. After tightening the propeller shaft-to-transmission-connection, drive the rollers in

position. Fig. 4.7 shows the difference between the types used for a fitted manual transmission and an automatic transmission.

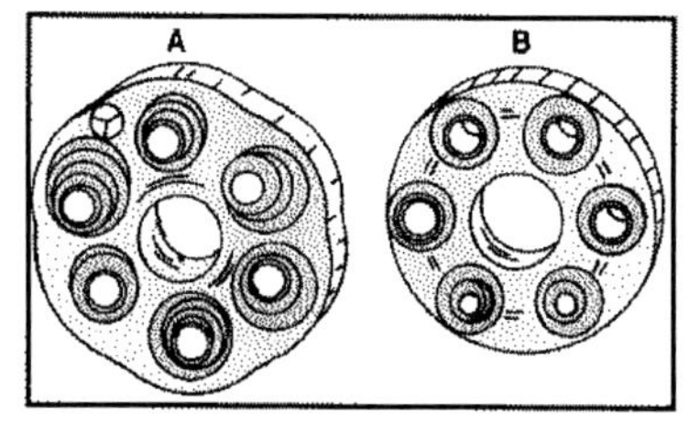

Fig. 4.7 – View of the of the different propeller shaft couplings. View A shows the soft, flexible coupling, front. Absorbs acceleration and deceleration shocks. View B shows a standard flexible coupling, fitted to the rear.

- Refit the rear mounting crossmember to the floor frame and tighten the bolts to 4.5 kgm (32 ft.lb.).
- Lower the transmission and fit the rear rubber mounting. Tighten to 4.5 kgm (32 ft.lb.).
- Lower the vehicle onto its wheels and push it backwards and forwards a few times. Jack up the vehicle once more and tighten the propeller shaft clamp nut to 3.0 – 4.0 kgm (21 – 29 ft.lb.), using a large open-ended spanner. If bolts with collar and centering point are used, tighten them to 3.0 kgm (22 ft.lb.).
- Refit the exhaust heat shield.
- Finally tighten the two intermediate bearing bolts to 2.5 kgm (18 ft.lb.). The remaining operations are carried out in reverse order.

4.2. Intermediate Bearing – Removal and Installation

The relationship of vibration damper, companion flange (coupling disc) and front propeller shaft is marked as shown in Fig. 4.1.

- Fully unscrew the nut (2) in Fig. 4.1 until the front propeller shaft can be separated from the rear shaft. Pull the rubber sleeves (3) over the splines.
- Pull the rubber mounting with the radial ball bearing and protective cap together from the yoke by means of a two-arm puller. Place the puller claws underneath the intermediate bearing, with the spindle pressing against the shaft. Also remove the rear protective cap from the universal joint fork.

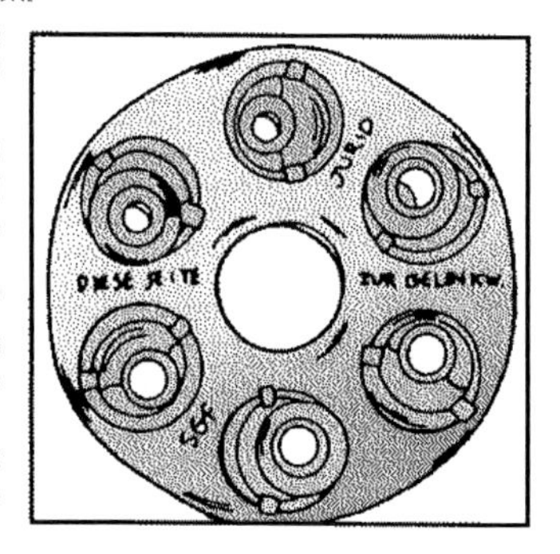

Fig. 4.8 – Correct installation of the companion plate for a vehicle with manual transmission. The side shown must face towards the propeller shaft.

- Press the radial ball bearing out of the rubber mount and slowly press a new ball bearing into the bearing seat until it rests against the contact surface.
- Pull the rubber sleeve over the splines of the yoke. Make sure that the sleeve is wells seated at the smaller diameter. Coat the splines with grease.
- Assemble the front and rear propeller shaft in accordance with Fig. 4.1 and in accordance with the marks in the shaft.

If the centering sleeves in the front or rear propeller shafts are worn there is no need to replace the propeller shaft. The sleeves can be replaced, but we recommend to have this carried out in a workshop.

NOTE: If the companion flange on a propeller shaft for a vehicle with manual transmission is replaced, arrange the plate as shown in Fig. 4.8, i.e. the side marked "Diese Seite zur Gelenkw". (this side towards the propeller shaft) must be attached to the propeller shaft – note to the transmission flange.

5 Front Axle and Front Suspension

5.0. Technical Data

Wheel bearing play:	0.01 – 0.02 mm
Max. permissible play:	0.05 mm
Wheel bearing grease:	Heat resistant grease No. 00 989 49 51
Amount of grease necessary:	
- In hub grease cap:	15 grams
- In wheel bearings and hub:	70 grams

Wheel Bearing and Hub Data:	
- Bore for inner taper roller bearing:	59.117 – 59.098 mm
- Bore for outer taper roller bearing:	45.220 – 45.195 mm
- Oil seal ring bore:	64.064 – 64.000 mm
Hub flange diameter:	150.00 mm
Max. Hub flange side clearance:	0.03 mm

Torsion Bar Diameter (all E-models listed):	
- Basic model:	24.0 mm
- Basic model, six cylinder:	26.0 mm
- Sport model:	24.0 mm
- Sport model, six-cylinder:	27.0 mm

5.1. Front Axle Half – Removal and Installation

Fig. 5.1 shows the assemble front suspension on one side. The front axle half consists of the steering knuckle together with the shock absorber strut and the upper and lower suspension arm. The whole assembly can be removed as follows:

- Jack up the front end of the vehicle and place chassis stands in position. Remove the front wheels. From underneath the vehicle remove the noise dampening panel. Make sure that the vehicle is adequately supported.

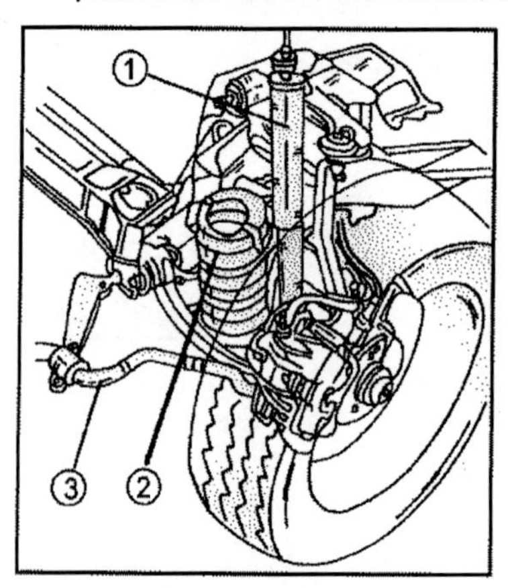

Fig. 5.1 – View of the assembled front suspension on one side.
1 Shock absorber strut
2 Coil spring
3 Stabiliser bar

- Remove the attachment of the stabiliser bar from the suspension arm. Refer to Fig. 5.2, remove the two nuts (1) and take off the clamp bracket. The nuts must always be replaced (self-locking) and are tightened to 2.0 kgm (14.5 ft.lb.).
- Withdraw the plug connectors from the wheel speed sensor for the ABS system and the brake pad wear indicator. The connectors must be located, depending which side is to be removed. Follow the cables and disconnect. Pull the

cable harness through the rubber grommets and free them from their retaining clamps.

- Disconnect the brake hose from the brake pipe and remove the hose/pipe connection from the attachment of the damper strut. Close the open ends of hose and pipe in suitable manner to prevent entry of dirt. **Note the following during installation.** The union nut is tightened to 1.5 kgm (10 ft.lb.) without allowing the brake hose to twist. After installation have the steering wheel turn from left to right and the car bounced up and down whilst checking the brake hose to make sure it cannot come near other parts of the front suspension.
- Remove the front spring as described later on.
- Remove the nut securing the track rod ball joint to the steering lever and separate the ball joint with a suitable puller. Immediately check the rubber dust cap and the ball joint for damage or excessive clearance. Fit new nuts during installation and tighten the nut to 5.0 kgm (36 ft.lb.).

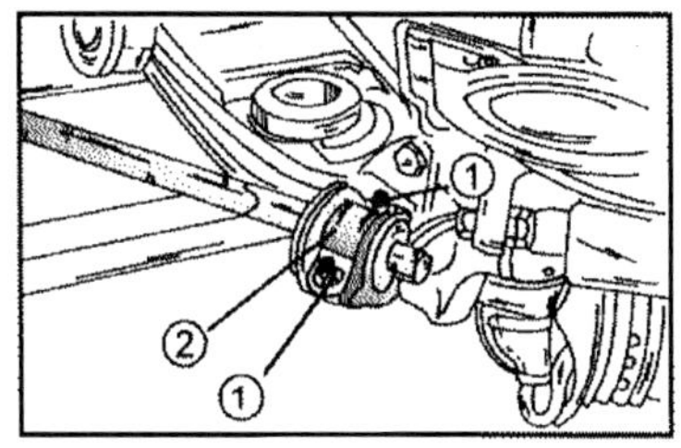

Fig. 5.2 – Attachment of the stabiliser bar to the suspension arm. Remove the bolts (1) and take off the mounting clamp (2).

- Place a trolley jack underneath the suspension arm and lift the arm until just under tension. In this position remove the nuts securing the upper end of the shock absorber strut. A locknut and a securing nut or a single nut are used, depending on the fitted type. The removal and installation of the shock absorber strut is described later on. Remove the metal washer from the end of the piston rod.
- Remove the nut securing the ball joint to the upper suspension arm and separate the connection with a suitable puller. Fit a new nut during installation. Tighten it to 4.5 kgm (32.5 ft.lb.).
- Mark the position of the two eccentric bearing pins for the lower suspension arm on the bolt head in relation to the bearing on the frame, remove the two nuts and drive out the bolts. If the vehicle is already of some age it may be possible that bearing bolts with hexagonal head are used. This signifies that the bolts are repair bolts which also will have different washers. Washers of normal bolts are centering washers, the other type is eccentric. The installation of the bolts is described later on. The attachment of the suspension arm is shown in Fig. 5.3.
- The front axle half can now be removed.

The front and rear bushes of the lower suspension arm can be replaced, but we recommend to take the suspension to a workshop to have the bushes replaced, if you find them worn beyond use.

The installation of the front axle half is a reversal of the removal procedure, but the following points should be noted:

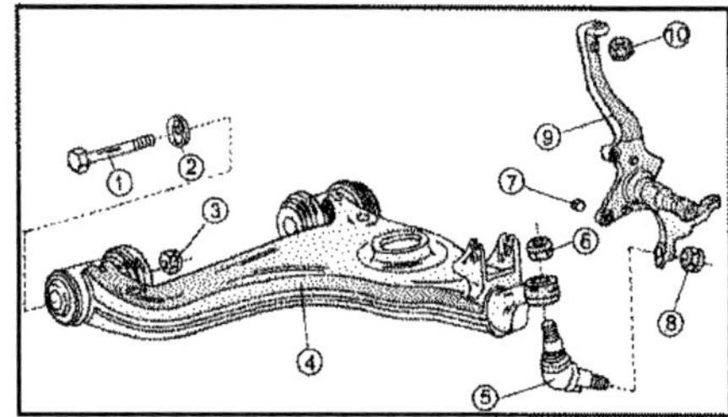

Fig. 5.3 – The lower suspension arm together with the steering knuckle.

1 Eccentric bearing bolt
2 Eccentric washer
3 Bolt, 15.0 kgm
4 Lower suspension arm
5 Suspension ball joint
6 Nut, 10.5 kgm
7 Dowel pin
8 Nut, 10.5 kgm
9 Steering knuckle
10 Nut, 4.5 kgm

- The tightening torques specified above must be observed.
- Place the front axle half onto a mobile jack and slowly lift it in position to guide the piston rod through the opening in the mounting bearing in the body.
- Place the metal washer over the end of the piston rod and fit the first nut. Tighten the mounting as described during the installation of the shock absorber strut.
- Fit the suspension arm to the frame crossmember. To do this align the eccentric bolts in accordance with the marks made before removal and hand-tighten the nuts. Insert the eccentric bolts from the front and push the towards the rear. As the eccentric bolts are used to adjust the camber angle (front bolt) and the castor (rear bolt) you will have to have the wheel alignment checked after completed installation.

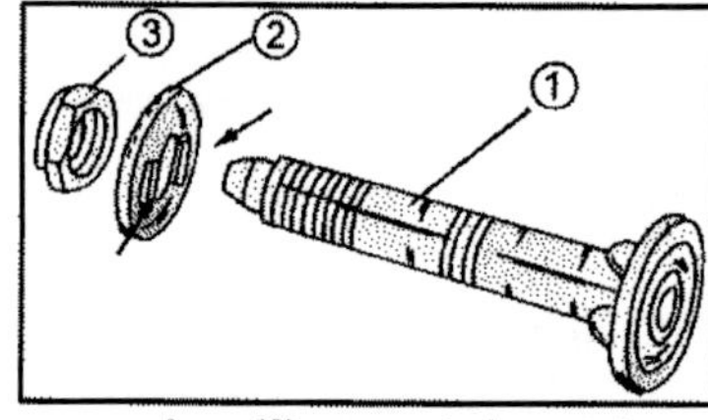

Fig. 5.4 – View of a normal bearing bolt. The arrows point to the cut-out where you can identify a centering washer. An eccentric washer only has one cut-out.

Also note the following in connection with the eccentric bolts:

- On a new vehicle the bolts have the shape shown in Fig. 5.4, i.e. you will recognise them on the bolt head. The washers (2) are centering washers.
- If the bolts have a hexagonal head repair bolts are fitted. In this case the centering washers are replaced by eccentric washers. If that type of bolt is fitted you must mark their installed position. Otherwise the front wheel alignment must be checked in a workshop.

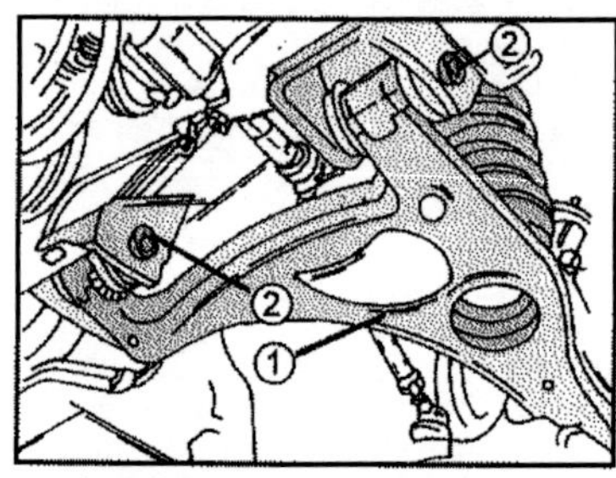

Fig. 5.5 – View of a suspension arm (1) from below. The bearing bolts (2) are fitted at the positions shown.

- The tightening torque of the nut (3) in Fig. 5.4 is 15.0 kgm (108 ft.lb.) at the front and the rear.
- Fig. 5.5 shows the fitted suspension arm from below with the location of the two bolts.
- Fit the stabiliser bar mounting clamp to the suspension arms. To facilitate the installation of the stabiliser bar you can lift the suspension arm on the opposite side with a jack.
- Re-connect the brake hose. Note the note already given earlier on.
- Re-connect all electrical leads. Push the rubber sleeve into place. Sharp metal edges could damage the cables.
- Re-connect the track rod ball joint to the steering lever and tighten the nut. If the ball joint pin rotates it will be possible to insert an Allen key into the end of the pin to prevent it from rotating.
- If removed, refit the two suspension ball joints and tighten them. The self-locking nuts must always be replaced. Observe the tightening torque.
- Fit the wheels and lower the vehicle to the ground. Check once more that the eccentric bolts are fitted in accordance with the marks made before removal and tighten the nut on the other side to 15.0 kgm (108 ft.lb.).
- Have the front wheel alignment and the headlamp adjustment checked in a workshop.

5.2. Front Springs – Removal and Installation

A spring compressor is necessary to remove the front springs. As the special compressor used in a workshop may not be available you can use a standard compressor, the claws of which are placed over three to four of the coils. Make sure the compressor has adequate strength for the springs. Remove a spring as follows:

- Place the front end of the vehicle on chassis stands and remove the front wheel. Also remove the noise encapsulation underneath the front end of the vehicle.
- Place the spring compressor over the spring coils as described above until the upper end lower ends are free of their spring seats.
- Place a mobile jack underneath the suspension arm and raise the arm slightly. The arm must not hang down under its own weight.
- Remove the shock absorber piston rod as described in Section 5.3 and slowly lower the jack until the spring is free and can be removed.

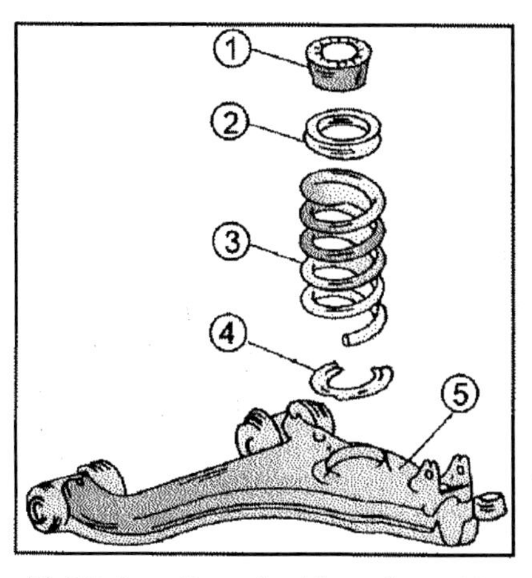

Fig. 5.6 – View of the coil spring as fitted to the suspension arm. The upper stop is not the same on all models.
1 Upper stop
2 Upper spring seat
3 Coil spring
4 Lower spring seat
5 Lower suspension arm

The installation is a reversal of the removal procedure. Make sure that the rubber seat in the floor frame is correctly positioned and the spring coil end is contacting the spring seat in the suspension arm. Tighten the nut at the upper end of the shock absorber as described in the next section.

Release the spring compressor slowly, checking that the spring is fitted properly at the upper end lower ends. Fit. 5.6 shows the parts of the spring and its position on the suspension arm.

The front wheel alignment and the headlamp adjustment must be checked after the completed installation.

Note: As the coil spring is clamped between the upper end lower spring seat it is also possible to remove a spring without removing the shock absorber at the upper end. In this case you will have to compress the spring as described above until it is free to be removed between the two spring seats.

5.3. Front Shock Absorbers – Removal and Installation

The shock absorbers are fitted between the body and the lower suspension arms. The attachment at the upper end is by means of a securing nut and a locknut, as shown in Fig. 5.7. The lower end is either secured with a bolt, nut and a washer or by means of two bolts. Figs. 5.8 and 5.9 show the two types of attachment.

As the piston rod may rotate when the lower nut is slackened the workshop will use a special wrench to prevent the piston rod from rotating. The tool is shown in Fig. 5.10, but we must mention that it may be sometimes difficult to slacken the lower nut without preventing the piston rod from rotating.

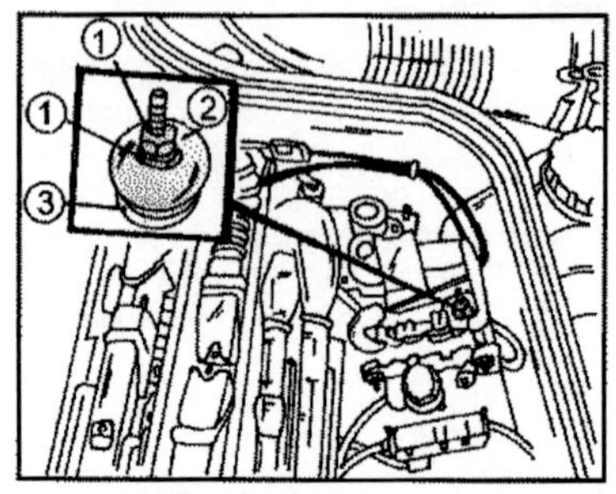

Fig. 5.7 – Attachment of the shock absorber at the upper end.
1 Securing nuts
2 Metal washer
3 Rubber bearing (bush)

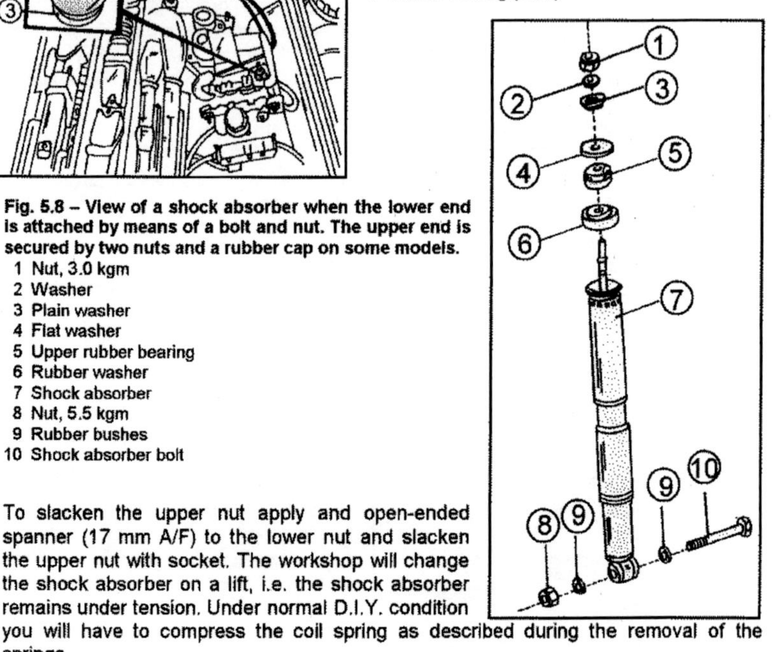

Fig. 5.8 – View of a shock absorber when the lower end is attached by means of a bolt and nut. The upper end is secured by two nuts and a rubber cap on some models.
1 Nut, 3.0 kgm
2 Washer
3 Plain washer
4 Flat washer
5 Upper rubber bearing
6 Rubber washer
7 Shock absorber
8 Nut, 5.5 kgm
9 Rubber bushes
10 Shock absorber bolt

To slacken the upper nut apply and open-ended spanner (17 mm A/F) to the lower nut and slacken the upper nut with socket. The workshop will change the shock absorber on a lift, i.e. the shock absorber remains under tension. Under normal D.I.Y. condition you will have to compress the coil spring as described during the removal of the springs.

- Place the front end of the vehicle on chassis stands and remove the front wheel.
- Attach the spring compressor as described in Section 5.2 until the spring can be removed from within the spring seats. You can also leave the spring in position.
- Remove the securing nuts from the shock absorber piston rod (Fig. 5.7 or 5.10) and remove the metal disc and the rubber bush. On the R.H. side it may be necessary to remove a relay to reach the mounting.
- Place a mobile jack underneath the suspension arm and slightly lift the arm without compressing the oil spring.
- At the lower end of the damper remove the nut and the washer and knock out the bolt. Mark where the bolt head is located. Fig. 5.11 shows the attachment of the shock absorber when a bolt with nut are used. On the other version remove the two bolts shown in Fig. 5.9.
- Slowly lower the jack until the upper end is free and the shock absorber can be disengaged from the suspension arm, push the damper slightly upwards and remove it out of the vehicle. If a new shock absorber is to be fitted check the marking as only a shock absorber with the same identification must be fitted. It is possible to replace a shock absorber on one side only.

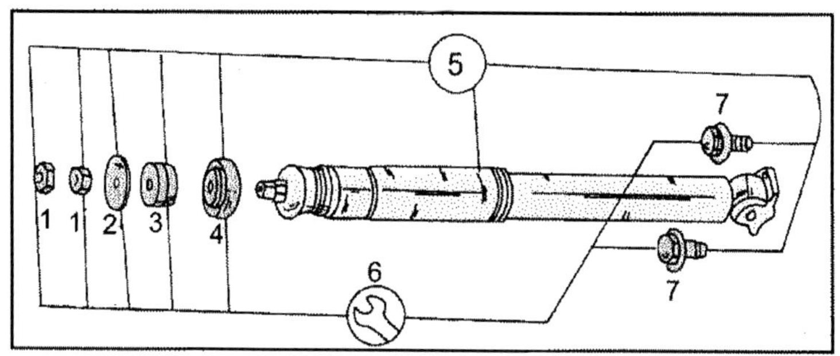

Fig. 5.9 – View of a shock absorber when the lower end is attached with two bolts. On this version the suspension arm shown in Fig. 5.6 is also slightly different, as the two mounting lugs on the outside are cast into the arm.

1 Shock absorber nuts, 3 kgm
2 Plain washer
3 Rubber bush
4 Rubber bearing
5 Shock absorber
6 Repair kit
7 Lower mounting bolts

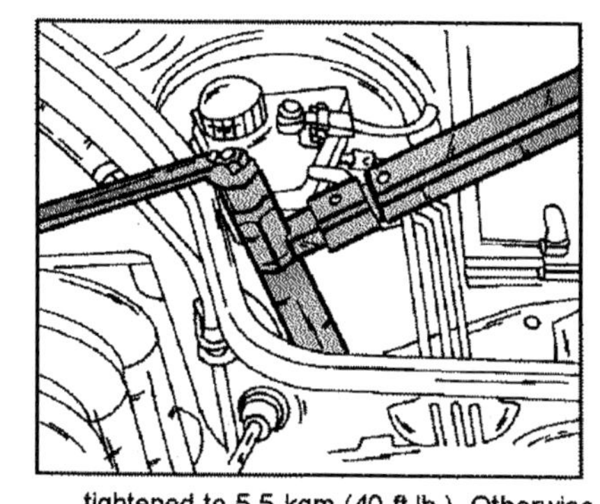

Fig. 5.10 – Special tool to slacken and tighten the upper shock absorber mounting. The tool prevents the piston rod from rotating.

The installation is a reversal of the removal procedure. The following points must be observed as you proceed:

- Place the parts over the upper end of the damper in the order shown in Fig. 5.12 and insert the shock absorber from below. Fit the bolt, washer and nut to the lower end. A new nut must always be used and is tightened to 5.5 kgm (40 ft.lb.). Otherwise attach the shock absorber with the two bolts.

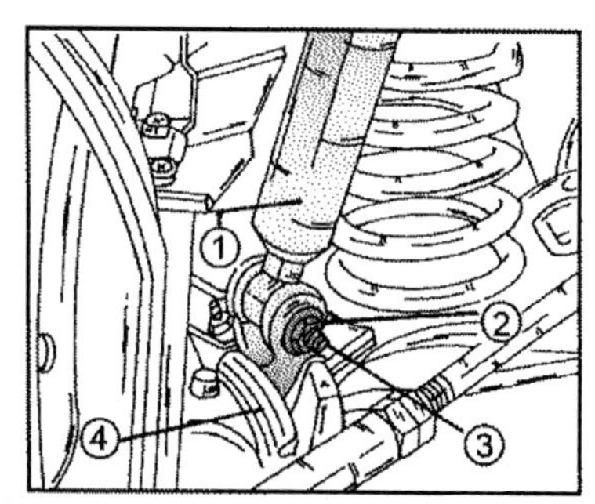

Fig. 5.11 – Attachment of the shock absorber at the lower end when a nut and bolt is used.

1 Shock absorber
2 Nut
3 Mounting bolt
4 Suspension arm

- Fit the rubber bush and the dished washer to the upper end and fit and tighten the first nut to 1.5 – 1.8 kgm (11 – 13 ft.lb.).
- Screw on the second nut, counterhold the lower nut with a 17 mm open-ended spanner and tighten the nut to 3.0 kgm.

- Release the front spring and reset it into its seats. After installation check that the shock absorber identification marks can be seen from the front.

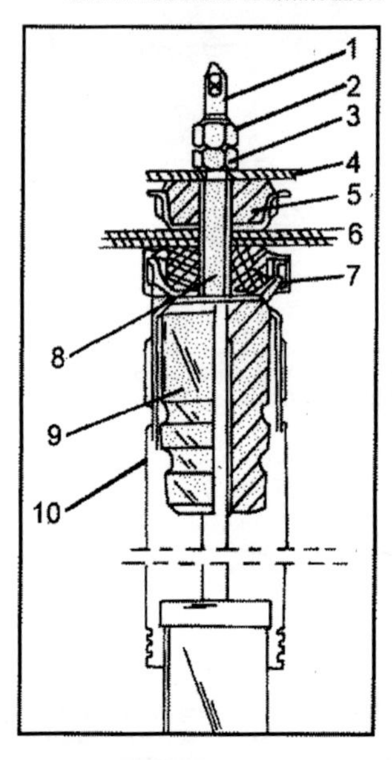

Fig. 5.12 – Sectional view of the upper shock absorber end with the location of the mounting parts.

1 Piston rod
2 Upper nut. 3.0 kgm
3 Lower nut, 1.5 – 1.8 kgm
4 Dished washer
5 Upper rubber bush
6 Body panel
7 Lower rubber bush
8 Spacer sleeve
9 Rebound rubber
10 Protective gaiter

5.4. Wheel Bearing Clearance – Adjustment

The max. clearance of the wheel bearing hub flange is 0.03 mm. Check as follows:

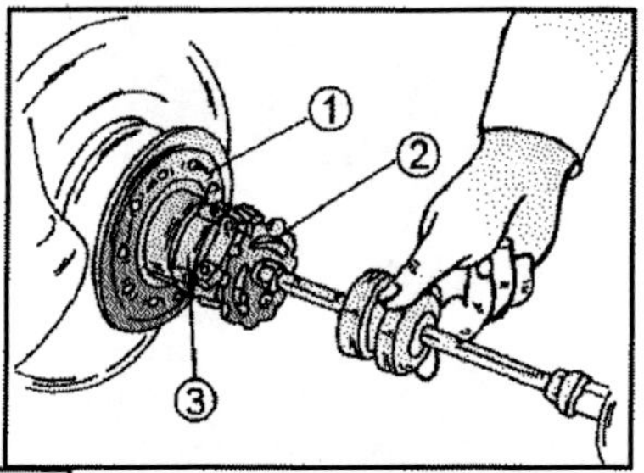

Fig. 5.13 – Removal of the hub grease cap with the special puller.

1 Front wheel hub
2 Special puller
3 Grease cap

- Place the front end of the vehicle on chassis stands and remove the wheel. Secure the brake disc to the wheel hub by using two of the wheel bolts and suitable spacers to eliminate any side play of the disc.

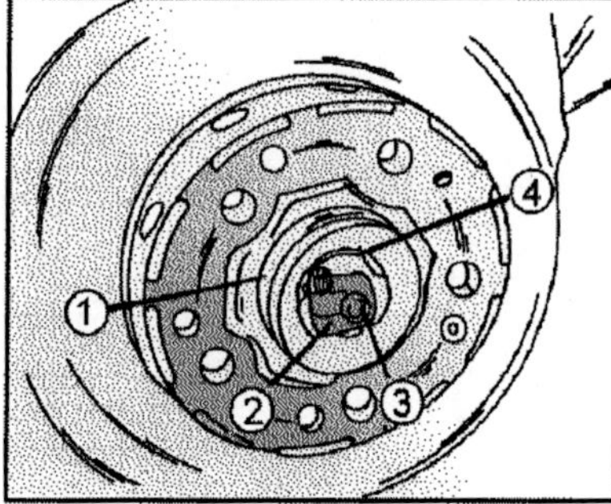

Fig. 5.14 – Details for the adjustment of the wheel bearings. The numbers are referred to in the text.

- Push the brake pads away from the disc. If difficult to move you can slacken the upper caliper bolt and push the caliper to one side.

- Remove the hub grease cap. Normally a special extractor with an impact hammer as shown in Fig. 5.13 is used. Otherwise try to knock off the cap by carefully inserting a screwdriver at different points of the circumference.

- Refer to Fig. 5.14 and slacken the socket head bolt securing the clamp nut (2). Slowly rotate the wheel hub (1) and at the same time tighten the clamp nut until the hub can just be rotated. From this position slacken the nut by 1/3 of a turn. Release any tension by hitting the end of the axle stump (3) with a plastic mallet.

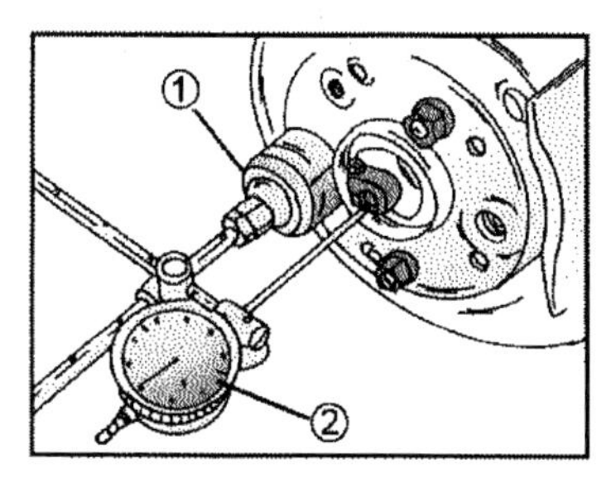

Fig. 5.15 – Checking the wheel bearing clearance. Attach the dial gauge holder (1) and the dial gauge (2) as shown.

- Attach a dial gauge as shown in Fig. 5.15 and set the dial gauge to a pre-tension of 2 mm. Grip the hub flange and move it in and out at the same time observing the dial gauge reading. The final reading should be between 0.01 and 0.02 mm. The wheel hub must be rotated before any new measurement takes place, but nor during the actual measurement.
- Tighten the socket head bolt of the clamp nut to 0.8 kgm (6.5 ft.lb.) and re-check the hub clearance. If the wheel bearing is correctly adjusted it must be possible to move the thrust washer (4) between the outer wheel bearing race and the clamp nut with the tip of a screwdriver.
- Fill the grease cap with a little grease and knock it over the wheel hub. Use the grease available from tour dealer (available in tubes of 150 grams). Only 15 grams are smeared into the grease cap.
- Refit the front wheel and lower the vehicle to the ground. Tighten the wheel bolts.

5.4. Front Wheel Hubs and Wheel Bearings

5.4.1. Removal and Installation

The wheel hub and wheel bearings can only be removed as a single item and must be replaced as such. Repair kits are available.

- Proceed as described above until the hub grease cap has been removed. Remove the radio interference contact spring.
- Remove the brake caliper and tie it with a piece of wire to the spring strut. Do not disconnect the brake hose. Remove the brake disc.
- Slacken the clamp nut (2) in Fig. 5.14, unscrew it and then remove the washer,
- Remove the front wheel hub. A tight hub can be removed as shown in Fig. 5.13. The puller is bolted to the wheel hub flange.
- If the inner wheel bearing race has remained on the axle stump you will have to remove it with a suitable puller. Finally remove the oil seal from the axle stump. Thoroughly clean all parts of the wheel hub and the axle stump and check for visible wear.

Refit the wheel hub as follows:

- Slightly cover the running area for the oil seal on the axle stump with wheel bearing grease, push the assembled wheel hub over the stump and drive it in position, using a plastic or rubber mallet.
- Fit the thrust washer and screw on the clamp nut. Adjust the wheel bearing clearance as described above.
- The remaining operations are carried out in reverse order. The caliper mounting bolts are tightened to 11.5 kgm (83 ft.lb.).

5.5. Steering Knuckle – Removal and Installation

- Place the front end of the vehicle on chassis stands and remove the front wheel.
- Remove the brake caliper and tie it to the front suspension, using a piece of wire. Remove the brake disc.
- Remove the front wheel hub as already described.
- Remove the wheel speed sensor (ABS) after unscrewing the Allen head bolts from the steering knuckle. New bolts must be used during installation. Tighten them to 2.2 kgm (16 ft.lb.).

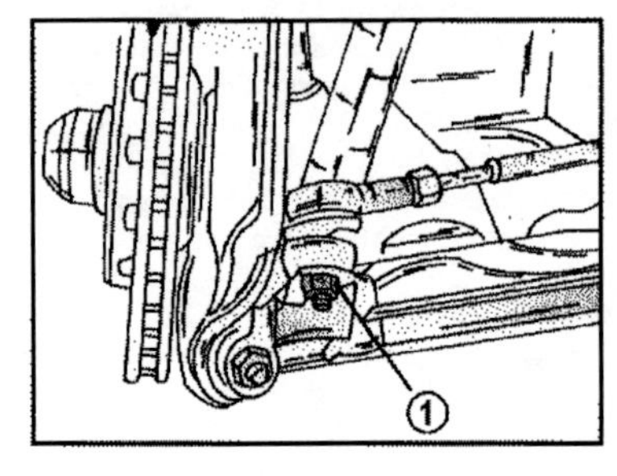

Fig. 5.16 – The nut (1) secures the track rod ball joint to the steering lever.

- Separate the track rod ball joint from the lever on the steering knuckle, using a suitable puller. Fig 5.16 shows the nut to be removed. The rubber dust cap must be checked for damage before re-connecting the joint. The self-locking nut must be replaced and it tightened to 5.5 kgm (40 ft.lb.). If the ball joint stud rotates during installation you can insert an Allen key into the end of the stud when the nut is tightened.
- At the bottom of the steering knuckle remove the nut securing the suspension ball joint and separate the joint with a suitable ball joint puller.
- Remove the bolts in the centre of the stub axle and take off the splash shield. Tighten the splash shield to 2.2 kgm (16 ft.lb.) during installation.
- In a similar manner as described for the lower suspension ball joint separate the upper joint from the steering knuckle. A new nut must in this case be tightened to 4.5 kgm (32.5 ft.lb.).
- Remove the steering knuckle from the front suspension. Check the suspension ball joints as described below and replace them if necessary.

The installation is a reversal of the removal procedure, observing the tightening torques already given. Also note the following points:

- Fit the wheel hub and adjust the wheel bearing clearance as described above.
- Check and adjust the front wheel alignment and the headlamp setting or have them checked professionally.

5.6. Suspension Ball Joints

Before replacing the upper or lower ball joint you can check them as follows:

- Place the front end of the vehicle on chassis stands without removing the wheels.
- First check the rubber dust caps. If cuts or other damage can be seen there is no need for further checks. If grease can be seen emerging from the rubber dust cap, replace the upper suspension arm as a complete item. The lower ball joint can be replaced separately.
- To check the ball joints for excessive wear ask a helper to move the wheel to and fro (inside to outside) and check for visible play. Again the upper suspension arm must be replaced if the upper joint shows excessive clearance (play). Fig. 5.17 shows the two joints in question with the items to be checked.

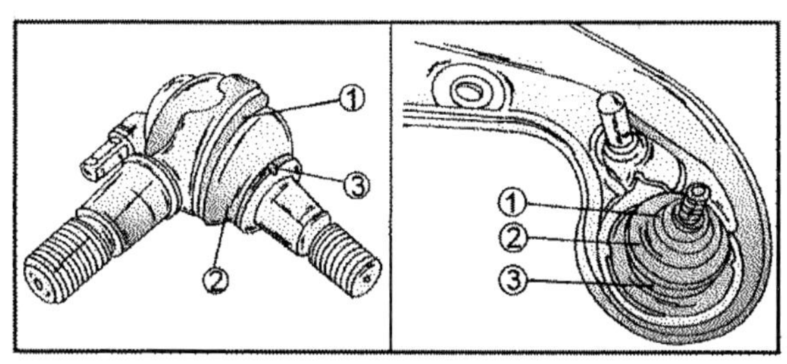

Fig. 5.17 – View of the lower suspension joint on the L.H. side and the upper joint on the R.H. side. The retaining clips (1) and (3) and the rubber dust cap (2) must be checked before re-installation.

5.6.1. Lower Suspension Ball Joints – Replacement

Suitable pullers are required to separate the ball joint from its seat in the steering knuckle and in the lower suspension arm. Both ends are secured in similar manner.

- Remove the nut at the lower end from the ball joint-to-steering knuckle connection and separate the joint with a suitable ball joint extractor (puller).
- Remove the nut securing the ball joint to the suspension arm and separate the joint with a suitable puller. Ball joints are greased for life, i.e. a damaged rubber dust cap requires the fitting of a new joint. New self-locking nuts must also be used.
- Clean the ball joint stud from oil or grease and refit it. Tighten the nut on the suspension arm side to 8.0 kgm (58 ft.lb.) and on the steering knuckle side to 105 kgm (76 ft.lb.).
- All other operations are carried out in reverse order.

5.7. Upper Suspension Arms – Removal and Installation

The upper suspension arm is secured by means of a bolt and nut in the inside of the wheel housing and with a ball joint to the upper end of the steering knuckle. A ball joint puller is required. Different operations apply to the L.H. and R.H. suspension arm.

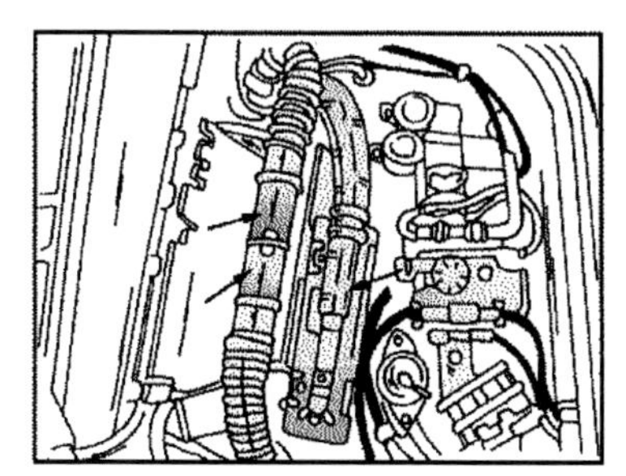

Fig. 5.18 – View of the connector plugs to be removed during the removal of the R.H. suspension arm.

On both sides

- Place the front end of the vehicle on chassis stands and remove the wheel.
- Remove the nut at the bottom of the ball joint and separate the joint from the steering knuckle. To prevent the steering knuckle from tilting towards the outside, tie it up with a piece of wire. The brake hose and the electrical cable must not be under tension. Tighten the ball joint nut to 4.5 kgm (32.5 ft.lb.).

On the R.H. side

- Disconnect the battery earth cable.

- Disconnect the cable connectors from the control elements shown in Fig. 5.18, locate the securing bolts and remove the control elements together with the support housing.
- Remove the self-locking nut at the inside of the suspension arm (noting where the bolt head is located) and knock out the bearing bolt. Hold the suspension arm as it will drop as soon as the bolt is free. Tighten bolt and nut to 5.5 kgm (40 ft.lb.) during installation.

On the L.H. side

- Remove the windscreen washer fluid reservoir.
- Remove the bolt and nut securing the suspension arm as described for the R.H. version. Nut and bolt are tightened as specified above.

The installation is a reversal of the removal procedure, noting the tightening torques already given. The ball joint must be free of oil or grease before installation.

5.8. Lower Suspension Arms – Removal and Installation

The suspension arms are secured at the inside by means of two bearing bolts to the chassis and at the outside with a ball joint to the steering knuckle. As the shock absorber must be detached at its lower end from the suspension arm, it will be necessary to undo the lower attachment when the wheels are resting on the ground or place a jack underneath the suspension arm and lift the arm until the shock absorber is slightly compressed. Before proceeding check the attachment of the shock absorber at the lower end.

- In the case of a new vehicle the bolts have the shape shown in Fig. 5.4, i.e. they can be recognised on the bolt head. The washers (2) are centering washers.
- If the bolts have a hexagonal head, they have been replaced previously. In this case the centering washers have been replaced by eccentric washers. The installed position of that type of bolt must be marked before removal to obtain the correct wheel alignment after installation.

Remove the suspension arm as follows:

- Place the front end of the vehicle on chassis stands and remove the front spring.
- Remove the stabiliser bar mounting from the suspension arm. To do this, remove two nuts and take off the clamp. Fig. 5.3 shows the attachment. The two nuts (new) are tightened to 2.0 kgm (14 ft.lb.).
- Remove the ball joint nut and separate the ball joint with a suitable puller. Check the rubber dust cap and the joint before installation. The nut (new) is tightened to 10.5 kgm (75 ft.lb.).
- Place a jack underneath the suspension arm and lift it until just under tension. The shock absorber can now be removed from the lower suspension arm. The securing nut (new) is tightened to 5.5 kgm (40 ft.lb.).
- Mark the installed position of the mounting bolts (eccentric bolts) on the chassis frame and the suspension arm, remove the two nuts and drive out the bolts. If a bolt with a hexagonal head is fitted, with different washers, the installed position MUST be marked. This is not too important on the other type. Note on which side the bolt heads are located. On all types remove the washers.

The installation is carried out as follows:

- Note the tightening torques already specified during the removal instructions.
- Fit the suspension arm to the frame crossmember. Align the eccentric bolts in accordance with the markings made before removal and fit and tighten the nuts finger-tight. Insert the eccentric bolts from the front and push them towards the

rear. As the eccentric bolts determine the wheel alignment it will be necessary to have it checked professionally.

- The tightening torque of the nuts at the front and rear is 15.0 kgm (108 ft.lb.),but the nuts must be tightened when the wheels are resting on the ground. Fig. 5.5 shows where you will find the two bolts and nuts.
- Refit the stabiliser bar to the suspension arm. To facilitate the installation of the bar it is possible to lift the opposite suspension arm slightly with a jack.

5.9. Front Wheel Alignment

The front wheel alignment cannot be measured by ordinary means when it comes to camber and castor checks as optical instruments are necessary and must therefore be carried out by a Mercedes dealer.

5.10 Front Suspension – Tightening Torques

Nuts, bearing bolts for suspension arms: .... 15.0 kgm (108 ft.lb.)
Nut, upper shock absorber mountings: .... 1.5 kgm (11 ft.lb.)
Lower shock absorber mounting: .... 5.5 kgm (40 ft.lb.)
Brake hose to brake pipe: .... 1.5 kgm (11 ft.lb.)
Stabiliser bar to suspension arms: .... 2.0 kgm (14 ft.lb.)
Track rod ball joint nuts: .... 5.0 kgm (36 ft.lb.)
Lower suspension arm ball joint nuts: .... 10.5 kgm (75.5 ft.lb.)
Upper suspension arm ball joint nuts: .... 4.5 kgm (32.5 ft.lb.)
Wheel hub bearing clamp nuts: .... 0.8 kgm (6 ft.lb.)
Brake disc splash shield bolts: .... 2.2 kgm (16.0 ft.lb.)
Brake caliper mounting bolts: .... 11.5 kgm (83 ft.lb.)
Wheel speed sensor to steering knuckle: .... 2.2 kgm (16.0 ft.lb.)
Brake disc splash shield bolts: .... 2.2 kgm (16.0 ft.lb.)
Wheel bolts: .... 11.0 kgm (80 ft.lb.)

6 Rear Axle and Rear Suspension

The so-called multi-link, independent rear suspension axle comprises the rear axle carrier, the rear axle centre piece, the rear axle shafts and the wheel carriers with double-row ball bearings. The wheel carriers are located by 5 specially arranged links, referred to as camber strut, pulling strut, pushing strut, track rod and spring links. Fig. 6.1 shows a view of the assembled rear suspension with the location of the individual struts and other parts. Fig. 6.3 shows the attachment of the links to the rear suspension.

6.0. Technical Data

Type: .... See above
Shock absorbers: .... Gas-filled, double-acting
Wheel bearing diameter: .... 45.020 – 45.011 mm
Max. run-out of drive shaft flange .... 0.12 mm (0.005 in.)
Oil capacity of differential: .... 0.7 litres
Oil grade: .... Hypoid oil

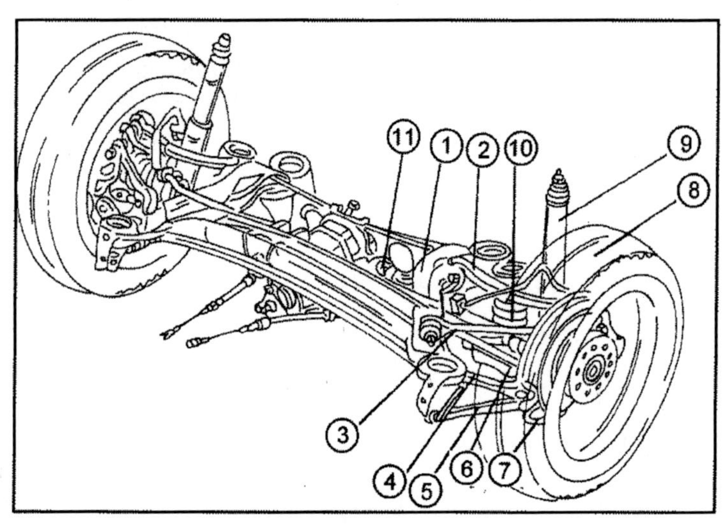

Fig. 6.1 – View of the assembled rear axle.

1 Rear axle carrier
2 Camber strut
3 Pulling strut
4 Track rod
5 Pushing strut
6 Spring link
7 Wheel carrier
9 Shock absorber
10 Rear spring
11 Rear drive shaft

6.1. Shock Absorbers – Removal and Installation

The shock absorbers act as rebound stop for the rear wheels. For this reason remove the shock absorber only when the wheels are resting on the ground or when the spring link is supported from underneath, i.e. the shock absorber is compressed. Shock absorbers are marked with colour lines and only a unit with the same marking must be fitted. Remove a damper as follows:

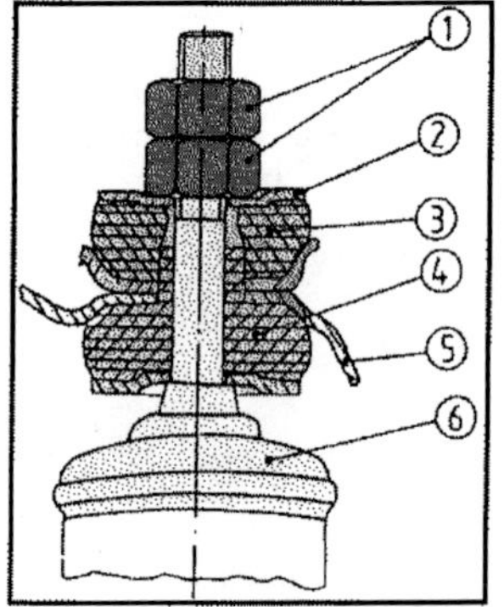

Fig. 6.2 – Sectional view of the upper shock absorber mounting.

1 Securing nuts
2 Metal plate
3 Upper rubber bearing
4 Lower rubber bearing
5 Body panel
6 Shock absorber

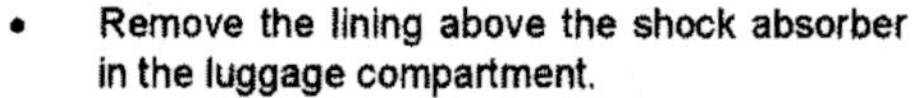

- Remove the lining above the shock absorber in the luggage compartment.
- Refer to Fig. 6.3 and remove the two nuts (1), lift off the washer (2) and remove the rubber bearing (3) from the piston rod. If the shock absorber tube is turning when the nuts are unscrewed, hold the tube manually in the wheel arch. A 17 mm open-ended spanner, which has been ground to a thickness of 5 mm, is required to slacken the locknut.

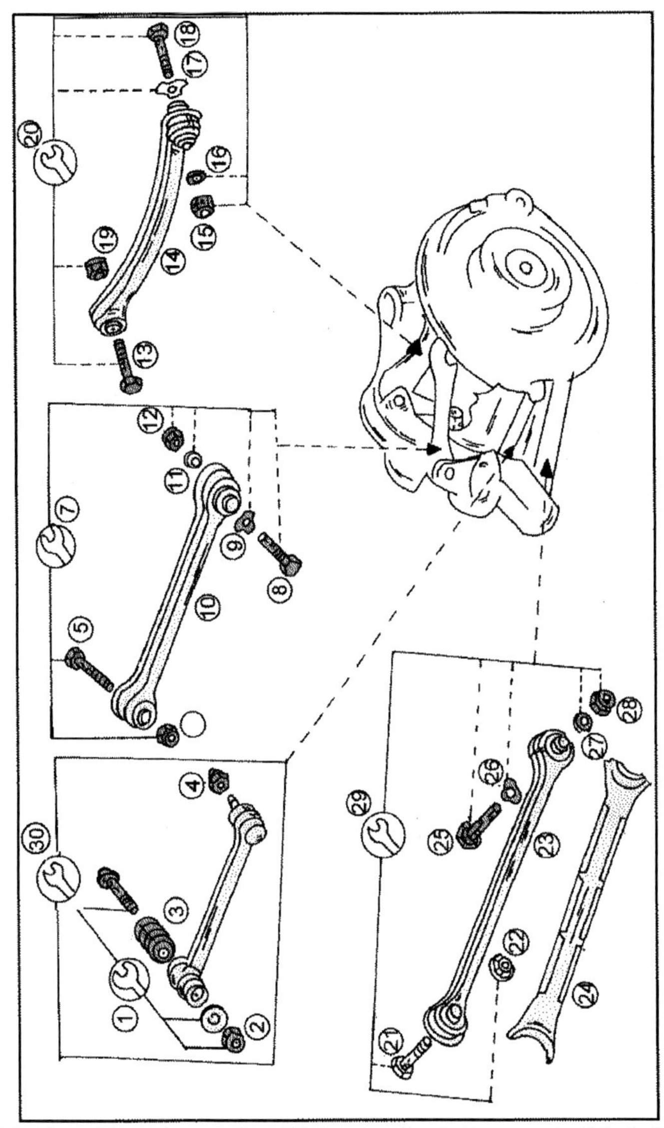

Fig. 6.3 – View of the various struts and their connection to the rear axle. Refer to the next page for the various items.

- Place a jack underneath the rear axle and lift up the assembly until the spring is compressed. Remove the spring link cover securing clamps and take off the cover.

Legend for Fig. 6.3.

1 Repair kit for track rod
2 Nut, 7.0 kgm
3 Rubber bush
4 Nut, 3.5 kgm
5 Bolt, pulling strut
6 Nut, 7.0 kgm, pushing strut
7 Repair kit for pushing strut
8 Bolt, pulling strut
9 Lock plate
10 Pulling strut
11 Plain washer
12 Nut, 4.5 kgm
13 Bolt for camber strut
14 Camber strut
15 Nut, 4.5 kgm
16 Washer
17 Lock plate
18 Bolt, pulling strut
19 Nut. 7.0 kgm, pulling strut
20 Repair kit for pulling strut
21 Bolt, pushing strut/rear axle
22 Nut, 7.0 kgm
23 Pushing strut
24 Reinforcement strut
25 Bolt, pushing strut
26 Lock plate
27 Washer
28 Nut, 4.5 kgm, pushing strut
29 Repair kit for pushing strut
30 Track rod and parts

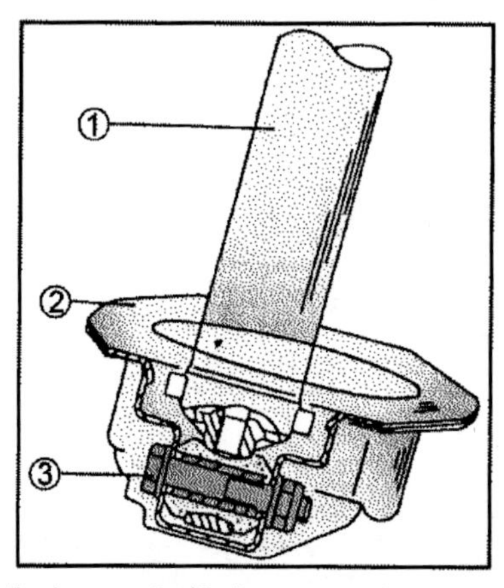

Fig. 6.4 – Attachment of the rear shock absorbers at the lower end.
1 Shock absorber
2 Spring link
3 Bolt with nut

- Refer to Fig. 6.4 and remove the bolt and nut securing the lower shock absorber mounting. Push the shock absorber downwards and out of the mounting bore in the vehicle floor and take out towards the rear.

The installation is a reversal of the removal procedure. Refer to Fig. 6.2 for the arrangement of the parts at the upper end. Lower the vehicle onto its wheels and tighten the lower of the two nuts at the upper end to the end of the thread to 1.5 – 1.8 kgm. Then screw on the upper nut, hold the lower nut with the open-ended spanner and tighten the nut to 3.0 kgm (22 ft.lb.). The lower shock absorber mounting is tightened to 5.0 kgm (36 ft.lb.).

6.2. Rear Springs – Removal and Installation

Fig. 6.5 shows a side view of the assembled rear suspension on one side. The following operations should be carried out by referring to this illustration. The removal of a spring requires, as for the front springs, a spring compressor or suitable compressor hooks. Proceed as follows:

- Place the rear end of the vehicle on chassis stands and remove the wheel. Remove the securing clamp for the spring link cover and take off the cover.
- Compress the spring in a suitable manner until it is free of the upper and lower spring seats. Remove the bolt and nut securing the lower shock absorber mounting (Fig. 6.4) and lift out the compressed spring. The rubber bearing can be removed from the spring.

The installation is a reversal of the removal procedure. Fit the rubber bearing with a rotating motion over the spring. Clean the inside of the spring link to ensure a clean seat and make sure that the water drain bore is clear. Fit the coil spring with the spring coil end into the recess of the spring link.

When releasing the spring compressor, observe the correct engagement of the spring into its upper and lower seats.

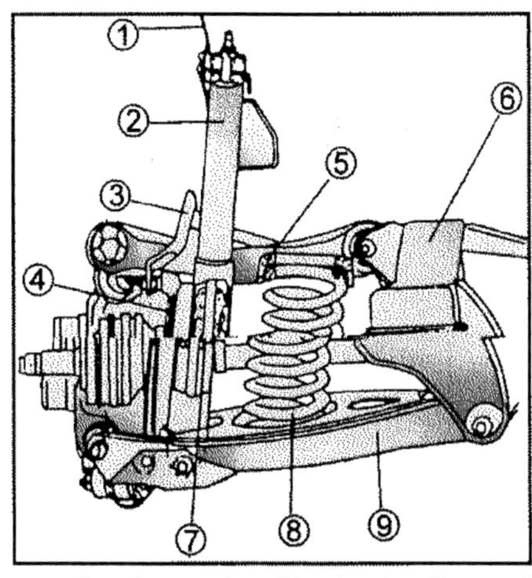

Fig. 6.5 – View of the rear suspension on one side.
1 Attachment to vehicle floor
2 Shock absorber
3 Stabiliser bar
4 Connecting rod for stabiliser bar
5 Rubber seat for rear spring
6 Rear axle carrier
7 Additional rubber spring
8 rear coil spring
9 Spring link

6.3. Stabiliser Bar – Removal and Installation

The stabiliser bar (3) in Fig. 6.5 is fitted by means of a connecting link (4) between the spring link (9) and the frame floor. The stabiliser is attached to the frame floor with a mounting clamp and a rubber bearing. Remove the stabiliser bar as follows:

- Place the rear end of the vehicle on chassis stands and remove the rear wheels.
- Remove the nut from the connecting link on both sides of the bar and disengage the connecting linkages from their locations.
- Remove the bolts securing the mounting clamps to the vehicle floor and take off the rubber bearings.
- Unscrew the propeller shaft centre bearing, separate the propeller shaft flange from the rear axle flange and carefully lower the shaft. Support the shaft from underneath. Do not allow it to hang down under its own weight.
- Remove the wheel speed sensor (ABS) together with the electrical lead after removal of the Allen head screw.
- If a level control system is fitted disconnect the connecting link to the stabiliser bar.
- Place a jack underneath the rear axle centre piece, jack up the axle slightly and remove the two bolts securing the front rear axle bearing. Slowly lower the axle on the jack until the stabiliser bar can be removed. Turn the bar a little towards the left to facilitate the removal.
- To remove the connecting links, take off the protective cover underneath the spring link and unscrew the bolt and nut of the lower mounting.

The installation is a reversal of the removal procedure, noting the following points:

- Insert the stabiliser bar so that the offset of the stabiliser bar ends is facing downwards and towards the rear, seen in the direction of drive.
- Fit the mounting clamps, if necessary with a new rubber bearing, to the frame floor. Tighten the nuts (new nuts) to 2.0 kgm (14 ft.lb.). The bushes should be coated with rubber grease before installation.
- Tighten the bolts for the front rear axle bearings to 7.0 kgm (52.5 ft.lb.).
- Re-connect the propeller shaft, fit the bolts and new nuts and tighten to 4.5 kgm (32.5 ft.lb.). Tighten the connecting linkages on both side to 3.0 kgm (22 ft.lb.) to the stabiliser bar.
- If the connecting links have been removed from the spring links, refit them with new nuts. Tighten the nuts to 2.0 kgm (14 ft.lb.). Finally fit the protective cover underneath the spring link with the securing clamps.

6.4. Rear Suspension Struts

Figs. 6.6 and 6.7 show views of the rear suspension on one side with the location of the individual struts that form the rear suspension. The attachment of the struts can be seen in Fig. 6.2.

6.4.0. CAMBER STRUT – REMOVAL AND INSTALLATION

The camber strut bushes cannot be replaced. If the rubber bushes are worn, replace the complete strut.

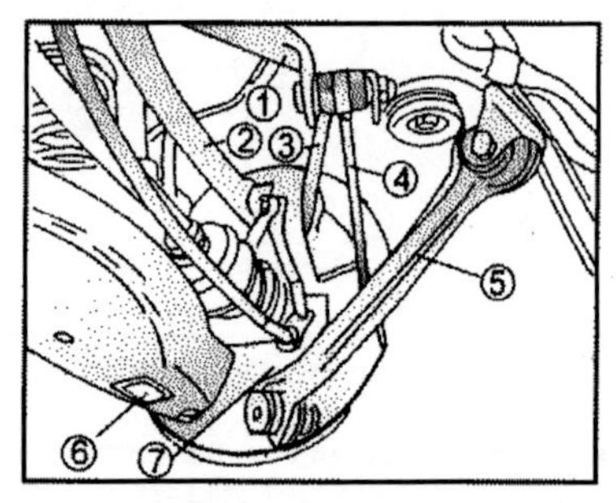

Fig. 6.6 – View of the various struts fitted to the rear suspension.
1 Rear axle carrier
2 Camber strut
3 Pulling strut
4 Track rod
5 Pushing strut
6 Spring strut and cover
7 Wheel carrier

- Place the rear of the vehicle on chassis stands and remove the wheel. Remove the self-locking nut securing the strut (2) and carefully drive out the bolt.
- Remove the self-locking bolt securing the strut to the wheel carrier and carefully drive out the bolt. Remove the metal plate and the washer.
- Push the camber strut clamp sleeve out of the wheel carrier. Withdraw the camber strut from the wheel carrier mounting towards the top, at the same time pulling the wheel carrier towards the outside. Then remove the camber strut towards the bottom.

The installation is a reversal of the removal procedure. Always replace self-locking nuts and bolts. Tighten the nut at the axle carrier end to 7.0 kgm (52.5 ft.lb.) and at the wheel carrier end to 4.5 kgm (23.5 ft.lb.). The rear axle drive shafts must be in horizontal position before the strut securing bolts and nuts are tightened.

6.4.1. PULLING STRUT – REMOVAL AND INSTALLATION

Again the rubber bushes cannot be replaced. A new strut must be fitted.

- With the rear end placed on chassis stands and the wheel removed, mark the position of the eccentric bolt for the pulling strut (3) in relation to the carrier, remove the nut and take it off together with the eccentric bolt and the eccentric washer. Before removal of the bolt push the wheel carrier slightly towards the front to release some of the tension.
- At the bottom of the strut remove the nut and washer. Remove the clamping sleeve for the pulling strut out of the wheel carrier and remove together with the metal plate and the bolt. Withdraw the strut.

The installation is a reversal of the removal procedure. Always replace self-locking nuts. Fit the metal plate to the rubber-bushed end and the washer at the wheel carrier end. Fit the eccentric bolts in their original position. Before tightening of the bolts and nuts position the rear axle drive shafts horizontally.

Tighten the nuts o 5.0 kgm (36 ft.lb.). A cranked 19 mm ring spanner is required to tighten the nut and the torque must be estimated. The rear wheel alignment must be checked after installation.

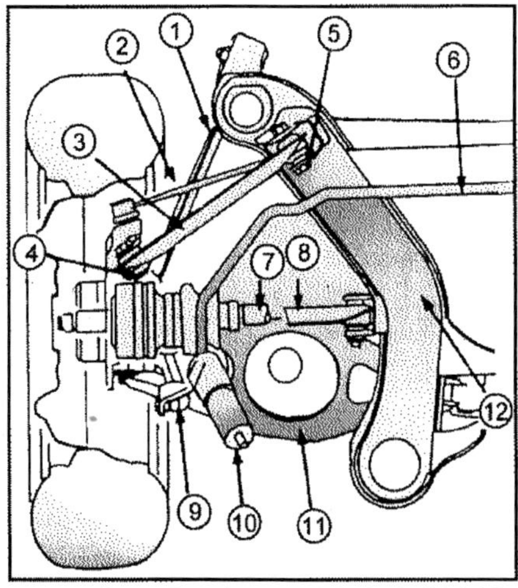

Fig. 6.7 – View of the rear suspension on one side from above with the location of the various struts.

1 Pushing strut
2 Track rod
3 Pulling strut
4 Wheel carrier
5 Eccentric bolt
6 Stabiliser bar
7 Rear axle drive shaft
8 Camber strut
9 Connecting linkage
10 Shock absorber
11 Spring link
12 Rear axle carrier

6.4.2. TRACK ROD – REMOVAL AND INSTALLATION

The rubber bush can be replaced separately. The track rod, must, however, be replaced as a unit if the ball joint is worn. A ball joint puller is required to remove the track rod.

- With the rear end of the vehicle on chassis stands and the rear wheel removed, remove the nut securing the track rod (4), separate the ball joint connection and withdraw the rod.
- At the wheel carried end mark the position of the eccentric bolt in relation to the carrier, using paint or a centre punch, and remove the nut, eccentric washer and eccentric bolt. Remove the track rod.

The installation is a reversal of the removal procedure. Replace the rubber bush, if necessary, under a press. Fit the eccentric bolt into its original position, bring the rear axle drive shafts into horizontal position and tighten the nut at the rear axle carrier side to 7.0 kgm (50.5 ft.lb.).

Thoroughly clean the ball joint stud bore, fit the ball joint and tighten the nut to 3.5 kgm (25 ft.lb.). If the ball joint is rotating, hold it with an Allen key. The camber adjustment must be checked after installation of the track rod.

6.4.3. PUSHING STRUT – REMOVAL AND INSTALLATION

The rubber bush cannot be replaced. A new strut must be fitted if worn.

- With the rear end of the vehicle on chassis stands and the wheel removed, remove the cover for the for the pushing strut (5). Remove the nut and bolt at the axle carrier side and carefully drive out the bolt.
- Remove the nut at the wheel carrier side and remove together with the washer, bolt and metal plate.
- Turn the pushing strut downwards at the rear axle carrier and force the clamping sleeve out of the wheel carrier. Remove the strut.

The installation is a reversal of the removal procedure. Arrange the rear axle drive shafts in horizontal position and tighten the bolt on the wheel carrier side to 4.5 kgm (32.5 ft.lb.) and at the axle carrier side to 7.0 kgm (52.5 ft.lb.).

6.4.4. SPRING LINKS – REMOVAL AND INSTALLATION

The spring links are the suspension arms of the rear suspension. The rear end of the vehicle must be resting on chassis stands and the wheel removed.

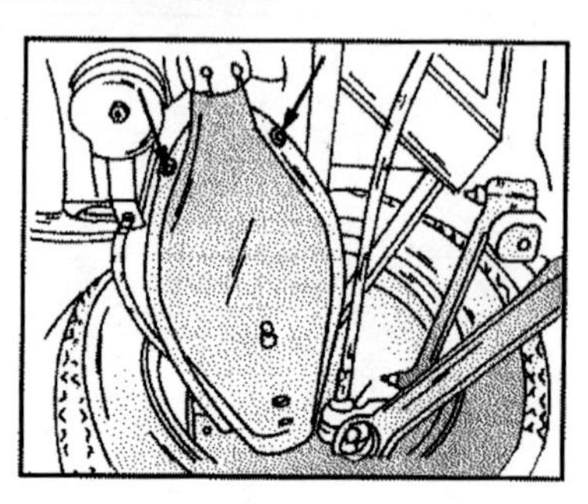

Fig. 6.8 – The two arrows show the location of the self-tapping screws which must be removed to take off the protective cover.

- Remove the four securing clips and the two self-tapping screws shown in Fig. 6.8 and take off the protective cover (6 in Fig. 6.6). Remove the nut securing the shock absorber to the bottom mounting and take off the washer (see Fig. 6.4). Lift the spring link until the shock absorber bolt can be withdrawn.
- Detach the stabiliser bar from the spring link and remove the rear spring as already described.

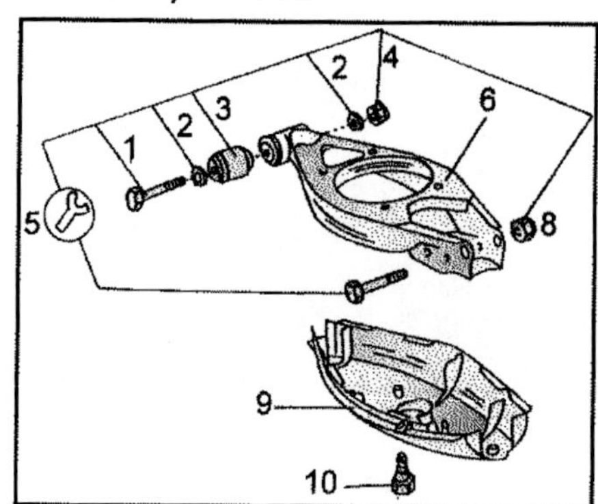

Fig. 6.9 – The attachment of a spring link (suspension arm).
1 Bolt, link to rear axle carrier
2 Washer
3 Rubber bearing
4 Nut, 7.0 kgm, spring strut to rear axle carrier
5 Repair kit for spring strut
6 Spring strut
7 Bolt, spring strut to wheel carrier
8 Nut, 12.0 kgm, spring strut to wheel carrier
9 Protective cover
10 Screw for protective cover

- Remove the nut securing the spring link fulcrum bolt to the rear axle and take off the washer. Carefully knock out the bolt.
- Remove the nut securing the spring link to the wheel carrier and again knock out the bolt. The spring link can now be removed. Fig. 6.9 shows the removed spring link together with the protective cover and the attachment at the inside and the outside.

The installation is a reversal of the removal procedure. Tighten the rear axle carrier end to 7.0 kgm (52.5 ft.lb.) and the wheel carrier end to 12.0 kgm (101 ft.lb.). The vehicle must be resting with its wheels on the ground.

6.5. Rear Axle Drive Shafts

Remove as follows:

- Slacken the axle shaft nut whilst the wheels are on the ground (12 point nut, 30 mm socket). Unscrew the shaft securing bolts from the intermediate flange and remove the washers. A splined Allen key insert is required. Insert a screwdriver between the flanges to separate the shaft flange. Push the shaft together and swing it towards the top, as soon as space is available.
- Using a puller, of the type shown in Fig. 6.10 and press the rear axle shaft out of the wheel hub. Lower the front end of the shaft and withdraw it from the rear suspension. Take care not to pull the shaft apart.

Fig. 6.10 – Removal of a rear wheel hub.

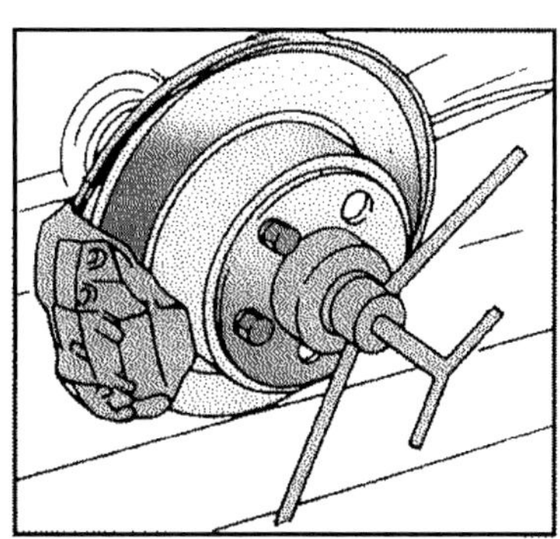

The rubber boots can be replaced with the shafts removed, however, as the joint must be dismantled, we recommend to have this operation carried out at your dealer.

Before installation of a shaft, check the C.V. joints and the rubber boots for wear or damage. Note the following points during installation:

- Always replace the flange bolts and the axle shaft nut. Thoroughly clean the flange faces. Coat the bolts and the bolt heads with engine oil.
- Tighten the flange bolts evenly to 7.0 kgm (52.5 ft.lb.).
- Lower the vehicle onto the ground and tighten the axle shaft nut to 28 – 32 kgm (202 – 230 ft.lb.). After tightening the nut, peen the nut collar into the groove of the axle drive shaft without splitting the metal.

6.6. Rear Hub Flange and Wheel Bearings - Removal and Installation

Certain extractors are required for this operation. Provided that these can be obtained, proceed as follows:

- Remove the rear axle drive shaft as described above and remove the brake caliper. Use a piece of wire to suspend the caliper to the rear suspension.
- Fig a puller, as shown in Fig. 6.10 to the hub flange and press the flange out of the wheel bearing.
- Use a pair of pliers and remove the circlip visible in the bearing bore. The wheel bearing must now be removed from the wheel carrier by means of the extractor 201 589 04 43 00. If this extractor cannot be obtained and the bearing must be replaced, remove the different struts and remove the wheel bearing under a press. A bearing race remains on the wheel hub flange and must be removed with a suitable puller.
- Check all parts for wear or damage. Bearings once removed should always be replaced, mainly if the bearing race has been removed from the hub flange.

Fit the new bearing until the circlip can be inserted. Check that the circlip has entered its groove fully. Draw the wheel hub flange in position with the special tool or carefully try to drive it in position with a soft-metal drift. The small thrust washer must press against the inner bearing race. There is no need to adjust the wheel bearings.

6.7. Rear Wheel Alignment

The checking and adjusting of the rear wheel alignment requires the use of special equipment and special wrenches and we recommend to have the alignment checked in a workshop. If you have removed parts of the suspension which will alter the alignment and have followed the advise to mark the eccentric bolts and washers there will be no problems to drive the vehicle to the wheel alignment centre to have the settings checked.

6.8. Level Control System

Certain models in the range can be optionally fitted with a level control system for the rear suspension. The system keeps the rear end of the vehicle at approximately the same level, irrespective of the load. This ensures additional driving safety and comport during trips with a full luggage compartment, when towing a trailer or at other occasions, when the rear of the vehicle carries more weight than usual. The pressure in the system is supplied by a pressure oil pump, fitted to the front of the engine. The oil under pressure is supplied from the level regulator (controller) on the rear axle to the shock absorbers (spring struts) when the rear of the vehicle is lowered due to the additional weight. The level regulator will open a port when the load is removed from the luggage compartment, and some of the oil will return to the oil reservoir in the engine compartment. The level regulator is operated by a lever, attached to the rear stabiliser (torsion) bar.

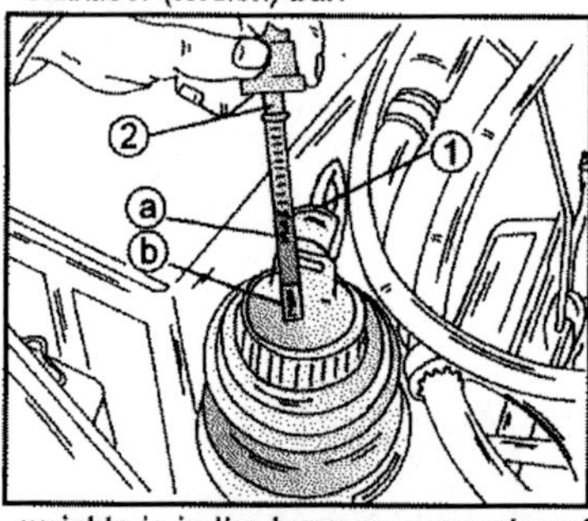

Fig. 6.11 – After withdrawing the fluid dipstick (2) out of the reservoir (1) the fluid level must be visible at the "Max" mark (a) or the "Min" mark (b). A cable plug is fitted to the screw cap.

The only operation recommended on the level control system is the checking of the fluid (oil) level in the reservoir. A cap with an attached dipstick is used to check the fluid level. With the vehicle in the normal laden condition, the fluid level should be between the MAX and MIN markings on the dipstick as shown in Fig. 6.11. If the fluid level is checked in the laden condition, i.e. additional weights is in the luggage compartment, the fluid level will be around the MIN mark or sometimes below and cannot be seen. Use only the fluid available from Mercedes-Benz (No. 000 989 85 03) to top up the system.

If the component parts of the rear suspension are removed it is necessary to disconnect the hydraulic pipes connecting the various parts. This mainly concerns the shock absorbers which have a pipe connection at the upper ends, as can be seen in Fig. 6.12. The hydraulic system must be drained before the pipes are disconnected.;

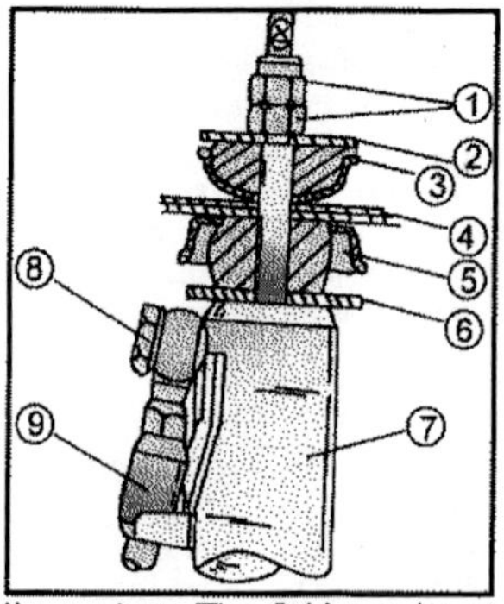

Fig. 6.12 – The attachment of a shock absorber at the upper end.

1 Hexagonal nut
2 Metal washer
3 Upper rubber ring
4 Chassis frame
5 Lower rubber ring
6 Plate
7 Shock absorber (damper)
8 Banjo bolt with sealing ring
9 Pressure hose, reservoir to damper

- Find the level regulator (controller) by referring to Fig. 6.13, push a hose over the drain plug connector, hold the hose into a container and open the drain plug on the level controller. Allow approx. 0.5 litre to drain from the system. The fluid can be re-used, but should be strained through a fine-mesh filter before it is filled into the container, using a funnel.

The hydraulic system must be refilled and bled of air as follows after the pipes have been re-connected:

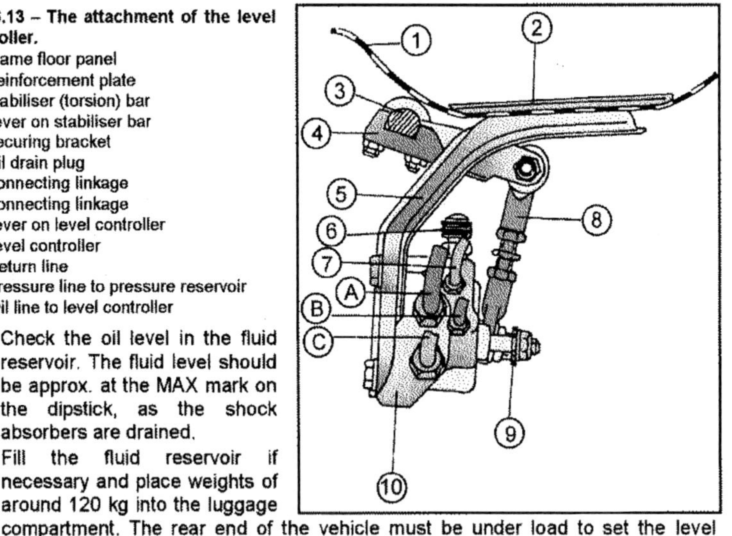

Fig. 6.13 – The attachment of the level controller.

1 Frame floor panel
2 Reinforcement plate
3 Stabiliser (torsion) bar
4 Lever on stabiliser bar
5 Securing bracket
6 Oil drain plug
7 Connecting linkage
8 Connecting linkage
9 Lever on level controller
10 Level controller
A Return line
B Pressure line to pressure reservoir
C Oil line to level controller

- Check the oil level in the fluid reservoir. The fluid level should be approx. at the MAX mark on the dipstick, as the shock absorbers are drained.
- Fill the fluid reservoir if necessary and place weights of around 120 kg into the luggage compartment. The rear end of the vehicle must be under load to set the level regulator into the "filling" position.
- Start the engine and check the fluid level, which must now be below the MIN mark.
- Switch off the engine and again check the level after the load has been removed from the luggage compartment. The fluid level on the dipstick must now be between the MAX and MIN marks.

6.9. Rear Axle and Suspension – Tightening Torques

Rear Axle

Lower shock absorber mounting: .... 6.5 kgm (47 ft.lb.)
Upper shock absorber mounting: .... Tighten nut to bottom of thread
Stabiliser bar mounting to frame floor: .... 2.0 kgm (14 ft.lb.)
Ball joints of linkage between stabiliser bar and spring link (suspension arm):
- Stabiliser bar end: .... 3.0 kgm (22 ft.lb.)
- Spring link end: .... 2.0 kgm (14.5 ft.lb.)
Rear axle centre piece rubber mountings to frame floor: .... 4.5 kgm, front, 12.0 kgm, rear
Propeller shaft intermediate bearing: .... 2.5 kgm (18 ft.lb.)
Propeller shaft to rear axle flange: .... 4.5 kgm (32.5 ft.lb.)
Propeller shaft clamp nut: .... 3.5 kgm (25.0 ft.lb.)
Speed sensor to rear axle (ABS): .... 0.8 kgm (6.0 ft.lb.)

Rear Suspension

Camber strut to rear axle carrier: .... 7.0 kgm (50 ft.lb.)
Camber strut to wheel carrier: .... 4.5 kgm (32.5 ft.lb.)
Pulling strut to rear axle carrier: .... 7.0 kgm (50 ft.lb.)
Pulling strut to wheel carrier: .... 4.5 kgm (32.5 ft.lb.)
Track rod to rear axle carrier: .... 7.0 kgm (50 ft.lb.)
Track rod to wheel carrier: .... 3.5 kgm (25.0 ft.lb.)
Pushing strut to rear axle carrier: .... 7.0 kgm (50 ft.lb.)
Pushing strut to wheel carrier: .... 4.5 kgm (32.5 ft.lb.)

Spring link to rear axle carrier: .... 7.0 kgm (50 ft.lb.)
Spring link to wheel carrier: .... 12.0 kgm (86.5 ft.lb.)
Stabiliser bar to spring link: .... 2.0 kgm (14.5 ft.lb.)
Shock absorber to spring link: .... 6.5 kgm (47 ft.lb.)
Flange nut for rear axle shaft: .... 28.0 – 32.0 kgm (202 - 230 ft.lb.)
Rear axle drive flange to connecting flange: .... 12.0 kgm (86.5 ft.lb.)
Axle centre piece to crossmember, front: .... 4.5 kgm (32.5 ft.lb.)
Wheel bolts: .... 11.0 kgm (80 ft.lb.)

7 Steering

From the introduction of the series W210 a rack and pinion steering with power assistance it fitted.

7.0. Technical Data

Type: .... See above
Opening Pressure of Steering Pump:
Filling capacity of steering system: .... approx. 1 litre
Fluid type: .... As for automatic transmissions

7.1. Checks on the Steering

Checking the Steering Play

Excessive play in the steering can be adjusted, but this should be left to a Mercedes dealer who has the necessary special tools. The steering can, however, checked as follows:

- Place the front wheels in the straight-ahead position and reach through the open window and turn the steering wheel slowly to and fro.
- The front wheels must move immediately as soon as you move the steering wheel.
- If there is no play in the straight-ahead position but the steering wheel is more difficult to move as you rotate the steering wheel further you can assume that the steering gear is worn and must be replaced.

Checking the track rod ball joints for excessive play

With the wheel fitted grip the track rod ball joint and move it up and down with a considerable force.

A worn track rod ball joint can be recognised by excessive "up and down" movement. If this exceeds 2 mm, replace the ball joint.

Checking the steering rubber gaiters

Check the rubber gaiter over its entire length and circumference for cuts or similar damage. Also check that the gaiters are securely fastened at both ends. Track rods with worn ball joints or damaged rubber gaiters must be replaced as described earlier on.

7.2. Steering Repairs

7.2.0. Removal and Installation

The removal and installation of the steering is a complicated operation. As it will be very rare to remove the steering, we recommend that the work is carried out in a workshop. Figs. 7.1 an 7.2 show the component parts of the steering for reference only.

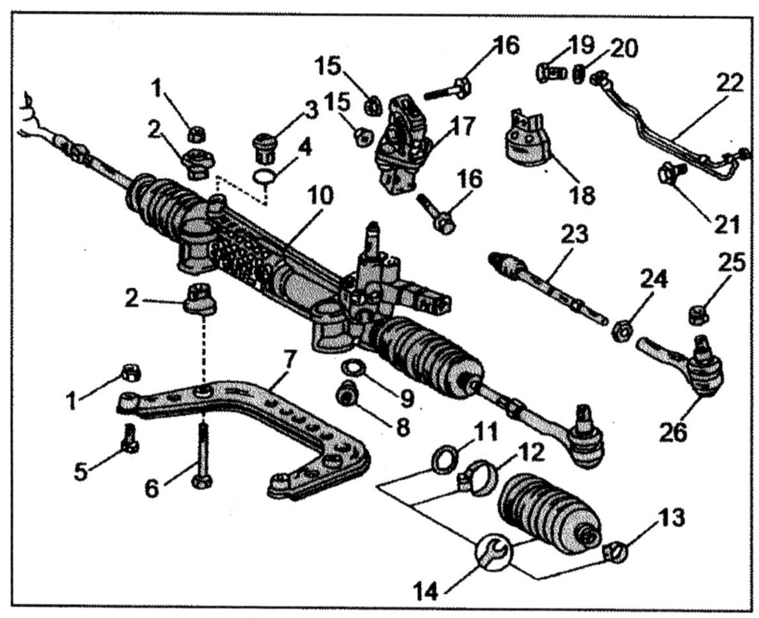

Fig. 7.1 – Steering gear installation.

1 Nut, steering to subframe
2 Rubber bearing
3 Connecting bolt
4 Sealing ring
5 Securing bolt for 7
6 Hexagonal bolt, steering gear
7 Steering mounting bracket
8 Connection
9 Sealing ring
10 Steering gear
11 Sealing ring to steering rack
12 Inner gaiter clamp
13 Outer gaiter clamp
14 Repair kit (gaiter)
15 Nut, universal joint bolt
16 Clamp bolt
17 Steering universal joint
18 Cover for universal joint
19 Banjo bolt
20 Sealing ring
21 Securing screw, pipe
22 Fluid pipe
23 Inner track rod
24 Locknut
25 Nut, track rod joint
26 Track rod ball joint

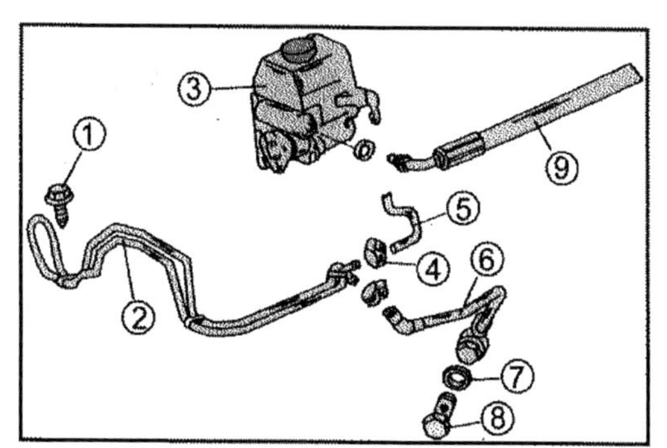

Fig. 7.2 – Steering fluid reservoir and connections.

1 Bolt for fluid pipe clamp
2 Fluid pipe
3 Fluid reservoir
4 Hose clamp
5 Connecting hose
6 Hose to steering gear
7 Sealing ring
8 Banjo bolt
9 Pressure hose

The following precautions must be followed if any work on the steering system is necessary:

- All operations must be carried out under the cleanest of conditions.
- After disconnecting any pipes or hoses from the steering system clean the connecting points immediately. All removed parts must be placed onto a clean surface and covered with clean paper or rags.
- Never use fluffy rags to clean any parts of the steering system.
- If fitting new parts take them out of the packing just before they are fitted. Only fit original parts.
- Never re-use fluid drained from the steering system.

7.2.1. Track Rods - Replacement

Before you decide to have a track rod ball joint or a track rod replaced, you can carry out the following checks:

- Check the track rod ball joints rubber dust caps for cuts or other damage,
- Have the steering wheel turned into one lock (helper required), grip the track rod ball joint with one hand and ask the helper to move the steering wheel to and fro. The engine should be started to facilitate the steering wheel movements. Excessive play requires the fitting of a new track do ball joint.
- Place the front end of the vehicle on chassis stands and grip the track rod with one hand. Move the track rod up and down. Excessive play in the ball joints requires the replacement of the joint.
- Similar checks can be carried out on the inner ball joint. In this case the rubber gaiter must be detached from the steering rack. Excessive clearance at the inner joints requires the replacement of the track rod. As the steering must be removed to replace a track rod a dealer will be the only solution.

7.2.2. Track Rod Ball Joints - Replacement

The ball joints at the ends of the track rods can be replaced with the steering fitted. After disconnecting the ball joint from the steering lever undo the locknut securing the ball joint end to the track rod and unscrew the end piece from the track rod, at the same time counting the number of turns necessary.

When fitting the new ball joint screw it onto the track rod by the same number of turns (also half-turns) and provisionally tighten the locknut. If the operations have been carried out properly there should be no need to check the toe-in setting.

7.3. The hydraulic System

7.3.1. Filling the System

If the steering system has been drained for any reason it must be refilled and bled of air. Fluid drained from the system must not be re-used to fill the reservoir. The filter must be replaced if the system has been completely drained or the fluid is changed for any reason. The following sequence must be adhered to:

- Remove the screw cap from the fluid reservoir. There is a seal inside the cap which could drop out.
- Fill the reservoir to the upper edge.
- Start the engine a few times and immediately switch it off again. This will fill the complete steering system. During this operation the fluid level in the reservoir will drop and must be corrected immediately. Never allow the reservoir to drain as otherwise fresh air will be drawn into the system. A helper is obviously required to start and switch off the engine.

- When the fluid level in the reservoir remains the same the system is filled and the fluid level must be within the markings on the fluid dipstick. As already shown the dipstick has an upper and a lower mark. The total capacity of the system is approx. 1 litre.
- Push the seal into the screw cap and refit the cap to the reservoir.

We recommend to check the fluid level during each check of the engine oil level. This will assure you not to overlook the check. During the level check or topping-up of the reservoir make sure that no foreign bodies or dirt can enter the system.

7.3.2. Bleeding the hydraulic System

After the fluid level remains the same after the engine has been started and switched off a few times bleed the system as described below. A helper is required to turn the steering wheel:

- Have the steering wheel moved from one lock to the other and back again to eject the air out of the steering cylinder. The steering wheel must be moved slowly, just enough for the piston inside the steering cylinder to contact its stop.
- Observe the fluid level in the reservoir during this operation. If the level drops, fill in additional fluid, as it must remain on the MAX mark. No air bubbles must be visible during the bleeding operation.

7.3.3. Checking the System for Leaks

Sometimes it is possible that fluid is lost for some unknown reason. A quick check may establish where the fluid is lost:

- Ask a helper to turn the steering wheel from one lock to the other, each time holding the wheel in the maximum lock. This will create the max. pressure in the system and any obvious leaks will be shown by fluid dripping on the floor.
- From below the vehicle (on chassis stands) check the area around the steering pinion. Slacken the rubber gaiters on the steering rack and check the ends of the rack. The rack seals could be leaking.
- Check the hose and pipe connections with reference to Fig. 7.2. These must be dry.

7.4. Tightening Torques – Steering

Union nut for high pressure hose: .... 2.5 – 3.0 kgm (18 – 22 ft.lb.)
Return pipe to pump: .... 3.5 – 4.5 kgm (25 – 32.5 ft.lb.)
Steering pump to bracket: .... 2.5 kgm (18 ft.lb.)
Track rod ball joint nuts: .... 5.0 kgm (36 ft.lb.)
Steering intermediate lever nut: .... 9.0 – 10.0 kgm (65 – 72 ft.lb.)
Clamp on track rod: .... 5.0 kgm (36 ft.lb.)

8 Brake System

8.0. Technical Data

Type of system	See description below
Front Brakes	
Caliper piston diameter, front brakes, one piston:	57.00 – 57.05 mm
Caliper piston diameter, front brakes, two pistons:	38 and 42 mm

Thickness of brake pads, incl. back plate:	19.6 or 20.5 mm
Min. thickness of linings:	3.5 mm
Brake disc thickness:	25,0 or 28.0 mm
Brake disc diameter:	288 or 300 mm
Min. thickness of brake disc:	23.0 or 26.0 mm
Max. run-out of brake discs	0.10 mm
Wear limit of brake discs, per side	max. 0.05 mm

Rear Disc Brakes

Caliper piston diameter:	Depending on fitted version 38,0 or 42,0 mm
Brake disc diameter:	278,0 or 290 mm
Brake disc thickness:	Between 9 and 22 mm
Min. brake disc thickness:	8,0 mm or 21,0 mm
Min. thickness of brake pads, incl. metal plate:	14.0 mm
Min. thickness of pad linings	2.0 mm

Handbrake

Diameter of handbrake drum:	164.0 mm (6.51in.)
Outer diameter of handbrake brake shoes:	164.0 mm (6.51in.)
Width of brake shoes:	20.0 mm (0.8 in.)
Width of brake shoes:	20.0 mm (0.8 in.)
Brake linings thickness:	2.65 mm (0.104 in.)
Min. width of brake linings:	1.00 mm (0.04 in.)
Number of notches required for fully engaging handbrake, using average force:	8
Number of notches until handbrake becomes effective:	1

Brake Master Cylinder

Cylinder Diameter of Pressure Piston:	
- Four- and five-cylinder engines:	23.81 mm or 25.40 mm
- Six-cylinder engine:	25.40 mm
Cylinder Diameter of Intermediate Piston:	
- Four-cylinder engines:	17.00 mm
- Six-cylinder engines:	19.05 mm

8.1. Short Description

All models covered in this manual are fitted with a hydraulic dual-circuit brake with vacuum-operated brake servo unit. The brake servo unit is supplied with vacuum from the vacuum pump fitted to the engine. Sliding (floating) or fixed brake calipers with 2 or 4 pistons, depending on the model, are fitted to the front wheels. Most W210 models covered in the manual are fitted with sliding calipers with 2 pistons. Some models in the series are, however, fitted with fixed calipers with 4 pistons and are also covered in the manual. Disc brakes, working on the same principle as the fixed front brake calipers are used on the rear wheels. The brake calipers are not the same on all models and again you can consult the technical data section for details.

Vehicles are either built with or without ABS system.

The brake system is diagonally split, i.e. one circuit serves one of the front brake calipers and the diagonally opposed rear brake caliper. The other circuit operates the other two brake assemblies accordingly. If one brake circuits fails, the brakes will operate as normal, but more brake force will be required.

Checking the Brake Fluid Level

The brake fluid reservoir is fitted above the master brake cylinder in the position shown in Fig. 8.1. At all times make sure that the brake fluid is between the "Min" and "Max" marks on the outside of the cylinder. If the brake fluid level sinks below the "Min" mark there are other reasons which must be investigated.

Fig. 8.1 – The brake fluid reservoir is fitted to the master brake cylinder. Note the "min" and "max" marks on the outside of the cylinder. The two buttons shown with the arrows can be depressed to check the brake fluid level.

1 Brake servo unit
2 Vacuum hose
3 Fluid reservoir
4 Screw cap
5 Master brake cylinder

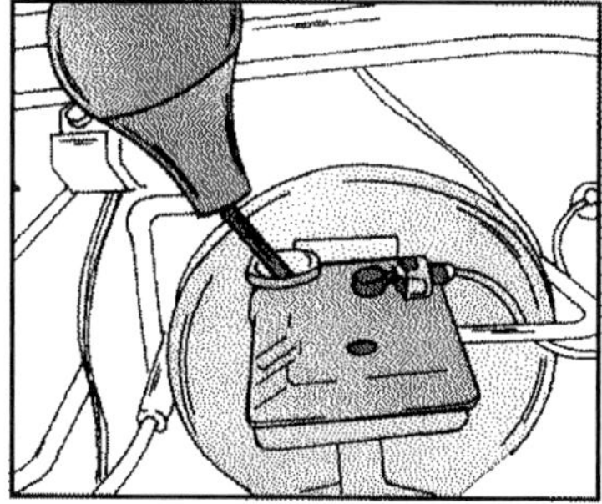

Fig. 8.2 – Brake fluid can be removed from the fluid reservoir as shown, when it is necessary to drain the brake system.

Brake fluid must sometimes be removed from the reservoir. If this is necessary we recommend the method shown in Fig. 8.2.

8.2. Front Disc Brakes

As already mentioned, sliding brake calipers with one piston are fitted to the front wheels. The assemblies consist of a caliper mounting bracket bolted rigidly to the front axle steering knuckle and a separate caliper cylinder, irrespective if the cylinder contains one or two pistons. When the brake is operated, the piston pressed first with its brake pad against the brake disc. The caliper cylinder then slides on glide bolts and moves against the direction of the pressure, until the other brake pad is pressed against the brake disc on the other side. Only the caliper cylinder must be removed to replace the brake pads, the mounting bracket remains on the steering knuckle. Only some models have fixed calipers with four pistons.

8.2.0. FLOATING CALIPERS

Checking and Replacing the Brake Pads

Fig. 8.3 shows this type of caliper in fitted position. Fig. 8.4 shows the component parts of a dismantled caliper.

To check the pad thickness without removing the pads, use a torch and shine through one of the slots in the wheel rim. If the thickness of the pad material appears to be less than 3.5 mm (0.14 in.), replace the brake pads on both sides.

Important: If it is possible that the brake pads can be re-used, mark them in relation to the side of the car and to the inside or outside position of the caliper. Never interchange brake pads from left to right or visa versa, as this could lead to unequal braking.

Fig. 8.3 – Fitted brake caliper with the location of some of the parts.
1 Brake caliper frame
2 Brake caliper
3 Hexagon
4 Securing bolt
5 Brake hose
6 Bleeder screw
7 Lower slide bolt

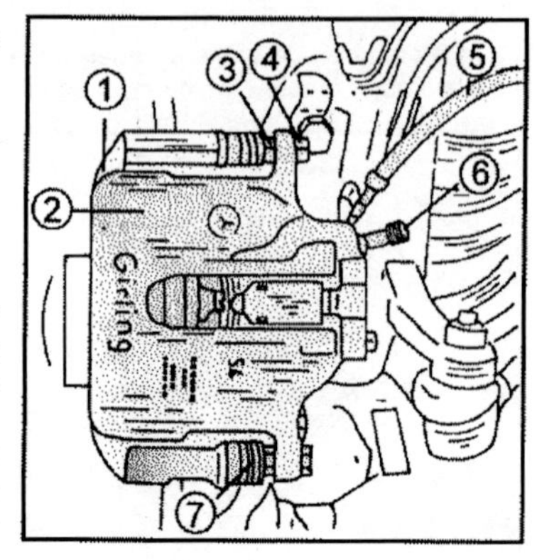

To remove the brake pads, either for examination or replacement, first jack up the front end of the vehicle and remove the wheel. Then proceed as follows:

- Insert a screwdriver into the retaining spring as shown in Fig. 8.5 and disengage it from the brake caliper housing.

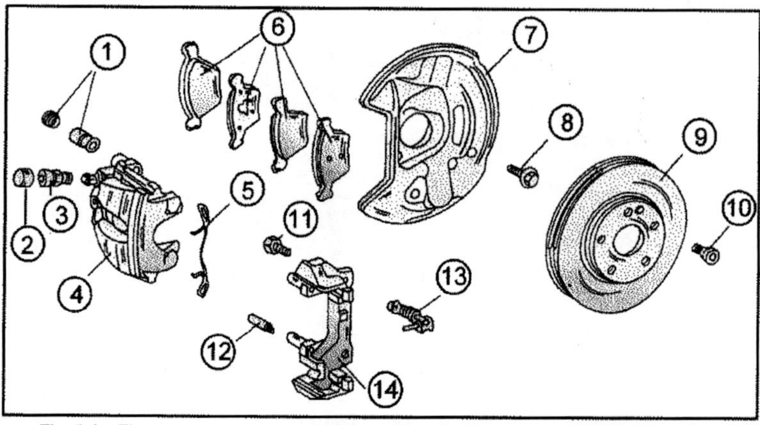

Fig. 8.4 – The component parts of a front caliper (Girling) together with the brake pads.

1 Slide bolt rubber boots
2 Bleeder valve rubber dust cap
3 Bleeder valve
4 Brake caliper housing
5 Tensioning spring
6 Brake pad set
7 Splash shield
8 Splash shield bolt
9 Brake disc
10 Brake disc bolt
11 Bolt, 11.5 kgm
12 Slide bolts
13 Brake pad wear indicator
14 Brake caliper bracket

Fig. 8.5 – Removal of the retaining (tensioning) spring from the caliper.

- Lift the two tabs at the side of the covering for the brake pad wear indicator plug and disconnect the plug from the socket on the brake caliper. Do not pull on the cable. Fig. 8.6 shows the attachment of the plug. Also disconnect the plug from the wheel speed sensor for the ABS system.
- Remove the lower hexagonal bolt. Counterhold the hexagon on the slide

bolt with an open-ended spanner, as shown in Fig. 8.7. The illustration shows the removal of the upper bolt, but this is no longer applicable as the lower bolt must be removed.

- Swing the brake caliper upwards and tie it in position with a piece of wire. Remove the brake pads from the sides of the caliper. The sensor for the brake pad wear indicator is fitted to the outer brake pad as shown in Fig. 8.8. The connector tab of the sensor can be withdrawn from the pad if new pads are fitted.

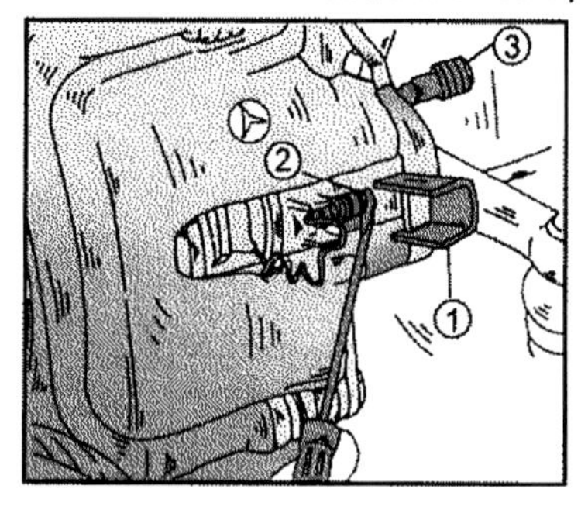

Fig. 8.6 – Removal of the brake pad wear indicator sensor.
1 Covering
2 Cable plug
3 Bleeder screw

Measure the thickness of the pad material. If the thickness is around 3.5 mm, fit new pads. Although the pad material can be worn down to 2.0 mm, you will find that the pad wear indicator lights up when a thickness of 3.5 mm is reached. Never replace one pad only even if the remaining pads look in good order.

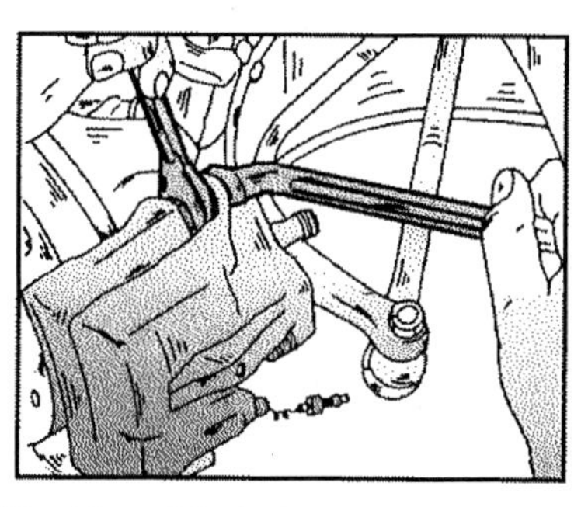

Fig. 8.7 – Slacken the lower brake caliper slide bolt with a ring spanner and an open-ended spanner at the inside. The illustration shows the removal of the upper bolt (necessary on previous models).

The sensors for the brake pad wear indicator should be replaced if the insulation shows signs of chafing or other damage.

Check the thickness of the brake disc and compare the dimensions given in Section 8.0. Replace the discs if the thickness is below the minimum permissible.

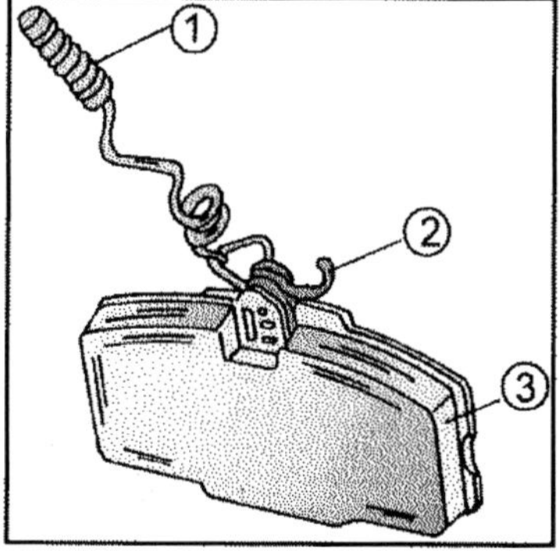

Fig. 8.8 – View of the brake pad with the pad wear indicator sensor.
1 Sensor
2 Spring clip
3 Brake pad

Before fitting the new pads clean the caliper opening with a brush and clean brake fluid or methylated spirit. Wipe off any spirit remaining and lubricate the exposed part of the piston with rubber grease. Fit the pads as follows:

- Open the fluid reservoir and draw off some of the fluid as shown in Fig. 8.2.
- Push the piston back into its bore. Either use the special pliers available for this purpose (see Fig. 8.9) or place a wooden block in position and lever back the piston carefully with a large screwdriver blade.

- Insert both brake pads into the caliper carrier and carefully lower the caliper housing carefully over the pads. Use new self-locking bolts and tighten the bolts to 3.5 kgm (25 ft.lb.), at the same time counterholding the hexagon of the slide bolt as shown in Fig. 8.7.
- Coil the brake pad wear indicator cable and connect to the terminal on the brake caliper. Fit the connector protective cover.

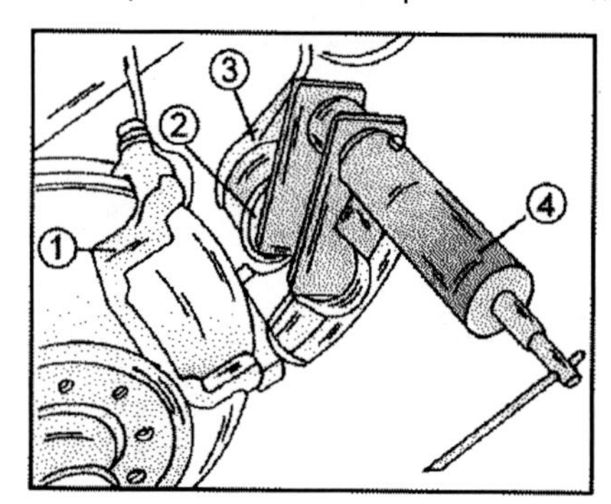

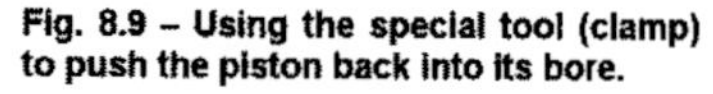
Fig. 8.9 – Using the special tool (clamp) to push the piston back into its bore.

1 Brake caliper
2 Piston
3 Cylinder housing
4 Special tool

- Apply the foot brake several times after installation of the brake pads. This is an essential requirement to allow the new pads to take up their position. Bleed the brake system if necessary and check and if necessary correct the fluid level in the master cylinder reservoir.

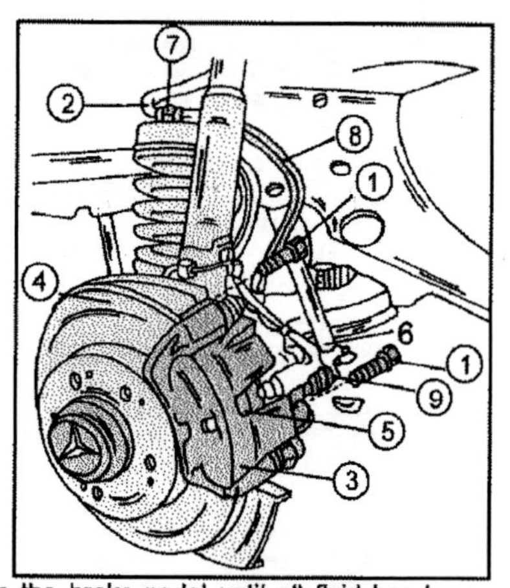

Fig. 8.10 – View of a fitted front brake caliper.
1 Brake caliper bolts
2 Brake tube
3 Cylinder body
4 Brake caliper bracket
5 Clip, wear indicator
6 Connector plug
7 Brake hose connection
8 Brake hose
9 Bolt

Removal and Installation of a Brake Caliper

The front end of the vehicle must be resting on chassis stands and the wheel removed.

- Place a bleeder hose over the bleeder screw (3) in Fig. 8.6 (remove the rubber cap first) and insert the other end of the hose into a container (glass jar). Open the bleeder hose and pump the brake pedal until all fluid has been drained from the system.
- Disconnect the brake pipe from the brake hose at the inside of the wheel arch housing by unscrewing the union nut. Knock out the spring plate to free the hose. Suitably close the hose and pipe ends to prevent entry of dirt.
- Lift the two lugs for the cover of the brake pad wear connector with a small screwdriver and pull off the plug (refer to Fig. 8.6). The brake hose can be unscrewed from the brake caliper, if desired, whilst the caliper is still fitted.
- Unscrew the brake caliper mounting bolts and lift off the unit. Discard the bolts as new bolts must be used during installation. The brake pads can now be removed from the caliper.

The installation of the caliper is a reversal of the removal procedure. Tighten the new caliper bolts to 11.5 kgm (83 ft.lb.). Finally bleed the brake system as described later on.

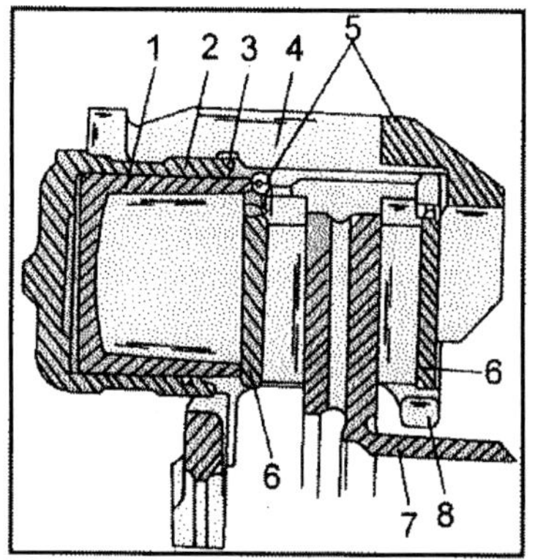

Fig. 8.11 – Sectional view of a front brake caliper.

1 Piston
2 Piston seal
3 Rubber dust seal
4 Cylinder housing
5 Spring clip
6 Brake pad
7 Brake disc
8 Brake caliper screw

Brake calipers - Overhaul

Remove the caliper from the steering knuckle and carry out the servicing on a bench, under the cleanest possible conditions. Before commencing any servicing, note the general points below, these being applicable to servicing of any of the brake hydraulic units.

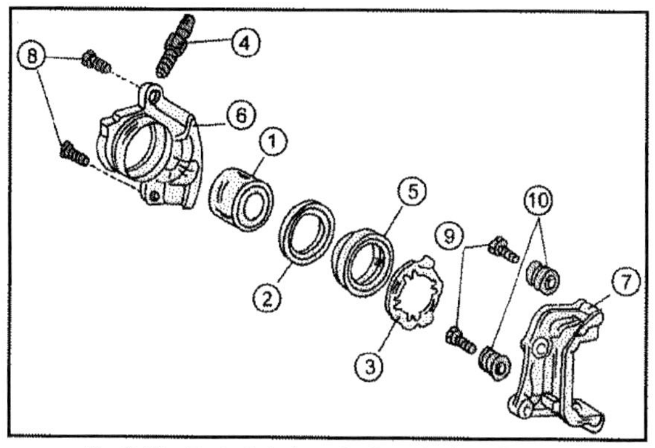

Fig. 8.12 – Exploded view of a front brake caliper.

1 Piston
2 Cylinder bore seal
3 Shield
4 Bleeder screw
5 Rubber dust seal
6 Brake caliper cylinder
7 Brake caliper carrier
8 Brake caliper bolt
9 Brake caliper carrier bolts
10 Rubber grommets

- Always replace all seals each time a unit is dismantled. These seals slowly deteriorate in service and even through they may look quite good, they should nevertheless be replaced in the interest of your own safety.
- Do nor **ever** attempt to re-use any piston or caliper if the surfaces are worn, scored or corroded. Obtain a complete new unit and discard the old one.
- Use only the hydraulic brake fluid of the recommended specification and **DO NOT** use fluid that has been kept over a long period.
- Lubricate the seals and pistons either with clean brake fluid or, better still, a proprietary brake paste or lubricant (ATE, Girling, etc.).

- Fig. 8.11 shows a sectional view of a caliper for your reference, with the location of the individual parts. Fig. 8.12 shows an exploded view of the caliper.

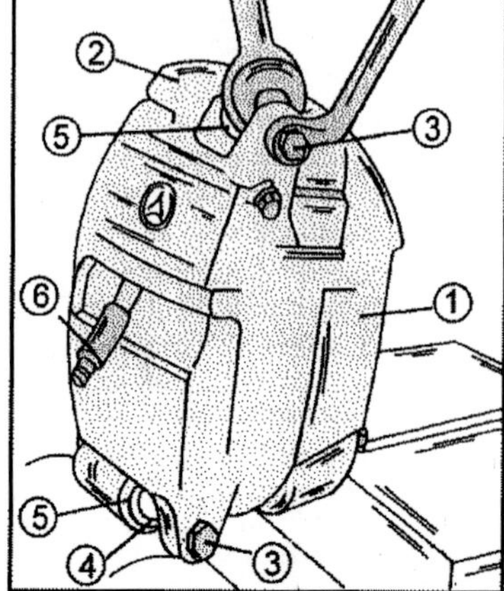

Fig. 8.13 - Dismantling a front brake caliper.
1 Brake caliper cylinder
2 Brake caliper carrier
3 Brake caliper bolts
4 Slide bolt
5 Slide bolt
6 Connector for pad wear indicator

A caliper can be overhauled as follows:

- With the caliper clamped carefully into a vice, unscrew the cylinder housing by removing the two slide bolts. Counterhold the hexagon with an open-ended spanner whilst the bolts are slackened as shown in Fig. 8.13.
- Using a screwdriver, carefully lever the rubber dust cap from the caliper housing and the piston. Apply an airline to the brake hose connection and blow out the piston. Insert a piece of wood or thick piece of rubber into the caliper opening. Keep the fingers away from the area when the piston is ejected. This operation can be carried out with an air line of a petrol station. Using a non-metal instrument, take out the cylinder sealing ring from the groove in the inside of the cylinder.

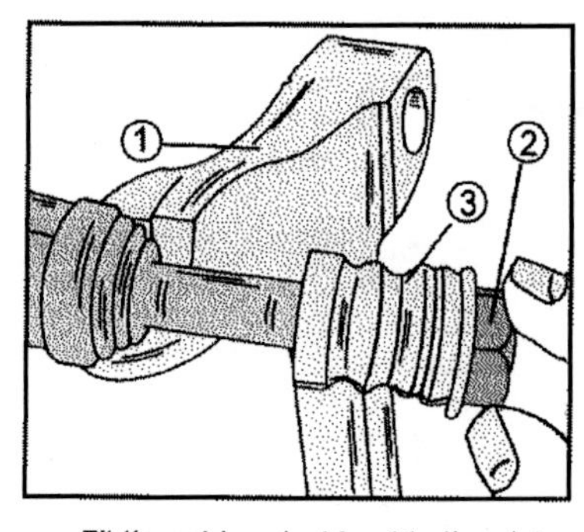

Fig. 8.14 – Fitting the slide bolts and the rubber boots into the brake caliper.
1 Brake caliper
2 Slide bolt
3 Rubber boot

Examine all parts for wear or damage as necessary. Use a repair kit to overhaul the caliper. Assemble in the reverse sequence to dismantling, noting the following points:

- Fit the cylinder seal into the bore, using the fingers only. Coat the piston with ATE brake paste and insert the piston into the cylinder.
- Fit the rubber dust boot to the piston and coat the piston face with brake paste or clean brake fluid.
- Push the piston into the cylinder bore and fit the rubber dust seal into the cylinder housing groove, after the piston is in position.

If the rubber boots for the slide bolts are damaged, remove them from the brake caliper mounting bracket and the slide bolt. Thoroughly clean the locating bores in the caliper bracket and the outer faces of the bolts, coat the parts with brake paste and refit the boots to the brake caliper mounting bracket. Fig. 8.14 shows the installation of the parts.

8.2.1. BRAKE CALIPERS – With 4 Pistons

As already mentioned calipers with four pistons are used. The piston diameters are, however, not the same (38 and 42 mm). Fig. 8.15 shows the component parts of this caliper type.

Checking and Replacing the Brake Pads

When the thickness of the brake pad linings has reached 3.5 mm, a warning light in the dashboard will light up, signalling that new brake pads must be fitted.

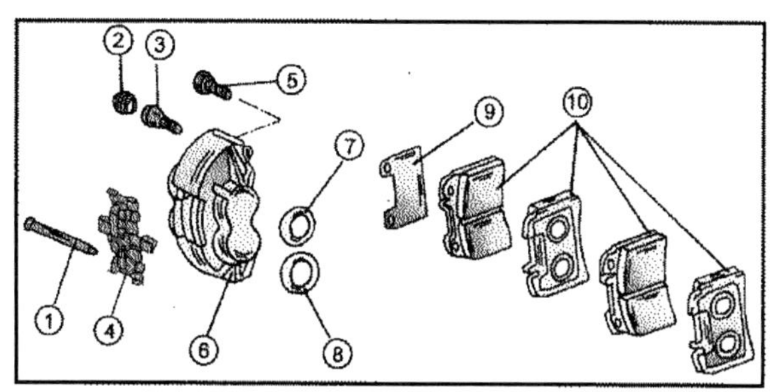

Fig. 8.16 – The component parts of a fixed caliper with four pistons.

1 Retaining pin
2 Rubber dust cap
3 Bleeder screw
4 Cross spring
5 Brake caliper bolt
6 Brake caliper
7 Cylinder seal, small piston
8 Cylinder seal, large piston
9 Shim
10 Brake pad set

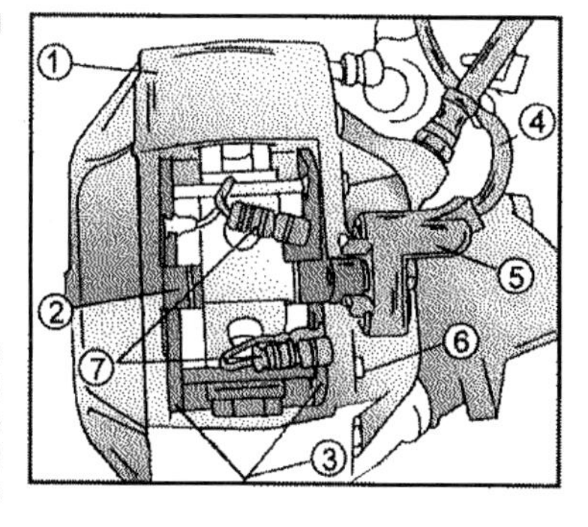

Fig. 8.16 – View of a brake caliper after removal of the wheel.

1 Brake caliper
2 Cross spring
3 Brake pads
4 Cable
5 Cable connector plug
6 Retaining pins
7 Contact sensor, brake pad wear indicator

- Place the front end of the vehicle on chassis stands and remove the front wheels. The caliper will now have the appearance shown in Fig. 8.16. First withdraw the contact sensor connectors (7).
- Using a drift of suitable diameter drive the retaining pins out of the caliper from the outside towards the inside in the manner shown in Fig. 8.17. Remove the retaining spring in the centre.
- Remove the brake pads. The workshop used a special tool to withdraw the pads. Otherwise hook a piece of wire through the two brake pad holes and withdraw the pads one after the other with a short, sharp pull.
- Push the pistons back into their bores, using a re-setting pliers as shown in Fig. 8.18. Otherwise use a piece of wood and carefully push the piston into the bore. It may be that the fluid reservoir overflows during this operation. Keep an eye on it. If necessary remove the fluid (Fig. 8.2).

- Withdraw the contact sensor out of the metal plate or the brake pad. The sensor must be replaced if the insulating layer on the contact plate is worn or any other part of the sensor or its cable is damaged.

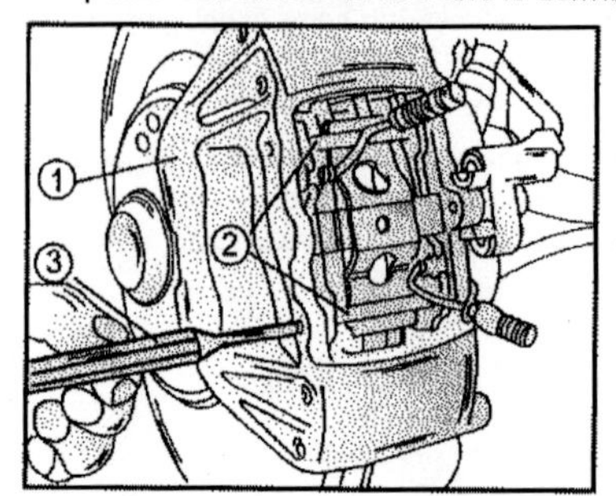

Fig. 8.17 – The retaining pins are removed from the outside towards the inside when the pads are removed.
1 Brake caliper
2 Retaining pins
3 Drift

- Measure the thickness of the pad material. If the thickness is around 3.5 mm, fit new pads. Although the pad material can be worn down to 2.0 mm, you will find that the pad wear indicator lights up when a thickness of 3.5 mm is reached. Never replace one pad only even if the remaining pads look in good order.

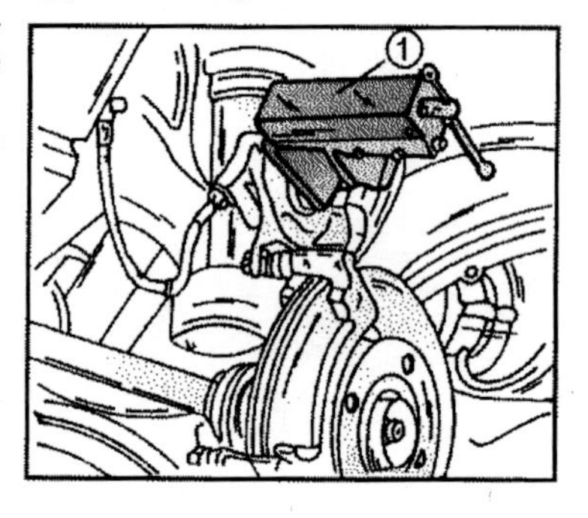

Fig. 8.18 – Piston can be pushed into their bores using a pair of re-setting pliers. Brake fluid could overflow from the fluid reservoir.

- Check the thickness of the brake disc and compare the dimensions given in Section 8.0. Replace the discs if the thickness is below the minimum permissible.

Before fitting the new pads clean the caliper opening with a brush and clean brake fluid or methylated spirit. Wipe off any spirit remaining and lubricate the exposed part of the piston with rubber grease. Fit the pads as follows:

- Coat the brake pad metal plates and their sides with Molycote paste and insert the pads into the brake caliper mounting brackets.
- Place the tensioning spring over the brake pads and insert the two retaining pins from the inside towards the outside into the caliper and through the brake pads. The pins have a clamping shape on one side which will keep them in position. Carefully drive them in position to their stop.
- Operate the brake pedal a few times to set the brake pads against the brake disc.
- Check the fluid level in the fluid reservoir and correct if necessary. Treat the new brake pads with feeling at the beginning before they are fully bedded in.

Removal and Installation of a Brake Caliper

The removal and installation of the caliper is carried out as described for the other type. One fitted bolt and a normal bolt are used to secure the caliper. Both are tightened to 11.5 kgm (83 ft.lb.) during installation.

Brake calipers - Overhaul

Refer to the points to be observed for the other caliper type. Additionally note that the two halves of a caliper must not be separated. A caliper can be overhauled as follows, but we would like to advise that a little experience with caliper units is helpful. Fig. 8.19 shows the component parts of a caliper.

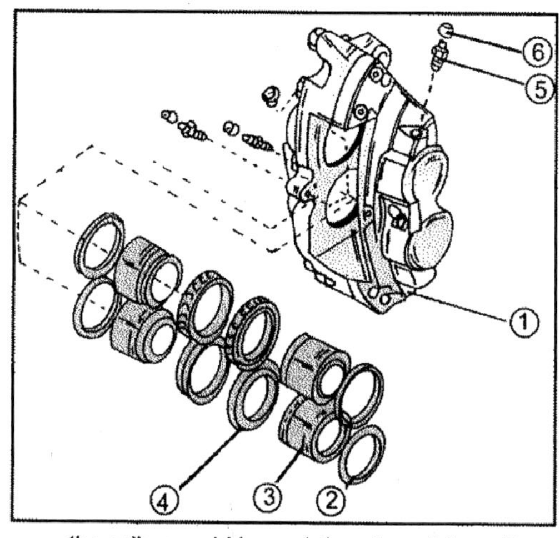

Fig. 8.19 – Exploded view of a fixed brake caliper with four pistons.
1 Brake caliper housing
2 Piston seal
3 Piston
4 Rubber dust seal
5 Bleeder screw
6 Rubber dust cap

- Clamp the caliper into a vice and remove the dust seal from the caliper housing and piston as shown in Fig. 8.20, using a screwdriver.
- Remove the first piston from the caliper. The workshop uses a special tool for this operation. As this will not be available, use a clamp to clamp one of the piston to the caliper and blow out the other piston with an airline held to the brake pipe connection bore. A piece of wood must be inserted into the caliper opening to prevent the piston from hitting the other side. Keep the fingers away from the area. The method described necessitates that each piston is overhauled on its own, i.e. removal and installation, before the second piston is dealt with.

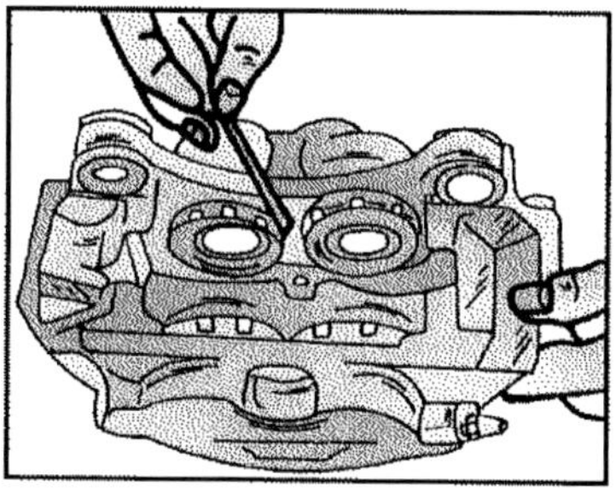

Fig. 8.20 – Remove the rubber dust seals with a screwdriver from the caliper housing.

- Remove the cylinder sealing ring out of the groove in the caliper bore. This operation is shown in Fig. 8.21. Do not tools with sharp edges.
- Clean all parts and use all parts supplied in the repair kit.

Fig. 8.21 – The cylinder seals are removed with a suitable tool from the inside of the bore.

- Fit the new sealing ring coated with brake paste or brake fluid into the groove of the cylinder bore.
- Fill half of the inside of the rubber dust seal with brake paste and fit it to the piston and insert the piston wit a rotating motion into the bore.
- Fit the rubber dust seals over the pistons, noting that both pistons must be fitted during this operation, to spread the pressure evenly. A pair of pliers working in the principle shown in Fig. 8.22 must be used. Finally refit the brake caliper and the brake pads as already described.

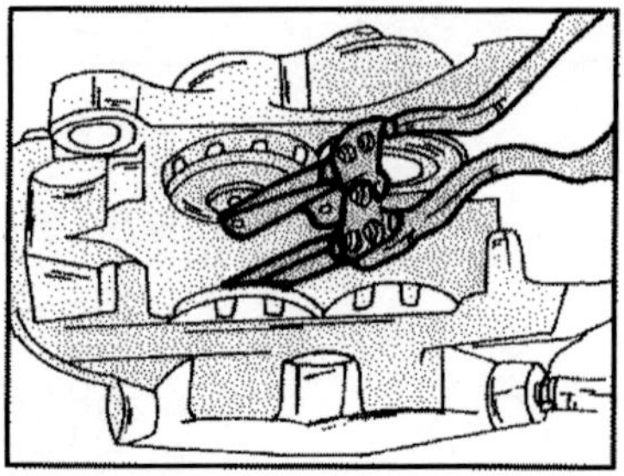

Fig. 8.22 – Rubber dust seals are pressed in position as shown.

8.3. Rear Disc Brakes

8.3.0. BRAKE PADS

The rear wheels must be removed to check the pad thickness. The remaining pad material can be checked by inspecting the distance between the edge of the cross spring and the metal plate of the brake pad, indicated by the arrows in Fig. 8.23. If this gap nearly disappeared, i.e. the pad material is less than 2.0 mm (0.8 in.), replace the brake pads of both rear calipers.

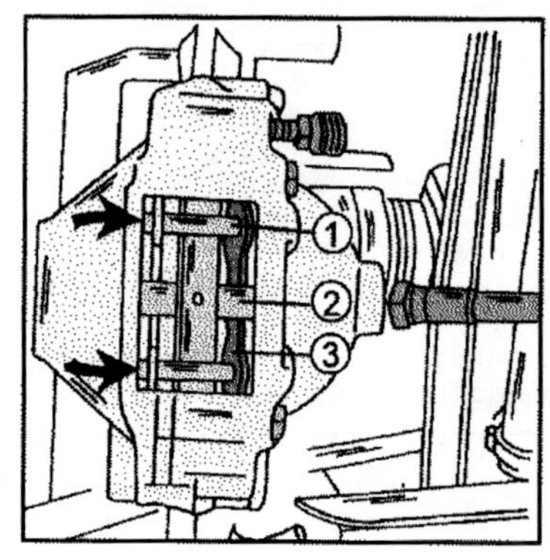

Fig. 8.23 – View of the rear brake caliper. The gap between the arrows gives an indication of the brake pad wear.

1 Retaining pins
2 Cross spring
3 Brake pad

ATTENTION: If it is possible that the brake pads can be re-used, mark them in relation to the side of the car and to the inside or outside position of the caliper. Never interchange brake pads from the inside to the outside, or visa versa, as this could lead to uneven braking.

Remove the brake pads as follows, referring to Fig. 8.24:

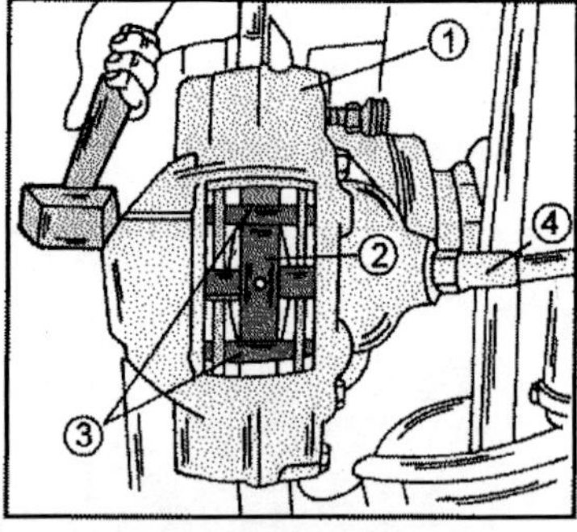

Fig. 8.24 – Removal of rear brake pads.

1 Brake caliper
2 Cross spring
3 Retaining pins
4 Brake hose
5 Drift

- Place the rear of the vehicle on chassis stands and remove the rear wheels.
- Drive out the retaining pins (3) at the top and bottom, with a suitable drift and take off the cross spring (2) from the caliper (1).
- If ASR is fitted, there is also a brake pad wear indicator fitted to the rear axle. Before disconnecting the plug note the routing of the cable.
- Remove the brake pads from the caliper. Sometimes the brake pads are difficult to remove. In this case use a strong screwdriver into one "ear" of the pad and lever the pad as shown in Fig. 8.25. Remove the second pad in a similar manner. The brake pedal must not be operated after the pads have been removed. Remember to mark where the pads have been located in case they are to be refitted.

Fig. 8.25 – Removal of the brake pads (1). Pull the screwdriver into the direction shown by the arrow.

Before fitting clean the caliper opening with a brush and clean brake fluid or methylated spirit. Wipe off any remaining spirit and lubricate the exposed parts of the piston with rubber grease. Push the pistons back into their bores. Either use the method described above (re-insert one of the old brake pads) or use a wooden block in position and lever back the piston carefully with a screwdriver blade. Note that one pad must be placed into the caliper as the other piston is pushed into the bore,. Otherwise the opposite piston will be pushed out when one of the pistons is pushed in. As this is done the level in the master cylinder reservoir will rise so either empty some of the fluid or alternatively release the bleeder screw to allow some fluid to escape as the piston is pushed in. The bleeder screw is only opened a little, and only whilst the piston is move. It should not be necessary to bleed the brake system.

Fig. 8.26 – Coat the faces shown by the arrows with the lubricant specified in the text.

Install the brake pads in reverse order to the removal, fitting the outer pad first and making sure that the inner pad does not jam as it is inserted. The pad faces shown by the arrows in Fig. 8.26 should be slightly coated with Molycote paste "U". You may be able to obtain some from your Mercedes dealer.

When the brake pads are replaced, it is essential to fit new cross-springs (spreader springs) and retaining pins that are supplied in the kit. Replace the pads as a complete set and **never** individually. Drive the retaining pins in position. First fit one pin with "feeling" to its stop, insert the cross spring underneath the fitted pin, push the spring towards the inside at the other end and fit the second pin so that the spring end comes underneath.

Apply the foot brake several times after installation of the brake pads. This is an essential requirement to allow the new pads to take up their proper position. Bleed the brake system as necessary and check the fluid level in the master cylinder reservoir and top-up if necessary.

8.3.1. BRAKE CALIPERS – REMOVAL AND INSTALLATION

The removal of a brake caliper is a simple operation.

- Jack up the rear end of the vehicle, place chassis stands in position and remove the wheel on the side in question.
- Place a bleeder hose over the bleeding screw in the caliper (remove the rubber dust cap first) and insert the other end of the hose into a container (glass jar). Open the bleeder screw and pump the brake pedal until the fluid has been drained from the system.
- Disconnect the brake pipe from the brake hose and free the brake hose from its bracket. Suitably close the hose and pipe ends to prevent entry of dirt. Slacken the

brake hose connection at the caliper whilst the caliper is till fitted. Then unscrew the caliper mounting bolts and lift off the unit. Discard the bolts as new ones must be fitted during installation. The brake pads can now be removed from the caliper. Figs. 8.27 and 8.28 show the caliper in fitted position (front view and rear view).

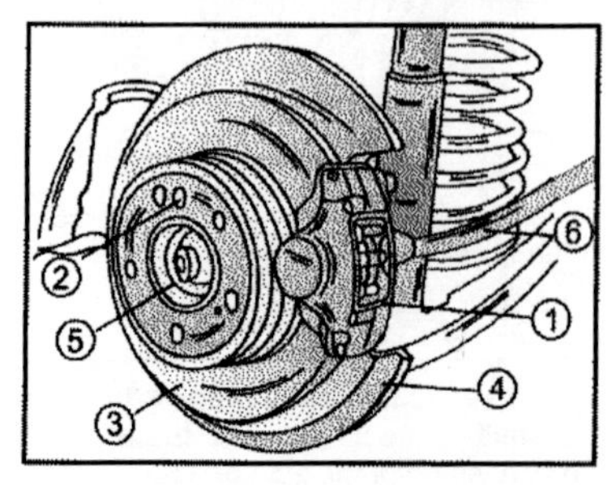

Fig. 8.27 – Rear caliper in fitted position.

1 Brake caliper
2 Brake disc screw
3 Brake disc
4 Splash guard
5 Wheel hub
6 Brake hose

The installation is a reversal of the removal procedure. Use new bolts and tighten them to 5.0 kgm (36 ft.lb.). In the case of a fitted 6 cylinder engine with ASR tighten the bolts to 7.0 kgm (50.5 ft.lb.). Tighten the brake hose with 1.8 kgm (13 ft.lb.) if it has been disconnected from the caliper. The brake system must be bled of air after installation of the caliper.

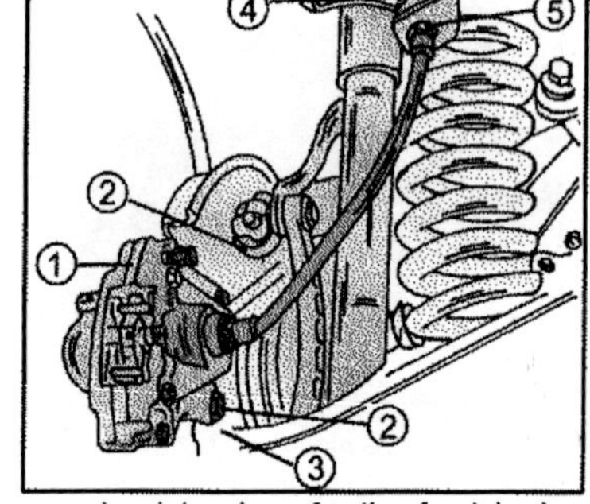

Fig. 8.28 – View of the fitted caliper from the rear.

1 Brake caliper
2 Caliper bolts, 5.0 kgm, 7.0 kgm (ASR)
3 Wheel carrier
4 Brake hose connection, 1.8 kgm
5 Brake hose bracket
6 Brake hose

8.3.2. BRAKE CALIPERS – OVERHAUL

Before commencing any servicing, note the general points given for the front brake calipers and the following points:

- Do not separate the two caliper halves since the mounting bolts have been tightened to a specific torque by the manufacturer.
- Brake calipers should only be dismantled if the piston setting gauge 000 589 35 23 00 can be obtained. This will ensure the correct position of the piston to the cylinder bore. If the gauge is not available have the caliper overhauled in a workshop or fit a new one.

The caliper can be dismantled and assembled as follows:

- With the caliper carefully clamped into a vice and using a screwdriver, carefully lever the rubber dust cap from the caliper housing. Fit a clamp to one of them pistons to keep it inside the bore.
- Apply an air line to the brake hose connection and blow out the piston. Keep the fingers away from this area when the piston is injected. Also place a piece of wood or rubber into the caliper opening to prevent the piston from hitting then other side. This operation can be carried out with an air line of a petrol station.
- Using a non-metal instrument, take out the cylinder sealing ring from the groove in the inside of the cylinder bore.

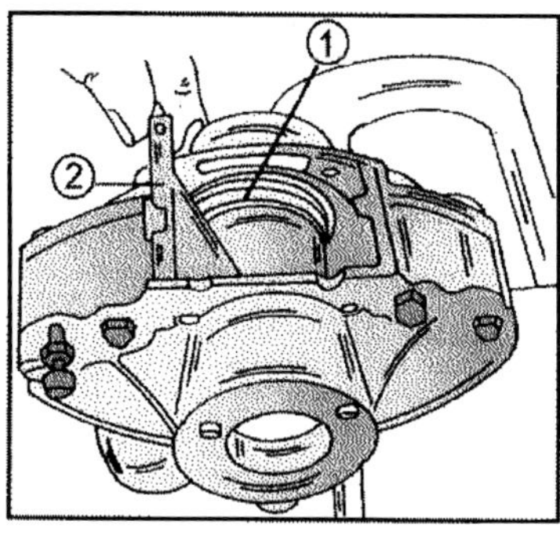

Fig. 8.29 – Aligning the piston (1) with the piston positioning gauge.

Examine all parts for wear or damage and replace as necessary. Use a repair kit to overhaul the caliper/ Assemble in the reverse order to dismantling, noting the following points:

- Fit the cylinder seal into the bore with the fingers only. Coat the piston with ATE brake paste and insert the piston into the cylinder. Use the piston positioning gauge, already mentioned, and check the position of the piston as shown in Fig. 8.29. The straight edge of the gauge must rest against the caliper and the angled edge must contact the cut-out in the piston. If necessary rotate the piston. A pair of special pliers is normally used for this purpose. Make sure that the piston is fitted the correct way round.
- Fit the rubber dust seal to the piston and to the outside of the caliper housing. Finally refit the brake pads as described earlier on.

8.4. Brake Discs

Brake discs can be re-machined, but below the thickness given in Section 8.0. Remove a brake disc as follows:

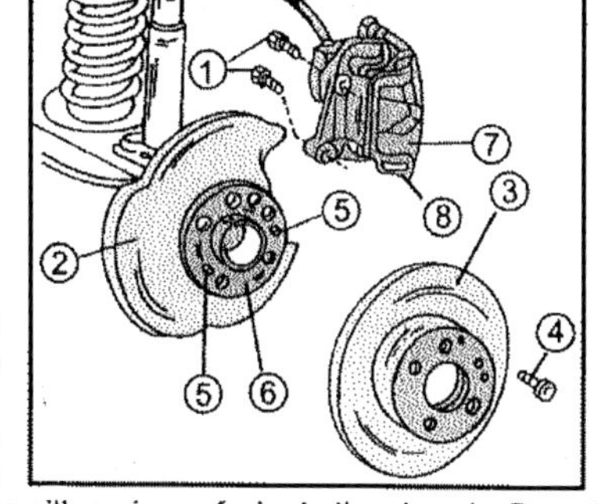

Fig. 8.30 – Details for the removal and installation of a front brake disc.

1 Caliper bracket bolt
2 Splash guard
3 Brake disc
4 Screw
5 Clamping sleeve
6 Front wheel hub
7 Cylinder housing
8 Brake caliper bracket
9 Brake hose

- Place the front or rear end of the vehicle on chassis stands and remove the wheel..
- Remove the two bolts securing the brake caliper (bolts must be replaced) and lift off the caliper. Attach the caliper with a piece of wire to the chassis. Do not allow the caliper to hang down on the brake hose.

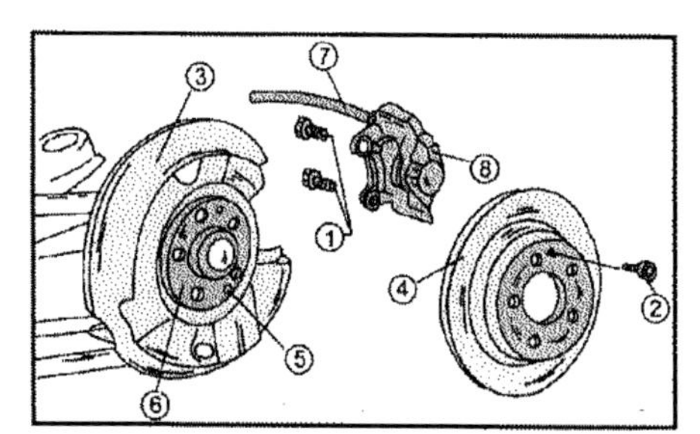

Fig. 8.31 – Details for the removal and installation of a rear disc brake.

1 Brake caliper bolts
2 Disc securing screw
3 Splash guard
4 Brake disc
5 Dowel pin
6 Wheel hub
7 Brake hose
8 Brake caliper

- The brake disc can now be removed after unscrewing the small securing screw. Use a rubber or plastic mandrel to knock off a sticking disc. Figs. 8.30 and 8.31 show how the discs are fitted.

New disc are coated for protection and s suitable solvent must be used to clean them. Refit the disc as follows:

- In the case of the front discs, check that the two clamping sleeves in the hub are properly fitted. Fit the disc and tighten the securing screw. Tighten the caliper mounting bolts to 11.5 kgm (83 ft.lb.).
- In the case of the rear discs, coat the surface of the wheel hub flange with long-term grease (Molykote paste "U" is recommended). This will facilitate removal at a later stage. Ensure that the centering dowel in the wheel hub engages properly into the brake disc, when the disc is pushed fully against the hub. Tighten the securing screw. Fit the brake caliper and tighten the bolts to 5.0 kgm (36 ft.lb.) – 7.0 kgm if ASR is fitted.
- After installation of the disc, place a dial gauge against the brake disc as shown in Fig. 8.32. Slowly rotate the disc and observe the reading of the dial gauge. If the run-out of the disc is more than specified in section 8.0, there could be two reasons:

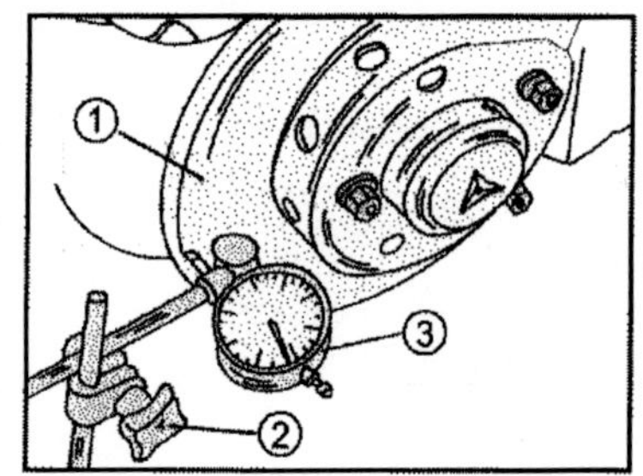

Fig. 8.32 – Checking a brake disc for run-out.

1 Brake disc
2 Dial gauge bracket
3 Dial gauge

- The front wheel bearings are not properly adjusted (adjust as described in section "Front Suspension").
- The brake disc is not fitted correctly to the hub. In this case remove the disc, move it around to the next fitting position and refit the disc.

Before driving off operate the brake pedal several times to establish the correct clearance between the brake pads and the brake disc. Check the fluid level in the reservoir and top-up if necessary.

8.5. Master Brake Cylinder

All vehicles use a tandem master cylinder with a twin reservoir, enabling the supply of brake fluid to the two circuits to the dual-line brake system. The brake pipes are split between the front and rear brakes. The piston nearest to the push rod operates the front brakes, the intermediate piston operates the rear brakes. The master cylinder should not be overhauled. Fit a new unit if the original one is worm beyond use.

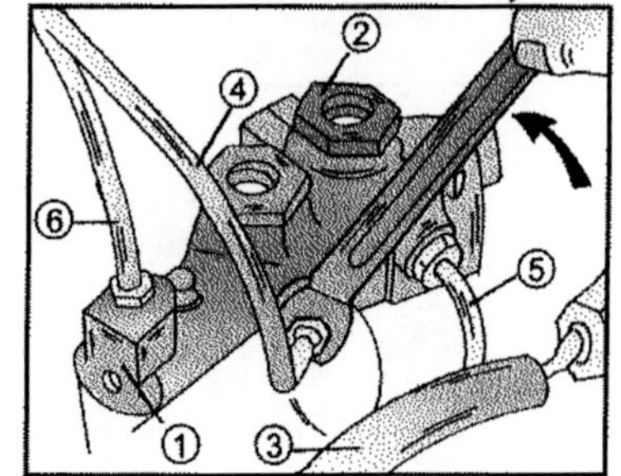

Fig. 8.33 – Details for the removal and installation of the master cylinder. The wrench (arrow) is shown removing one of the pipe union nuts.

1 Master cylinder
2 Connection for reservoir
3 Hose to clutch (M/T)
4 Pipe to R.H. front caliper
5 Pipe to L.H. front cylinder
6 Pipe to rear brakes

The fluid level in the reservoir is monitored by means of a warning light in the instrument panel. The operation of the light must be checked when the reservoir is topped-up. To do this, switch on the ignition, release the handbrake and, using the thumb, press down the two rubber caps, shown by the arrows in Fig. 8.1.
To remove the cylinder proceed as follows:

- Place a bleeder hose over one of the bleeder screws of one of the front calipers and another one over a bleeder screw of one of the rear calipers (remove the rubber dust cap first). Insert the other end of the hoses into a container (glass jar). Open both bleeder screws and operate the brake pedal until the system is empty.
- Referring to Fig. 8.33, remove the plug from the brake fluid warning switch, next to the rubber cap. If a manual transmission is fitted, disconnect the hose (3) for the hydraulic clutch operation from the reservoir. If ASR is fitted the hose is also connected. Remove the reservoir cap.
- Disconnect the brake lines from the master cylinder. An open double-box wrench should be used in order not to damage the union nuts. This is a wrench similar to a ring spanner, with a cut-out which can be placed over the pipe ends with rubber caps (from the bleeder screws) – which must of course be clean.
- Rock the fluid reservoir to and fro to remove it from the rubber grommets. Carefully lift out the reservoir without spilling brake fluid over painted area.
- Remove the cylinder from the brake servo unit. An "O" sealing ring is fitted between brake servo unit and cylinder, fitted into a groove in the cylinder flange.

The installation is a reversal of the removal procedure. The "O" sealing ring must always be replaced as the connection must be vacuum-tight. Insert the sealing ring into the groove of the cylinder. Tighten the cylinder securing nuts to 1.5 kgm (11 ft.lb.). There is no need to adjust the master cylinder push rod. Fill the brake system and bleed the complete system as described later on.

8.6. Parking Brake

The parking brake (handbrake) is a "duo-servo-type" brake shoe system. "Duo" indicates that the brake is effective in both directions of brake disc rotation, "servo" indicates the transmission of the brake shoe movement from one shoe to the other.

8.6.0. PARKING BRAKE SHOES – REMOVAL AND INSTALLATION

The special installation tool 112 589 09 61 00 is required to remove the brake shoes with the rear wheel hub fitted. Otherwise the rear hub must be removed to gain access to the hold-down springs. Fig. 8.34 shows a sectional view of the brake system and should be referred to locate the individual parts. The vehicles are fitted with an automatic handbrake cable compensating mechanism which must be tensioned before the following operations can be carried out and must de-tensioned after the operations are completed. The operation is described later on. Remove the brake shoes as follows:

- Remove the rear seat bench.
- Tension the automatic handbrake cable compensator (Section 8.6.2).
- Remove the brake caliper and the brake disc as already described.
- Turn the rear axle shaft flange until one of the threaded holes is opposite the spring (1) in Fig. 8.35. Then slightly turn the spring by 90° and disconnect the spring from the cover plate. Remove the spring in the same manner from the second brake shoe.

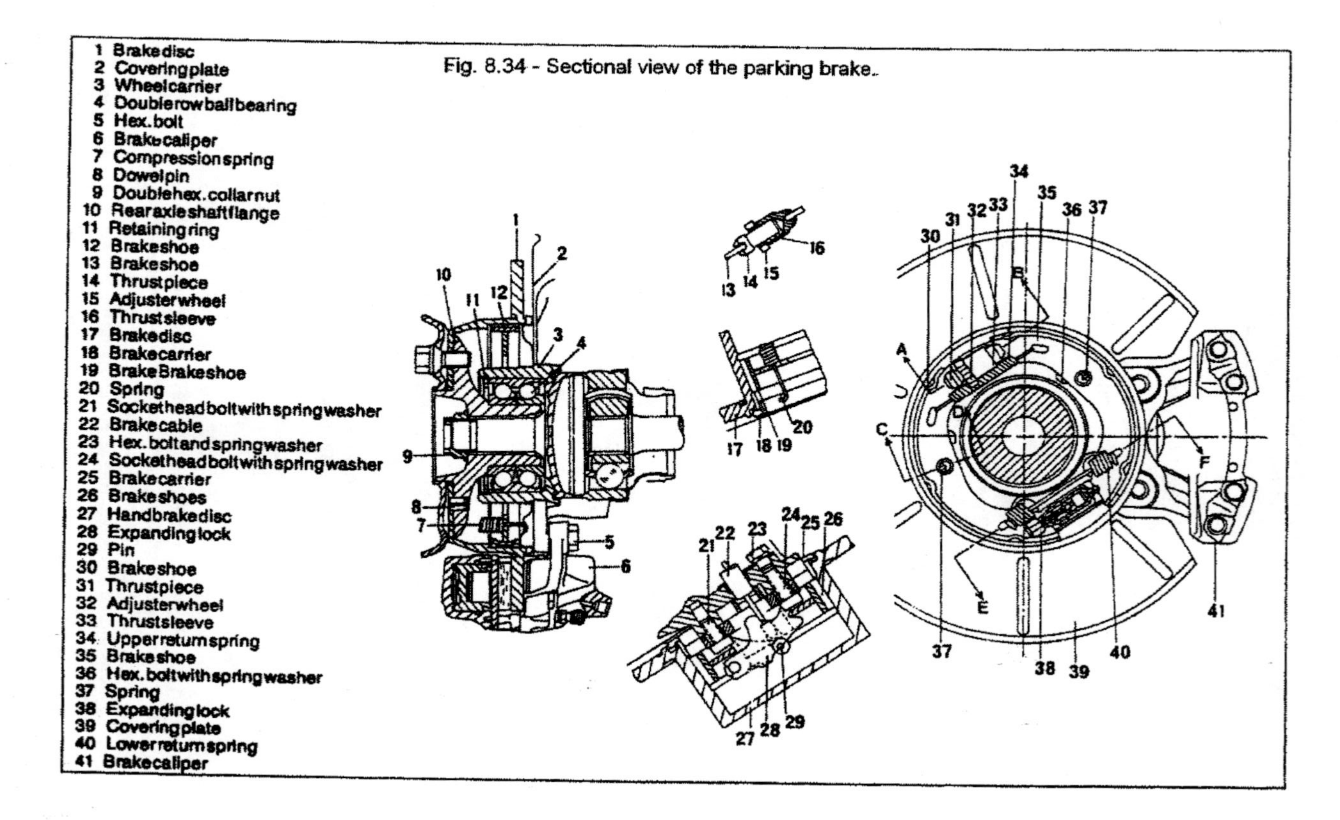

Fig. 8.34 - Sectional view of the parking brake.

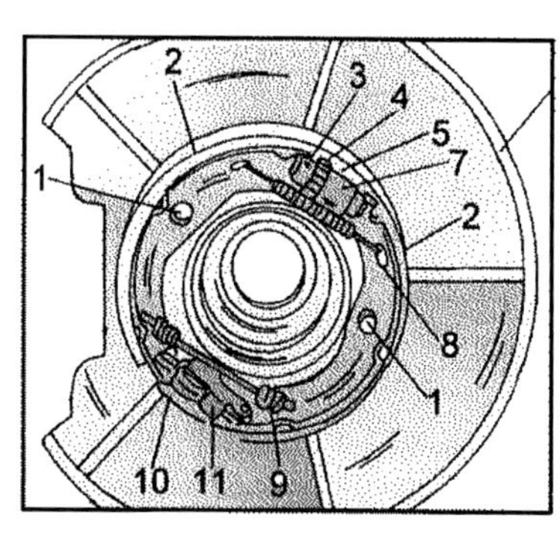

Fig. 8.35 – Removal and installation of brake shoes.
1 Hold-down spring
2 Brake shoes
3 Thrust piece
4 Adjuster wheel
5 Brake back plate
6 Cover plate
7 Thrust sleeve
8 Upper shoe return spring
9 Lower shoe return spring
10 Brake carrier
11 Expanding lock

- Remove the lower return spring with a screwdriver or a pair of pliers. Pull the two brake shoes apart until they can be lifted over the axle shaft flange towards the top. Disconnect the upper return spring (34) and remove the adjusting device (21 to 29). Remove the pin from the expanding lock and from the parking brake cable. Fig. 8.34 shows the items of the brake expanding lock.

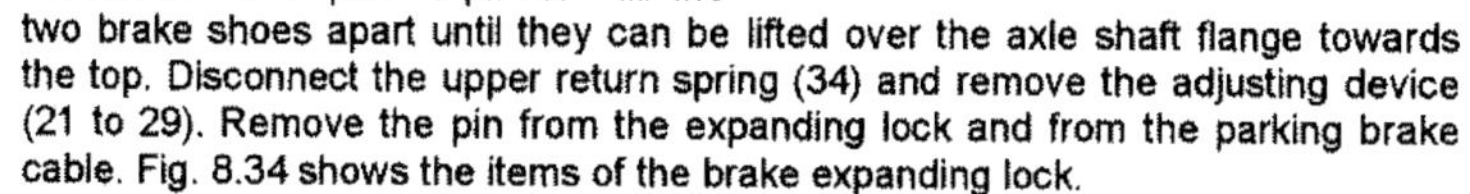

The installation of the new brake shoes is carried out as follows:

- Coat all bearing and sliding faces on the expanding lock with Molykote paste and fit the parking brake cable (22) with the pin (29) to the expanding lock (28). Then push the expanding lock against the cover plate.
- Coat the threads of the thrust piece (31) and the cylindrical portion of the adjuster wheel (32) with long-term grease and assemble the adjusting device. Turn the adjuster completely back.
- Insert the adjuster between the two brake shoes, with the adjuster wheel nearest the front of the vehicle. Fit the upper return spring (34) to the two brake shoes.
- Pull the brake shoes apart at the bottom, lift them over the drive shaft and attach them to the expanding lock.
- Fit the hold-down spring (37) to one of the brake shoes, insert the installation tool through one of the threaded bores in the drive shaft flange, compress the spring slightly and ten turn it by 90° to attach it to the cover plate. Check that the spring is correctly fitted and fit the other hold-down spring in the same manner.
- Fit the lower return spring with the smaller hook to one of the brake shoes and expand the spring until it can be engaged into the other brake shoe. This can be accomplished by means of a wire hook and a screwdriver to guide the spring into the anchor hole.
- Finally fit the brake disc and the caliper as described earlier on and adjust the braking brake as described in the next section.

8.6.1. PARKING BRAKE - ADJUSTMENT

The parking brake must be adjusted if it can be operated by more than 4 "clicks" of a total of 6 without locking the rear wheels. To adjust the handbrake proceed as follows:

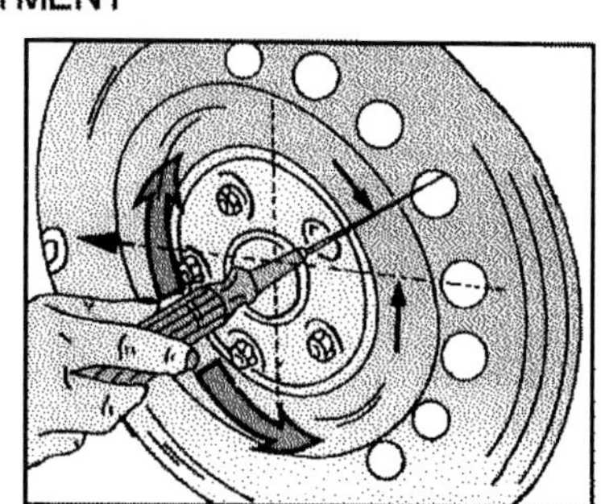

Fig. 8.36 – Adjusting the parking brake. The arrows show the adjusting direction (see text).
1 Wheel
2 Shaft flange
3 Screwdriver

- Place the rear end of the vehicle on chassis stands.

- On each wheel remove one of the wheel bolts and turn one of the wheels until the threaded hole (where the wheel bolt was fitted) is in the position shown in Fig. 8.36.

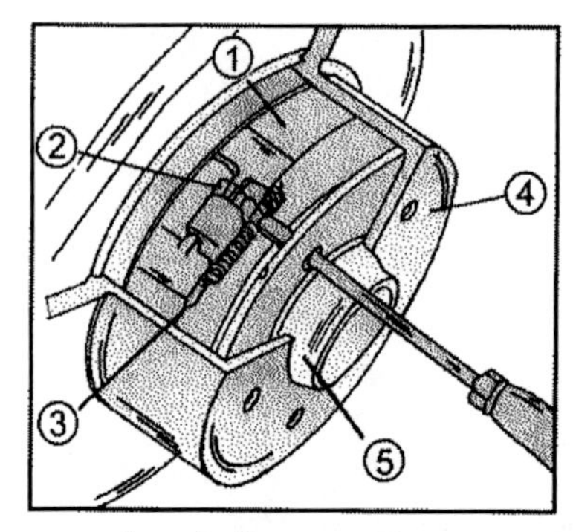

Fig. 8.37 – Adjusting the parking brake.
1 Brake shoe
2 Adjuster wheel
3 Upper return spring
4 Brake disc/drum
5 Wheel shaft flange

- Insert a screwdriver blade of 4.5 mm in diameter into the hole. The screwdriver will pass through the brake disc/drum and the drive shaft flange and engages with the adjuster wheel inside the brake drum. Fig. 8.37 shows where the engagement takes place. The drum has been cut-away to give a better view.
- Operate the screwdriver up and down in the correct direction, referring to the arrows in Fig. 8.36 until the wheel is locked. Operate the screwdriver from bottom to top in the case of the L.H. wheel or from top to bottom in the case of the R.H. wheel.
- With the wheel locked, turn back the screwdriver by 5 to 6 "clicks" of the adjuster wheel until the wheel once more is free to rotate.

8.6.2. THE AUTOMATIC PARKING BRAKE COMPENSATOR

As already mentioned the compensator must be pre-tensioned bore the removal of the parking brake shoes and de-tensioned after installation. Fig. 8.38 shows the compensator. Read the instructions carefully as it is rather complicated.

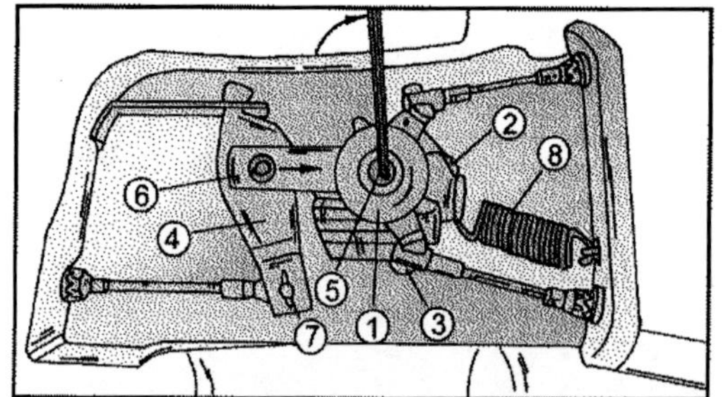

Fig. 8.38 – View of the handbrake compensating mechanism.
1 Spring housing with locking eccentric
2 Spring clip
3 Compensating lever
4 Intermediate lever
5 Locking eccentric
6 Cable length compensator
7 Securing bolt
8 Return spring

Underneath the rear seat bench there is a cover which becomes visible after removal of the seat. The cover has openings, shown in Fig. 8.39 with the arrows, through which an Allen key is inserted to pre-tension the compensator. Insert the key into the locking eccentric (5) and turn it by half a turn towards the right, as shown in Fig. 8.38 with the arrow.

Consider that the compensator is shown from above in the illustration. The key is inserted into the centre of the spring housing (1). As the Allen key and the bolt is rotated you will have to push back the bolt into the elongated hole, until the locking eccentric in the spring housing (1) engages with the spring clip (2). As mentioned above, the operation is not easy to understand if you do it the first time. The best way is perhaps to remove the cover to have a better idea what the assembly looks like.

Fig. 8.39 – The cover underneath the rear seat bench covers the handbrake cable compensating mechanism. The arrows show where the Allen key is inserted. We recommend to remove the cover for better access.

To de-tension the compensator after completing the work on the brake shoes use a screwdriver and lift the spring clip (2). The compensator will then be de-tensioned and the length of the handbrake cable will be automatically compensated. Operate the handbrake a few times to set the mechanism into operation.

8.6.3. PARKING BRAKE CABLES – REPLACEMENT

The replacement of the front handbrake cable and the rear handbrake cables is a very complicated operation, which we cannot recommend to carry out under DIY conditions. If the front cable or the rear cables require replacement you will have to seek the help of a Mercedes dealer.

8.7. Brake Servo Unit

Brake servo units should not be dismantled, as special tools are required to dismantle, assemble and test the unit. Different servo units are fitted in the range covered in this publication, manufactured by either Teves, Bendix or Girling. Always make sure to fit the correct part if the servo unit is replaced. Remember that a failure of the servo unit to act will not affect the efficiency of the braking system but, of course. additional effort will be required for the same braking distance to be maintained.

ATTENTION! If you coast downhill, for whatever reason, with a vehicle equipped with a brake servo unit, remember that the vacuum in the unit will be used up after a few applications of the brake pedal and the brake system will from then onwards operate without power-assistance. Be prepared for this.

The master brake cylinder must be removed to gain access to the brake servo unit (Section 8.5.0.). Then proceed as described:

- Disconnect the vacuum hose from the brake servo unit. Tighten the hose connection to 3.0 kgm (22 ft.lb.).
- Disconnect the brake servo push rod from the brake pedal after removal of the covering for the dashboard in the foot well. To do this, remove the lock and pull out a flange bolt. Unscrew the brake servo unit from the mounting bracket (nuts).

The installation is a reversal of the removal procedure. Tighten the securing nuts to 1.5 - 25 kg m (11 -18 ft.lb.).

8.8. Bleeding the Brakes

Bleeding of the brake system should be carried out at any time that any part of the system has been disconnected, for whatever reason. Bleeding must take place in the order left-hand rear side, right-hand rear, left-hand front and right-hand front. If only one of the brake circuits has been opened, either bleed the front or the rear circuit. The procedure given below should be followed and it should be noted that an assistant will be required, unless a so-called "one-man" bleeding kit is available.

Always use clean fresh brake fluid of the recommended specification and never re- use fluid bled from the system. Be ready to top up the reservoir with fluid (a brake bleeding kit will do this automatically) as the operations proceed. If the level is allowed to fall below the minimum the operations will have to be re-started.

Obtain a length of plastic tube, preferably clear, and a clean container (glass jar). Put in an inch or two of brake fluid into the container and then go to the first bleed point. Take off the dust cap and attach the tube to the screw, immersing the other end of the tube into the fluid in the container.

Open the bleed screw about three quarters of a turn and have your assistant depress the brake pedal firmly to its full extent while you keep the end of the tube well below the fluid level in the container. Watch the bubbles emerging from the tube and repeat the operation until no more are seen. Depress the brake pedal once more, hold it down and tighten the bleed screw firmly.

Check the fluid level, go to the next point and repeat the operations in the same way. Install all dust caps, depress the brake pedal several times and finally top up the reservoirs.

8.9. ABS System

Work should not be carried out on the ABS system, with the exception of the removal and installation of the speed sensors on front and rear wheels. The sensors are inserted into the steering knuckles, in the case of the front wheels and near the differential, in the case of the rear wheels. The final drive pinion has the function of measuring the average speed of the two rear wheels.

The following precautions must be taken after carrying out any work on a vehicle with ABS system:

- When carrying out electric welding work, withdraw the plug from the electronic control unit.
- If the vehicle is re-sprayed outside a Mercedes body shop, point out that the electronic control unit must not be subjected to temperatures of more than 95° C for a short while or more than 85° C for more than 2 hours.
- Tighten the battery terminal clamps well when re-connecting the battery.
- If a rear axle centre piece is replaced, make sure to fit the correct tooth wheel with the correct ratio for the speed sensor.
- After work has been carried out on the brake system check the ABS system as follows:
- Start the engine and check if the warning light, marked "ABS" goes off immediately and does not come on after a speed of 3 to 5 mph is reached.

8.10. Tightening Torques – Brakes

Brake caliper to knuckle, front: .... 11.5 kgm (83 ft.lb.)
Cylinder housing to brake caliper: .... 3.5 kgm (18 ft.lb.)
Brake caliper to wheel carrier, rear: .... 5.0 kgm (36 ft.lb.)
Brake hose to caliper: .... 1.5 kgm (11 ft.lb.)
Master cylinder to brake servo unit: .... 1.5 kgm (11 ft.lb.)
Brake pipe union nuts: .... 1.0 kgm (7 ft.lb.)
Brake servo unit: .... 1.5 – 2.5 kgm (14.5 - 18 ft.lb.)
Vacuum pipe to brake servo unit: .... 3.0 kgm (22 ft.lb.)
Wheel bolts: .... 115 kgm (83 ft.lb.)

9. ELECTRICAL EQUIPMENT

9.0. Battery

Voltage: .... 12 volts
Polarity: .... Negative earth (ground)
Condition of Charge:
Well charged: 1.28
Half charged : 1.20
Discharged: 1.12

To check the voltage of the battery, use an ordinary voltmeter and apply between the two battery terminals. A voltage of 12.5 volts or more should be obtained.
If a hydrometer is available, the specific gravity of the electrolyte can be checked. The readings of all cells must be approximate by the same. A cell with a low reading indicates a short circuit in that particular cell .Two adjacent cells with allow reading indicates a leak between these two cells.
A battery can be re-charged, but the charging rate must not exceed 10% of the battery capacity, i.e. 7.2 amps. The battery must be disconnected from the electrical system. Charge the battery until the specific gravity and the charging/voltage are no longer increasing within 2 hours. Add distilled water only. Never add acid to the battery.
The level of the battery electrolyte should always be kept above the top of the plates.

9.1. Alternator

9.1.0. ROUTINE PRECAUTIONS

The vehicle covered in this manual employs an alternator (90 or 150 amps) and control unit. This equipment contains polarity-sensitive components and the precautions below must be observed to avoid damage:

- Check the battery polarity before connecting the terminals. Immediate damage will result to the silicon diodes from a wrong connection — even if only momentarily.
- Never disconnect the battery or alternator terminals whilst the engine is running.
- Never allow the alternator to be rotated by the engine unless ALL connections are made.
- Disconnect the alternator multi-pin connector before using electric welding equipment anywhere on the vehicle.
- Disconnect the battery leads if a rapid battery charger is to be used.
- If an auxiliary battery is used to start the engine. take care that the polarity is correct. Do not disconnect the cables from the vehicle battery.

9.1.1. DRIVE BELT TENSION

Always tension the drive belt whenever the alternator, water pump or drive belt have been removed or slackened for any reason. The single drive belt is properly tensioned when the operations described in Section "Cooling System" are followed. The alternator runs at higher speed than the older D.C. dynamo generators and the belt tension should be maintained accurately for the best results. When a new belt has been fitted, it is as well to re-check the tension after a few hundred miles have been covered.

9.1.2. ALTERNATOR – REMOVAL AND INSTALLATION

The alternator is rigidly attached to the engine. Fig. 9.1 shows details of the attachment of the alternator. Remove as follows:

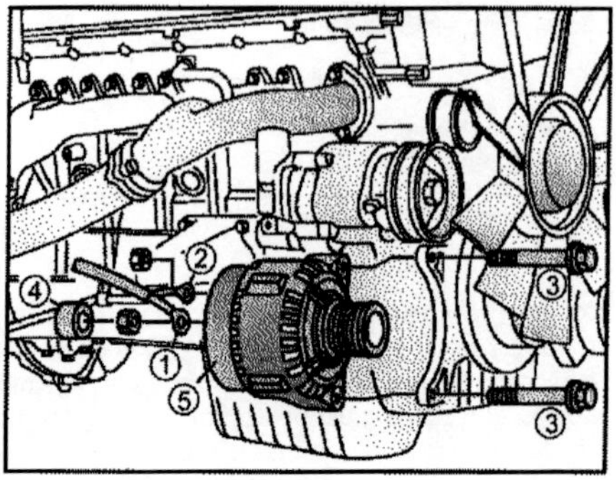

Fig. 9.1 – Fitted alternator.

1 Terminal 30 cable (B+)
2 Terminal 61 cable (D+)
3 Mounting bolts, 2,o kgm
4 Contact protection
5 Alternator

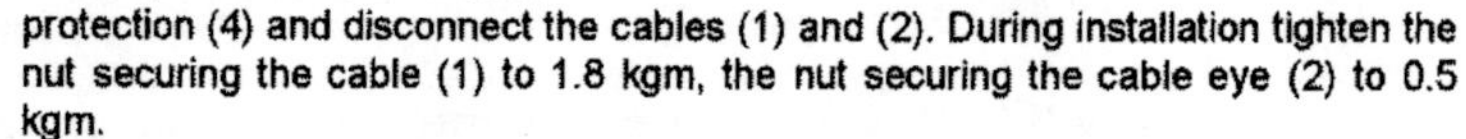

- Disconnect the battery.
- Slacken the Poly V-belt as described in section "Cooling System" and lift it off the alternator pulley.
- Remove the bottom parts of the engine compartment panelling.
- If fitted, remove the terminal protection (4) and disconnect the cables (1) and (2). During installation tighten the nut securing the cable (1) to 1.8 kgm, the nut securing the cable eye (2) to 0.5 kgm.

Unscrew the mounting bolts (3) and remove the alternator downwards.

The installation is a reversal of the removal procedure. Tighten the alternator mounting bolts (3) to 2.0 kgm (14.5 ft.lb.).

9.1.3. SERVICING

A Bosch alternator is used on the engines dealt with in this manual, having an output of 70, 90 or 120 amps if an air conditioning system or automatic climate control system is fitted.

We do not recommend that the alternator or control unit should be adjusted or serviced by the owner. Special equipment is required in the way of test instruments and the incorrect application of meters could result in damage to the circuits.

The alternator is fitted with sealed-for-life bearings and no routine attention is required for lubrication. Keep the outside of the alternator clean and do not allow it to be sprayed with water or any solvent.

Fig. 9.2 – Th dimension "a" gives the remaining brush length.

The alternator is fitted with sealed-for life bearings and no routine attention is required for lubrication. Keep the outside of the alternator clean and do not allow it to be sprayed with water or any solvent.

The alternator brush gear runs in plain slip rings and the brushes have a long life, requiring inspection only after a high mileage has been **covered.** To inspect the brushes, we recommend the removal of the alternator. Take out the two screws from the brush holder assembly and withdraw for inspection. Measure the length of the brushes, shown by "a" in Fig. 9.2. If the protruding length is less than 5.0 mm (0.2 in.) or approaching this length, replace the brushes. New brushes will have to be soldered in position. We would like to point out that it is not an easy operation to guide the brushes over the slip rings when the slip ring cover is being fitted.

9.2. Starter Motor

9.2.0. REMOVAL AND INSTALLATION

Disconnect the battery earth (ground) cable. The bottom parts of the noise encapsulation must be removed to gain access to all attached parts. Disconnect two cables from the starter motor connections.
Remove the starter motor support bracket from starter motor and cylinder block. Remove the starter motor mounting bolts.
Withdraw the unit from the car in a downward direction. The starter motor may have to be turned to get it passed the various parts.
Install in the reverse sequence to removal. Tighten the starter motor bolts to 4.2 kgm. Make sure that the mating faces are clean before bolting up. Re-connect the wires and the battery terminals.

9.2.1. SERVICING

It may be of advantage to fit an exchange starter motor if the old one has shown fault. Exchange starter motors carry the same warranty as a new unit are therefore a better proposition.

9.3. Headlamps - Replacement

As the replacement of a headlamp requires the adjustment of the headlamp beams, which should be carried out at a Mercedes dealer or a workshop dealing with headlamp adjustments we will not describe the operations for the removal and installation of the units. Fig. 9.3 shows the individual parts of a headlamp, which can be referred to when you intent to remove a lamp unit. Remember, always have the headlamp alignment checked after a headlamp is replaced.

9.4. Bulb Table

Main and dipped beam:	Halogen H4, 60&55 watts, ECE H4
Fog lamps:	Halogen H3, 55 watts, ECE H3
Indicator lamps, front end rear:	ECE P, 21 watts
Reversing light, rear fog lamp:	ECE P, 21 watts
Parking lamps:	ECE T, 4 watts
Tail lights:	ECE R, 10 watts
Number plate lights:	5 watts
Luggage compartment, interior lamps:	10 watts

9.5. Headlamp Adjustment

The adjustment of the headlamps is carried out from the rear of the units. Optical equipment is required to adjust the headlamp beams and only a workshop equipped with such equipment can carry out the adjustment. Provisional adjustments can be carried out as follows:

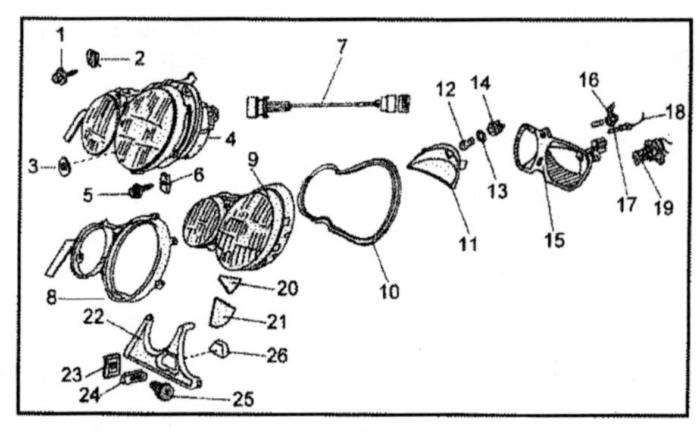

Fig. 9.3 – The component parts of a headlamp unit.

1 Combi screw, upper attachment
2 Metal nut, upper attachment
3 Securing nut
4 L.H. headlamp unit
5 Combi screw, lower attachment
6 Metal nut
7 Connecting cable for headlamp unit
8 L.H. headlamp frame
9 L.H. headlamp insert
10 Inner headlamp sealing ring
11 Reflector, flasher lamp
12 Flasher lamp bulb, 21 watts
13 Sealing ring, flasher lamp carrier
14 Lamp socket, flasher lamp
15 Headlamp reflector
16 Double filament bulb, 55 watts
17 Parking lamp bulb, 5 watts
18 Parking lamp bulb socket
19 Dipped beam bulb, 55 watts
20 Insert (R.H. drive only)
21 As 20
22 Embellisher for headlamp
23 Metal nut
24 Headlamp support
25 Securing screw
26 Insert on headlamps without cleaning system

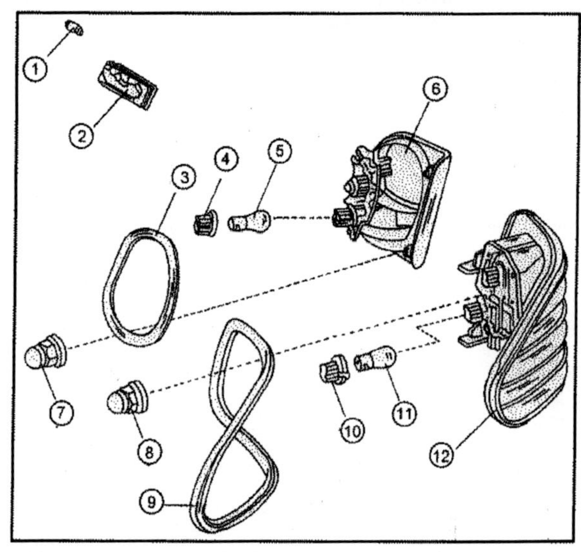

Fig. 9.4 – The component parts of a rear light cluster. The explanation of the numbers is given on the opposite page.

Fig. 9.4 – The component parts of a rear light cluster.

1 Bulb, number plate light, 5 watts
2 Number plate light
3 Sealing ring for 6
4 Bulb socket for reversing light and rear fog light
5 Bulb, reversing and rear fog light, 21 watt
6 Tail light assembly
7 Nut, tail light attachment
8 Nut, tail light attachment
9 Sealing ring for 12
10 Bulb socket for brake and tail light
11 Bulb, brake and tail light, 21 watts
12 Tail light cluster cover

- Drive the vehicle in front of the garage door and position it so that the front of the vehicle is approx. 5 metres away from the door.
- Switch on the headlamp and adjust the beams of the new headlamps to be level with the beams of the original headlamp on the other side. The vehicle can new be driven to a workshop to have the headlamps professionally adjusted.

9.6. Tail Light Units

The tail light unit consists of the items shown in Fig. 9.4. A bulb carrier, a reflector and a coloured plastic lens cover make up the light unit. The lens cover is attached from the inside together with the reflector by means of nuts. The bulbs of the unit can be replaced by referring to the illustration.

10. EXHAUST SYSTEM

10.1. Exhaust System

10.1.1. REMOVAL AND INSTALLATION

Fig. 10.1 shows the exhaust systems which is fitted to the vehicle with engines 604, 605 and 606 without turbo charger. Fig. 10.2 shows the parts of the system if a turbo charger is fitted (602 and 606 engine, E290 and E300 TD).

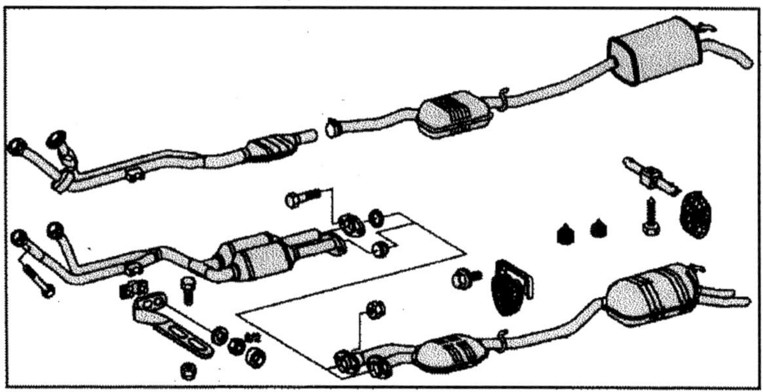

Fig. 10.1 – The component parts of the exhaust system as fitted to vehicles with 604, 605 and 606 engine, the latter without turbo charger.

The removal and installation of the exhaust system presents no major problems, but the following points must be noted:

- The vehicle must be resting on chassis stands to work underneath the vehicle.

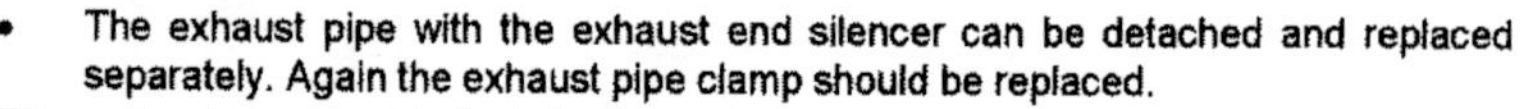

Fig. 10.2 – **The component parts of the exhaust system as fitted to vehicles with 602 and 606 engine with turbo charger. Letters "a" and "b" are explained below.**

- All self-locking nuts must be replaced during installation.
- Inspect the condition of the rubber hangers before re-using them.
- The exhaust pipe and the catalytic converter Fig.10.1 can be separated and replaced separately. The exhaust pipe clamp should always be replaced.
- The exhaust pipe with the exhaust end silencer can be detached and replaced separately. Again the exhaust pipe clamp should be replaced.

The system is removed in the following order:

- Remove the bottom section of the sound insulation capsule.
- Remove the exhaust pipes from the exhaust manifold. Tighten the bolts to 2.5 kgm (18 ft.lb.).
- Remove the bracket securing the exhaust system to the transmission. Again use self-locking nuts during installation (2.0 kgm).
- Detach the bracket from the front exhaust pipe. Again use self-locking nuts during installation (2.0 kgm). The bracket is attached to a threaded plate.
- Take off the complete exhaust system.

Note the following when the exhaust system of a turbo charged engine, shown in Fig. 10.2 is removed:

- The flange connection of the front exhaust pipe or the exhaust pipe with converter is connected by means of disc springs. To separate the connection use a pair of water pump pliers and press together the disc springs "a" and "b" in the illustration and press of the metal tongue with a screwdriver. Fig. 10.3 shows the operation. During installation press together the two springs (a) and (b) until the metal tongue engages.

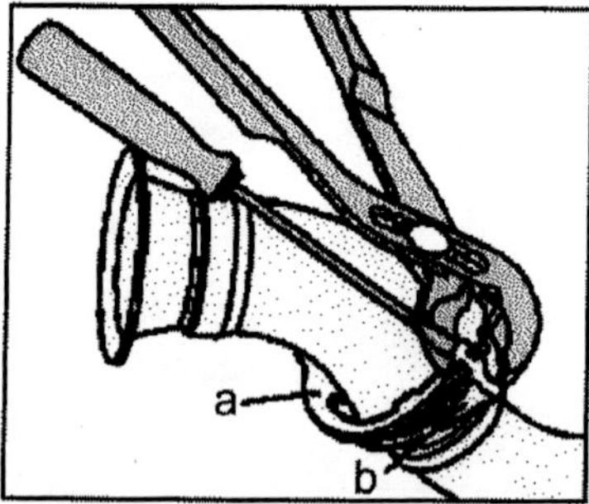

Fig. 10.03 – Releasing the spring connection between the primary and the three-way converter. The two spring discs (a) and (b) must be pressed together as shown during removal and installation.

10.3. Exhaust Manifold – Removal and Installation

The exhaust manifold can be removed after the removal of the turbo charger. Fig. 10.4 and Fig. 10.5 show details of the attachment of the manifold on the various engines. The turbo charger must be removed if one is fitted.

After removal of the self-locking nuts take off the exhaust manifold. The nuts must always be replaced. The same applies to the gasket or gaskets. The nuts are tightened to 3.0 kgm (22 ft.lb.) during installation.

Fig. 10.4 – Details for the removal and installation of the exhaust manifold. On the left for the 605 engine, on the right for the 606 engine without turbo charger.

Inspect the threads of the studs and replace them if necessary. To remove a stud, screw two nuts onto the stud and lock them against each other. Apply an open-ended

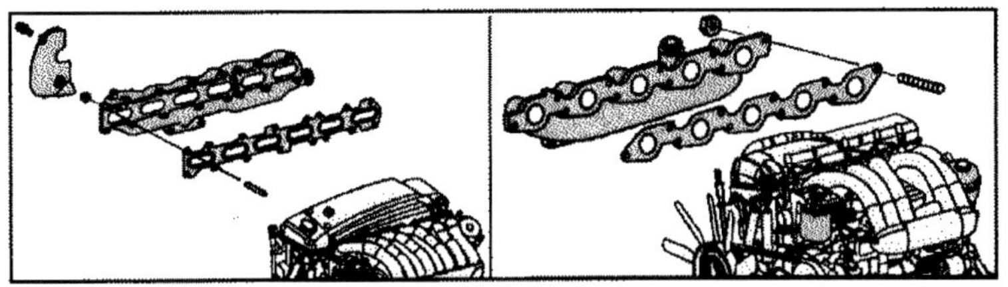

Fig. 10.5 – Details for the removal and installation of the exhaust manifold. On the left for the 606 engine with turbo charger, on the right for the 606 engine with turbo charger.

spanner to the lower stud and unscrew it. The new stud is fitted in the same manner, but in this case apply the spanner to the upper nut.

Install in reverse order. Noting the point given above.

11 Wheels and Tyres

Tyre pressure and the condition of the tyres should be checked once a week. Remember that the tyres are the only contact with the road surface.

Inspect the tyre walls for cracks, splits or bad damage. If the tyres are worn on one side, in most cases on the outside, check the front wheel alignment or have it checked professionally. Normally the toe-in/toe-out setting of the front wheels will require adjustment.

Excessive wear on both sides of the tyre(s) could indicate driving with an under-inflated tyre. Excessive wear in the centre of the tread indicates an over-inflated tyre.

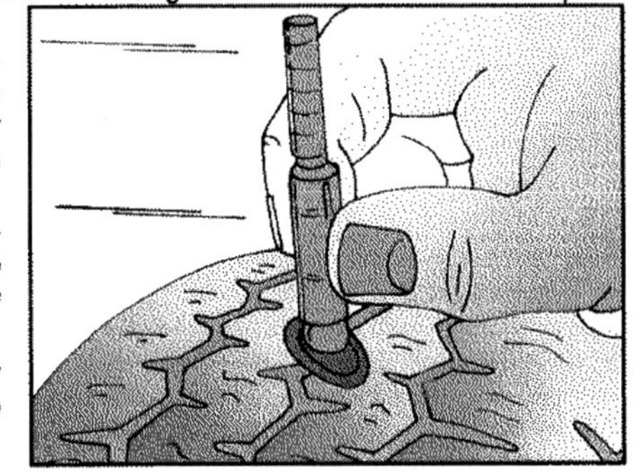

Fig. 11.1 – A gauge can be placed over the surface of the tyre as shown to measure the remaining depth of the tyre tread.

Damage can also be caused by sharp objects or contact with kerb stones.

A clear thread pattern should always be visible. Do not drive with tyres if the depth of the tread is less than 1.6 mm. Tyre manufacturers have gauges which you can use as shown in Fig. 11.1 to check the remaining tread depth of your tyre.
Check the tyre pressures once a week in accordance with the pressures given in your Operators Manual, noting the differences for different tyres. If the tyres lose more than 2 psi. a week, then the tyre has a slow puncture or the seal on the wheel rim is damaged. Take the faulty wheel to a tyre specialist, Always keep the valve caps in place as these will prevent leakage of air from the valves. Do not forget to replace them after you have checked the tyre pressure.
Punctures may occur slowly over night or suddenly whilst driving. If the latter is the case do not panic. Try to keep the steering wheel straight and if possible come to rest without braking. Drive onto the hard shoulder on the road (or as far as possible to the road side if there is no hard shoulder). Sudden movement will cause a skid. When you have stopped the vehicle, change the damaged wheel with the spare. Remember the wheel bolt tightening torque – 10.5 kgm (76 ft.lb.).

Note: Use a warning triangle a good distance behind the vehicle when changing a wheel on a public road. Also switch on the hazard warning lights. Do no forget to collect the warning triangle. Here is good advice. If you have a passenger make him responsible for the triangle, if you are by yourself, place something over the steering wheel. This will remind you that you have forgotten something.

12 Automatic Transmission

A five-speed automatic transmission is fitted to certain models, but not all models have the same transmission, as the gear ratios have been selected for the fitted engine.
The removal and installation of the transmission is a very comprehensive operation. Due to the electronic functions of some component parts, which can only be carried out by a dealer, we cannot recommend to remove the transmission yourself.

Fluid Level and Fluid Change.

The fluid level in the transmission changes with the temperature of the transmission fluid. The fluid dipstick is marked with two levels, one for a temperature of around 30° C (cold) and one for around 80° C (hot). These are shown in Fig. 12.1 with "A" and "B". On the other side of the dipstick you will find the temperatures in Fahrenheit (F). If the fluid level is correct you will find the fluid between the "Min" and "Max" marks. The transmission is filled with automatic transmission fluid (ATF). Dexron II fluid is recommended. The total capacity of the transmission is 6.3 litres, but only 3.3 litres will be used during a fluid change..

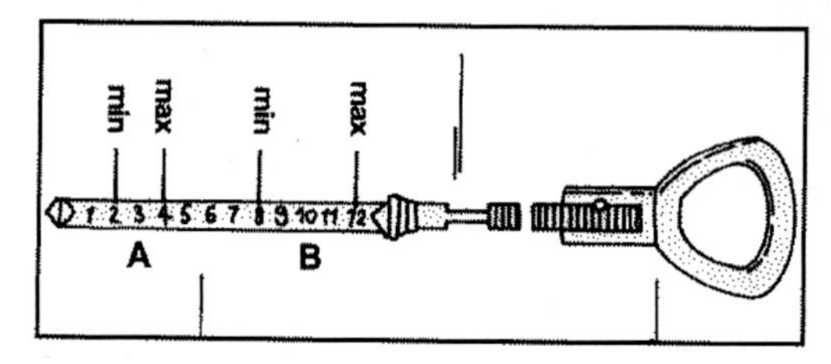

Fig. 12.1 – View of the fluid dipstick as fitted to an automatic trans-mission. The fluid must be within the area "A" when the fluid is cold or the area "B" when the fluid is hot.

Absolute cleanliness is to be observed during a fluid level check that even small particles entering the transmission can lead to malfunctions. Do not use fluffy cloth to wipe the dipstick. Tissue paper is best.

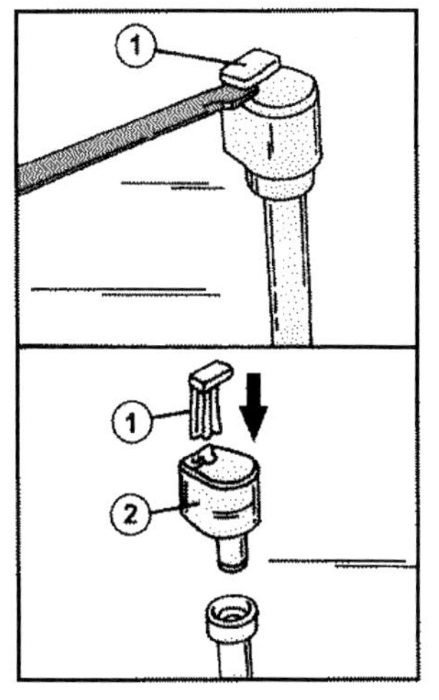

Fig.. 12.2 – Break off the lug on the plate (1) and pus the pin downwards in the direction of the arrow. The cap (2) can then be removed.

Filling in additional fluid is not straight forward. As you can see in Fig. 12.2, the upper end of the filler tube is fitted with a locking pin, which must be removed. To do this brake off the plate (1) of the pin with a screwdriver as shown and press out the remaining pin in the cap downwards. Remove the cap (2). The pin must, of course, be replaced.

The engine must be running at idle speed when fluid is filled in through the filler tube (a funnel is required). Apply the handbrake and depress the brake pedal and change through all gears. Finally leave the gear selector lever in position "P" and re-check the fluid level.

Finally refit the cap to the filler tube ands press in a new locking pin until it locks in position.

The fluid can only be changed at a dealer as a diagnostic system is used to carry out the operation.

13. SERVICING AND MAINTENANCE

Most of the maintenance operations can be carried out without much difficulties. In many cases it is, however, better to have certain maintenance operations carried out in a workshop as experience and special equipment, for example test instruments, are required to carry out a certain job. Most important are the regular inspections and checks which are described below. Operations to be carried out after a certain mileage are described later on in this section and the text will advise when specific jobs should be left to a Mercedes Dealer.

13.0. Regular Maintenance

Oil Level Check: Check the engine oil level every 500 miles. With the vehicle standing on level ground, remove the oil dipstick and wipe it clean with a clean rag or a piece of tissue paper. Re-insert the oil dipstick and remove once more. The oil level must be visible between the upper and the lower mark on the dipstick. If the oil level is below the lower mark, top-up with engine oil of the correct viscosity. The oil quantity between the two marks is approx. between 5.0 and 3.5 litre and from the actual level indicated you will be able to tell how much oil is missing. Never overfill the engine - the level must never be above the upper dipstick mark.

Checking the Brake Fluid Level: - The brake fluid reservoir is in the engine compartment on the drivers side. The reservoir is transparent and it is easy to check whether the fluid level is between the "Min" and "Max" mark. If necessary, top-up to the "Max" mark with the correct brake fluid.

Checking the Brake Lights: The operation of the brake lights can either be checked with the help of another person or you can check it by yourself by driving the vehicle backwards near the garage door. Operate the brake pedal and check if the reflection of the brake lights can be seen on the garage door by looking through the rear view mirror.

Checking the Vehicle Lights: In turn check every vehicle light, including the horn and the hazard warning light system. Rear lights and reversing lights can be checked in the dark in front of a garage door, without leaving the vehicle.

Checking the Tyre Pressures: Check the tyre pressures at a petrol station. Pressures are different for the various models. Either your Operators Manual or tyre charts will give you the correct pressures.

If continuous speeds of more than 100 mph are anticipated, increase the tyre pressure by 0.2 kg/sq.cm. (3 psi.).

Checking the Coolant Level: See Section "Cooling System". Never open the radiator filler cap when the engine is hot.

Checking the Fluid Level in the Automatic Transmission: The fluid level should be checked at regular intervals to ensure the correct operation of the transmission:

- Apply the handbrake and place the gear selector lever into the "P" position. Start the engine and allow to idle for 1 to 2 minutes.
- Remove the oil dipstick from the transmission and read off the fluid level. The level must be between the "Min" and "Max" mark when the transmission is at operating temperature; the level may be up to 10 mm (0.4 in.) below the "Min" mark if the transmission is cold.
- If necessary top-up the transmission with ATF fluid through the fluid dipstick tube. A funnel is required. Only use the fluid recommended for the transmission.

13.1. Service every 6000 Miles

Changing the Engine Oil and Oil Filter-: Some petrol stations will carry out an oil change free of charge – You only pay for the oil. The same applies to the oil filter (there may be a small extra charge), but not every petrol station will be able to obtain a Mercedes filter. To change the filter yourself, refer to the relevant page.

Lubrication Jobs: Apart from the engine lubrication there are further lubrication points which should be attended to. These include the throttle linkage and shafts (only grease the swivel points), the engine bonnet catch and the hinges (use a drop of engine oil) and perhaps the door mechanism.

13.2. Additional Service Every 12,000 Miles

Checking the Idle Speed: If the engine no longer idles as expected, have the idle speed checked and if necessary adjusted at your Dealer.

Air Filter Service: Remove the air filter element for cleaning.

Checking the Brake System: If no trouble has been experienced with the brake system, there is little need to carry out extensive checks. To safeguard for the next 6000 miles, however, follow the brake pipes underneath the vehicle. No rust or corrosion must be visible. Dark deposits near the pipe ends point to leaking joints. Brake hoses must show no signs of chafing or breaks. All rubber dust caps must be in position on the bleeder valves of the calipers. Insert a finger underneath the master

cylinder, where it is fitted to the brake servo unit. Moisture indicates a slightly leaking cylinder.
The brake pads must be checked for the remaining material thickness as has been described in Section "Brakes" for the front and rear brakes.

Adjusting the Parking Brake: Adjust the parking brake as described in Section "Brakes" under the relevant heading.

Brake Test: A brake test is recommended at this interval. Your will decide yourself if the brakes perform as you expect them to. Otherwise have the brakes tested on a dynamometer. The read-out of the meter will show you the efficiency of the brake system on all four wheels.

Checking the Wheel Suspension and Steering: In the case of the front suspension remove both wheels and check the shock absorbers for signs of moisture, indicating fluid leaks.
Check the free play of the steering wheel. If the steering wheel can be moved by more than 25 mm (1 in.) before the front wheels respond, have the steering checked professionally.
Check the rubber dust boots of the track rod and suspension ball joints. Although rubber boots can be replaced individually, dirt may have entered the joints already. In this case replace the ball joint end piece or the suspension ball joint.
Check the fluid level in the reservoir for the power-assisted steering. Refer to the "Steering" section for details. If steering fluid is always missing after the 12,000 miles check, suspect a leak somewhere in the system - See your dealer.

Tyre Check: Jack up the vehicle and check all tyres for uneven wear. Tyres should be evenly worn on the entire surface. Uneven wear at the inner or outer edge of front tyres points to misalignment of the front wheel geometry. Have the geometry measured at your dealer. Make sure that a tread depth of 1.6 mm is still visible to remain within the legal requirements. Make sure to fit tyres suitable for your model, mainly if you buy them from an independent tyre company.

Re-tighten Wheel Bolts:- Re-tighten the wheel bolts to 9.0 – 11.0 kgm (65 – 80 ft.lb.). Tighten every second bolt in turn until all bolts have been re-tightened.

Checking the Cooling System: Check all coolant hoses for cuts, chafing and other damage. Check the radiator for leaks, normally indicated by a deposit, left by the leaking anti-freeze. Slight radiator leaks can be stopped with one of the proprietary sealants available for this purpose.

Checking the Clutch: Check the clutch operation. The fluid reservoir should be full. If it is suspected that the clutch linings are worn near their limit, take the vehicle to a dealer.

Checking the Anti-freeze: The strength of the anti-freeze should be checked every 12,000 miles. Petrol stations normally have a hydrometer to carry out this check. Make sure that only anti-freeze suitable for Mercedes engines is used.

Checking the Manual Transmission Fluid Level: Refer to Section "Manual Transmission".

Checking the Rear Axle Oil Level: Remove the filler plug at the side of the rear axle centre piece, just above one of the drive shafts (L.H. side). The oil level should be to the lower edge of the filler hole. If necessary top-up with differential oil. Refit the plug.

13.3. Additional Service every 36,000 Miles

Automatic Transmission Oil and Filter Change: These operations should be carried out by a Dealer.

Air Cleaner Element Change: Refer to Section "Diesel Fuel Injection" for details.

Clutch: The wear of the clutch driven plate should be checked by a dealer with the special gauge available.

Propeller Shaft: Check the two shaft couplings for cuts or other damage. The sleeves must not be loose in the couplings. Check the intermediate bearing for wear by moving the shaft up and down in the bearing.

13.4. Once every Year

Brake Fluid Change: We recommend to have the brake fluid changed at your dealer. Road safety is involved and the job should be carried out professionally. If you are experienced with brake systems, follow the instructions in the "Brakes" section to drain, fill and bleed the brake system.

13.5. Once every 3 Years

Cooling System: The anti-freeze must be changed. Refer to Section "Cooling System" to drain and refill the cooling system.

FAULT FINDING SECTION

The following section lists some of the more common faults that can develop in a motor car. References to diesel engines are not aimed at the CDI system as this can only be checked out properly in a workshop dealing with this type of injection system, but many of the faults will be the same in all types of diesel engines. The section is divided into various categories and it should be possible to locale faults or damage by referring to the assembly group of the vehicle in question.

The faults are listed In no particular order and their causes are given a number. By referring to this number it is possible to read off the possible cause and to carry out the necessary remedies, if this is within the scope of your facilities.

ENGINE FAULTS

Engine will not crank:	1, 2, 3, 4
Engine cranks, but will not start:	5, 6, 7, 8
Engine cranks very slowly:	1, 2, 3
Engine starts, but cuts out:	5, 6, 9, 10
Engine misfires in the lower speed ranges:	5, 6, 9, 11
Engine misfires in the higher speed ranges:	5, 6, 11, 12
Continuous misfiring:	5, 6, 7, 10 to 15, 21, 22
Max. revs not obtained:	5, 6, 12, 22
Faulty idling:	5, 6, 8 to 11, 13, 15, 16, 21 and 22
Lack of power:	3, 5 to 11, 13 to 15, 22
Lack of acceleration:	5 to 8, 12, 14 to 16
Lack of max. speed:	5 to 8, 10, 12, 13 to 15 ,22
Excessive fuel consumption:	3, 5, 6, 15 ,16
Excessive oil consumption:	16 to 19
Low compression:	7, 11 to 13, 16, 20 to 22

CAUSES AND REMEDIES

1. Fault in the starter motor or its connection. Refer to "Electrical Faults".
2. Engine oil too thick. This can be caused by using the wrong oil, low temperatures or using oil not suitable for the prevailing climates. Depress the clutch whilst starting (models with manual transmission). Otherwise refill the engine with the correct oil grade, suitable for diesel engines.
3. Moveable parts of the engine not run-in. This fault may be noticed when the engine has been overhauled. It may be possible to free the engine by adding oil to the fuel for a while.
4. Mechanical fault. This may be due to seizure of the piston(s), broken crankshaft, connecting rods, clutch or other moveable parts of the engine. The engine must be stripped for inspection.
5. Faults in the glow plug system. Refer to "Glow Plug Faults".
6. Faults in the fuel system. Refer to "Fuel Faults".
7. Incorrect valve timing. This will only be noticed after the engine has been re-assembled after overhaul and the timing belt has been replaced incorrectly. Re-dismantle the engine and check the timing marks on the timing gear wheels.
8. Compression leak due to faulty closing of valves. See also under (7) or leakage past worn piston rings or pistons. cylinder head gasket blown.
9. Entry of air at inlet manifold, due to split manifold or damaged gasket.
10. Restriction in exhaust system, due to damaged exhaust pipes, dirt in end of exhaust pipe(s), kinked pipe(s), or collapsed silencer. Repair as necessary.

11. Worn valves or valve seats, no longer closing the valves properly. Top overhaul of engine is asked for.
12. Sticking valves due to excessive carbon deposits or weak valve springs. Top overhaul is asked for.
13. Cylinder head gasket blown. Replace gasket and check block and head surfaces for distortion.
14. Camshaft worn, not opening or closing one of the valves properly, preventing proper combustion. Check and if necessary fit new camshaft.
15. Incorrect valve (tappet) clearance. There could be a fault in the hydraulic tappets.
16. Cylinder bores, pistons or piston rings worn. Overhaul is the only cure. Fault may be corrected for a while by adding "Piston Seal Liquid" into the cylinders, but will re-develop.
17. Worn valve guides and/or valve stems. Top overhaul is asked for.
18. Damaged valve stem seals. Top overhaul is asked for.
19. Leaking crankshaft oil seal, worn piston rings or pistons, worn cylinders. Correct as necessary.
20. Loose glow plugs, gas escaping past thread or plug sealing washer damaged. Correct.
21. Cracked cylinder or cylinder block. Dismantle, investigate and replace block, if necessary.
22. Broken, weak or collapsed valve spring(s). Top overhaul is asked for.

GLOW PLUG FAULTS

Check a suspect glow plug as follows:

- Remove the glow plug lead from the rear glow plug and from the remaining plugs the bus bars.
- Connect a 12 volts test lamp to the plus terminal of the battery and with the other lead of the lamp touch in turn the connecting threads of each glow plug. The faulty plug is detected when the test lamp does not light up.

Further faults in the glow plug system can develop in the glow plug relay. Check as follows:

- Disconnect the electrical lead from the glow plug on the flywheel end and connect a test lamp between the lead and a good earthling point.
- Disconnect the electrical lead from the coolant temperature sender unit and move away from earth.
- Turn the ignition key to the "glowing" position and observe the test lamp. The lamp should light up for approx. 25 - 30 sec. and then switch off.
- Turn the ignition switch' 'off' and then again to the '4glowing" position. The test lamp must light up once more.
- Hold the disconnected lead from the temperature sender unit against earth. The test lamp must switch off.
- Turn the ignition switch "off' and then again to the "glowing" position. The test lamp should light up.
- Operate the starter motor and check that the test lamp remains "on".
- If the above tests cannot be carried out satisfactory, see your dealer.

LUBRICATION SYSTEM FAULTS

The only problem the lubrication system should give is excessive oil consumption or low oil pressure, or the oil warning light not going off.

Excessive oil consumption can be caused by worn cylinder bores, pistons and/or piston rings, worn valve guides, worn valves stem seals or a damaged crankshaft oil seal or leaking gasket on any of the engine parts. In most cases the engine must be dismantled to locate the fault.

Low oil pressure can be caused by a faulty oil pressure gauge, sender unit or wiring, a defective relief valve, low oil level, blocked oil pick-up pipe for the oil pump, worn oil pump or damaged main or big end bearings, In most cases it is logical to check the oil level first. All other causes require the dismantling and repair of the engine. If the oil warning light stays on, switch off the engine IMMEDIATELY, as delay could cause complete seizure within minutes.

COOLING SYSTEM FAULTS

Common faults are: Overheating, loss of coolant and slow warming-up of the engine:

Overheating:

1. ***Lack of coolant:*** Open the radiator cap with care to avoid injuries. Never pour cold water in to an overheated engine. Wait until engine cools down and pour in coolant whilst engine is running.
2. ***Radiator core obstructed by leaves, insects, etc.***: Blow with air line from the back of the radiator or with a water hose to clean.
3. ***Cooling fan not operating:*** Check fan for proper cut-in and cut-out temperature. If necessary change the temperature switch or see your Dealer.
4. ***Thermostat sticking:*** If sticking in the closed position, coolant can only circulate within the cylinder head or block. Remove thermostat and check as described in section "Cooling".
5. ***Water hose split:*** Identified by rising steam from the engine compartment or the front of the vehicle. Slight splits can be repaired with insulation tape. Drive without expansion tank cap to keep the pressure in the system down, to the nearest service station.
6. ***Water pump inoperative:*** Replace water pump.
7. ***Cylinder head gasket blown:*** Replace the cylinder head gasket.

Loss of Coolant:

1. ***Radiator leaks:*** Slight leaks may be stopped by using radiator sealing compound (follow the instructions of the manufacturer. In emergency a egg can be cracked open and poured into the radiator filler neck.
2. ***Hose leaks:*** See under 5, "Overheating".
3. ***Water pump leaks:*** Check the gasket for proper sealing or replace the pump.

Long Warming-up periods:

1. Thermostat sticking in the open position: Remove thermostat, check and if necessary replace.

DIESEL FUEL SYSTEM FAULTS

Engine is difficult to start or does not start	1 to 13
Engine starts, but stop soon afterwards:	14 to 20
Engine misfires continuously:	1 to 13
Bad idling:	14 to 20
Black, white or blue exhaust smoke:	21 to 29
Lack of power:	30 to 39
Excessive fuel consumption:	40 to 47

CAUSES AND REMEDIES

1. Fuel tank empty. Refuel.
2. Pre-glowing time too short. Operate until warning light goes "off".
3. Cold starting device not operated. Pull cable and push in after approx. 1 mm.
4. Glow plug system inoperative. Refer to "Glow Plug Faults".
5. Electro-magnetic cut-off device, loose or no current. Check cable to cut-off at top of injection pump. Ask a second person to operate ignition key and check if a "click" is heard. Either interrupted current supply or defective cut-off device.
6. Air in fuel system. Operate starter motor until fuel is delivered.
7. Fuel supply faulty. Slacken the injection pipes at injectors, and check if fuel is running out. Other faults: kinked, blocked or leaking injection pipes, blocked fuel filter, tank breathing system blocked. Wrong fuel for cold temperatures.
8. Injection pipes refitted in wrong order after repair.
9. Injection timing of pump out of phase: Have the adjustment checked and corrected.
10. One or more injectors faulty, dirty or incorrect injection pressure. Have injectors repaired or replace them.
11. Injection pump not operating properly. Fit an exchange pump or have it repaired.
12. Valves not opening properly.
13. Compression pressures too low. See item "8" under "Engine Faults".
14. Idle speed not properly adjusted. Adjust.
15. Throttle cable not properly adjusted or sticking. Re- adjust or free-off.
16. Fuel hose between filter and pump not tightened properly. Tighten connections.
17. Rear mounting of injection pump loose or cracked. Tighten or replace.
18. See items 6, 7, 9, 11, 12 and 13
19. Engine mounting not tightened properly or worn. Tighten or replace.
20. Sticking accelerator pedal. Free-off pedal.
21. Engine not at operating temperature. Check exhaust smoke colour again when engine is warm.
22. Too much acceleration at low revs. Use individual gears in accordance with acceleration.
23. Air cleaner contaminated. Clean or replace.
24. Fuel filter contaminated. Replace.
25. Max. speed adjustment incorrect. Re-adjust.
26. Injectors are dripping. Have them checked or replace faulty ones.
27. Injector nozzles sticking or broken. Replace injector.
28. Injection pressure too low. Have injectors checked and adjusted.
29. See items 9, 11, 12 and 13
30. Throttle cable travel restricted. Re-adjust. Check that floor mats cannot obstruct pedal movement.
31. Throttle cable not correctly adjusted. Re-adjust.
32. Operating lever loose on pump. Re-tighten.
33. Max. speed not obtained. Re-adjust max. speed or have it adjusted.
34. Injector pipes restricted in diameter (near connections). Disconnect pipes and check that diameter is at least 2.0 mm (0.08 in.).
35. Heat protection sealing gaskets under injectors not sealing or damaged. Remove injectors and check. Replace if necessary. Fit the washers correctly.
36. Injection pressure of injectors wrong. Have them re-adjusted.
37. See items 6, 7, 9, 11 and 13

38. See item 20.
39. See items 23, 24, 26 and 27.
40. Road wheels dragging. Brakes seized or wheel bearings not running freely.
41. Engine not running "free". Refers to new or overhauled engine.
42. Fuel system leaking. Check hoses, pipes, filter, injection pump, etc. for leaks.
43. Fuel return line blocked. Clean with compressed air if possible.
44. Idle speed too high. Re-adjust.
45. Max. speed too high. Re-adjust.
46. See items 10, 11, 12 and 13.
47. See items 24, 26, 27 and 28.

CLUTCH FAULTS

Clutch slipping:	1, 2, 3, 4, 5
Clutch will not disengage fully:	4, 6 to 12, 14
Whining from clutch when pedal is depressed:	13
Clutch judder:	1, 2, 7, 10 to 13
Clutch noise when idling:	2, 3
Clutch noise during engagement:	2

CAUSES AND REMEDIES

1. Insufficient clutch free play at pedal.
2. Clutch disc linings worn, hardened, oiled-up, loose or broken. Disc distorted or hub loose. Clutch disc must be replaced.
3. Pressure plate faulty. Replace clutch.
4. Air in hydraulic system. Low fluid level in clutch cylinder reservoir.
5. Insufficient play at clutch pedal and clutch release linkage. Rectify as described.
6. Excessive free play in release linkage (only for cable operated clutch, not applicable). Adjust or replace worn parts.
7. Misalignment of clutch housing. Very rare fault, but possible on transmissions with separate clutch housings. Re-align to correct.
8. Clutch disc hub binding on splines of main drive shaft (clutch shaft) due to dirt or burrs on splines. Remove clutch and clean and check splines.
9. Clutch disc linings loose or broken. Replace disc.
10. Pressure plate distorted. Replace clutch.
11. Clutch cover distorted. Replace clutch.
12. Fault in transmission or loose engine mountings.
13. Release bearing defective. Remove clutch and replace bearing.
14. A bent clutch release lever. Check lever and replace or straighten, if possible.

- The above faults and remedies are for hydraulic and mechanical clutch operation and should be read as applicable to the model in question, as the clutch fault finding section is written for all types of clutch operation.

STEERING FAULTS

Steering very heavy:	1 to 6
Steering very loose:	5, 7 to 9, 11 to 13
Steering wheel wobbles:	4, 5, 7 to 9, 11 to 16

Vehicle pulls to one side:	1, 4, 8, 10, 14 to 18
Steering wheel does not return to centre position:	1 to 6, 18
Abnormal tyre wear:	1, 4, 7 to 9, 14 to 19
Knocking noise in column:	6, 7, 11, 12

CAUSES AND REMEDIES

1. Tyre pressures not correct or uneven. Correct.
2. Lack of lubricant in steering.
3. Stiff steering linkage ball joints. Replace ball joints in question.
4. Incorrect steering wheel alignment. Correct as necessary.
5 Steering needs adjustment. See your dealer for advice.
6. Steering column bearings too tight or seized or steering column bent. Correct as necessary.
7. Steering linkage joints loose or worn. Check and replace joints as necessary.
8. Front wheel bearings worn, damaged or loose. Replace bearing.
9. Front suspension parts loose. Check and correct.
10. Wheel nuts loose. Re-tighten.
11. Steering wheel loose. Re-tighten nut.
12. Steering gear mounting loose. Check and tighten.
13. Steering gear worn. Replace the steering gear.
14. Steering track rods defective or loose.
15. Wheels not properly balanced or tyre pressures uneven. Correct pressures or balance wheels.
16. Suspension springs weak or broken. Replace spring in question or both.
17. Brakes are pulling to one side. See under "Brake Faults".
18. Suspension out of alignment. Have the complete suspension checked by a dealer.
19. Improper driving. We don't intend to tell you how to drive and are quite sure that this is not the cause of the fault.

BRAKE FAULTS

Brake Failure: Brake shoe linings or pads excessively worn, incorrect brake fluid (after overhaul), insufficient brake fluid, fluid leak, master cylinder defective, wheel cylinder or caliper failure. Remedies are obvious in each instance.

Brakes Ineffective: Shoe linings or pads worn, incorrect lining material or brake fluid, linings contaminated, fluid level low, air in brake system (bleed brakes), leak in pipes or cylinders, master cylinder defective. Remedies are obvious in each instance.

Brakes pull to one side: Shoes or linings worn, incorrect linings or pads, contaminated linings, drums or discs scored, fluid pipe blocked, unequal tyre pressures, brake back plate or caliper mounting loose, wheel bearings not properly adjusted, wheel cylinder seized. Rectify as necessary.

Brake pedal spongy: Air in hydraulic system. System must be bled of air.

Pedal travel too far: Linings or pads worn, drums or discs scored, master cylinder or wheel cylinders defective, system needs bleeding. Rectify as necessary.

Loss of brake pressure: Fluid leak, air in system, leak in master or wheel cylinders, brake servo not operating (vacuum hose disconnected or exhauster pump not operating). Place vehicle on dry ground and depress brake pedal. Check where fluid runs out and rectify as necessary.

Brakes binding: Incorrect brake fluid (boiling), weak shoe return springs, basic brake adjustment incorrect (after fitting new rear shoes), piston in caliper of wheel cylinder seized, push rod play on master cylinder insufficient (compensation port obstructed), handbrake adjusted too tightly. Rectify as necessary. Swelling of cylinder cups through use of incorrect brake fluid could be another reason.

Handbrake ineffective: Brake shoe linings worn, linings contaminated, operating lever on brake shoe seized, brake shoes or handbrake need adjustment. Rectify as necessary.

Excessive pedal pressure required: Brake shoe linings or pads worn, linings or pads contaminated, brake servo vacuum hose (for brake servo) disconnected or wheel cylinders seized. Exhauster pump not operating. Rectify as necessary.

Brakes squealing: Brake shoe linings or pads worn so far that metal is grinding against drum or disc. Inside of drum is full of lining dust (from handbrake brake shoes). Remove and replace, or clean out the drum(s). Do not inhale brake dust.

NOTE: Any operation on the steering and brake systems must be carried out with the necessary care and attention. Always think of your safety and the safety of other road users. Make sure to use the correct fluid for the power-assisted steering and the correct brake fluid.

Faults in the ABS system should be investigated by a dealer.

ELECTRICAL FAULTS

Starter motor failure:	2 to 5, 8, 9
No starter motor drive:	1 to 3, 5 to 7
Slow cranking speed:	1 to 3
Charge warning light remains on:	3, 10, 12
Charge warning light does not come on:	2, 3, 9. 11, 13
Headlamp failure:	2, 3, 11, 13, 14
Battery needs frequent topping-up:	11
Direction indicators not working properly:	2, 3, 9, 13, 14
Battery frequently discharged:	3, 10, 11, 12

CAUSES AND REMEDIES

1. Tight engine. Check and rectify.
2. Battery discharged or defective. Re-charge battery or replace if older than approx. 2 years.
3. Interrupted connection in circuit. Trace and rectify.
4. Starter motor pinion jammed in flywheel. Release.
5. Also 6, 7 and 8. Starter motor defective, no engagement in flywheel, pinion or flywheel worn or solenoid switch defective. Correct as necessary.
9. Ignition/starter switch inoperative. Replace.
10. Drive belt loose or broken. Adjust or replace.
11. Regulator defective. Adjust or replace.
12. Generator inoperative. Overhaul or replace.
13. Bulb burnt out. Replace bulb.
14. Flasher unit defective. Replace unit.

WIRING DIAGRAMS

Wiring Diagram Index

Cable Colour Code

bl	=	blue	nf	=	natural colour
br	=	brown	rs	=	pink
ge	=	yellow	rt	=	red
gn	=	green	sw	=	black
gr	=	grey	vi	=	violet
			ws	=	white

Cable identification: a = size, square mm, b = basic colour, c = second colour

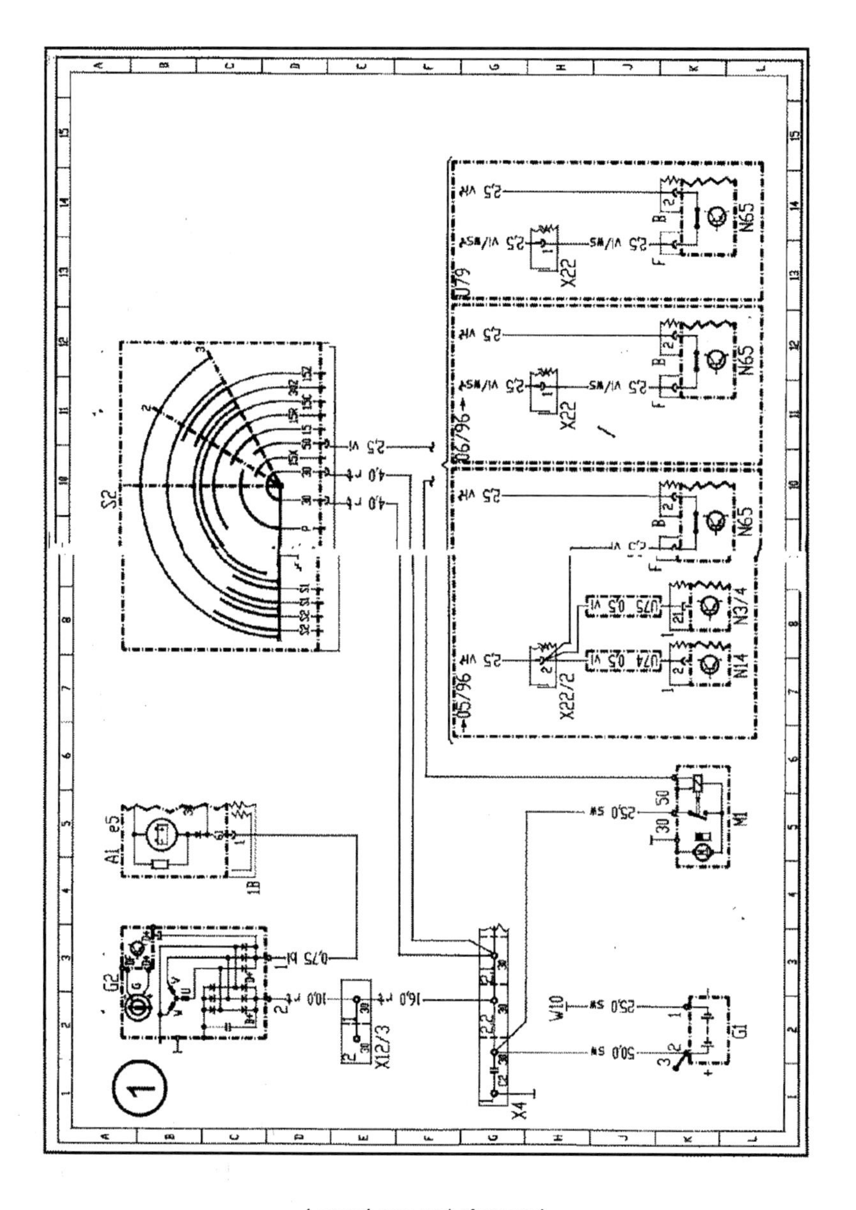

Legend, see end of manual

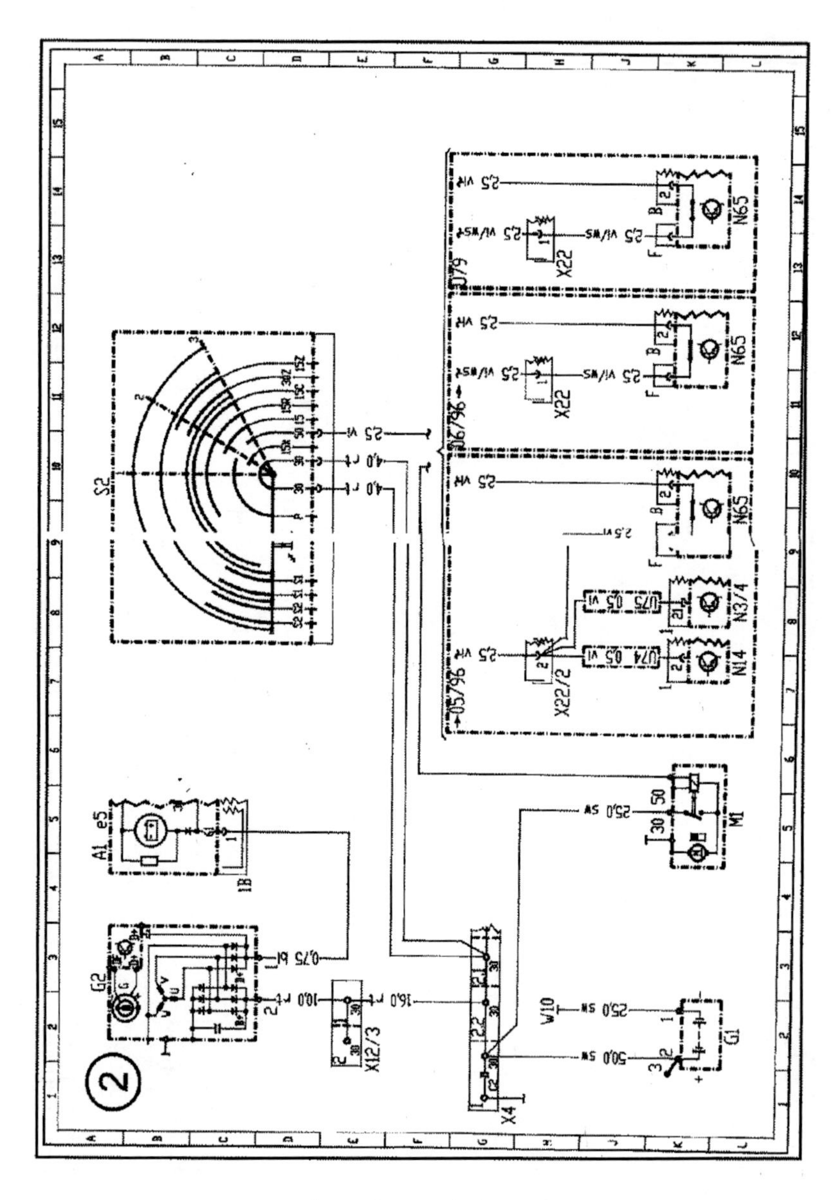

Legend, see end of manual

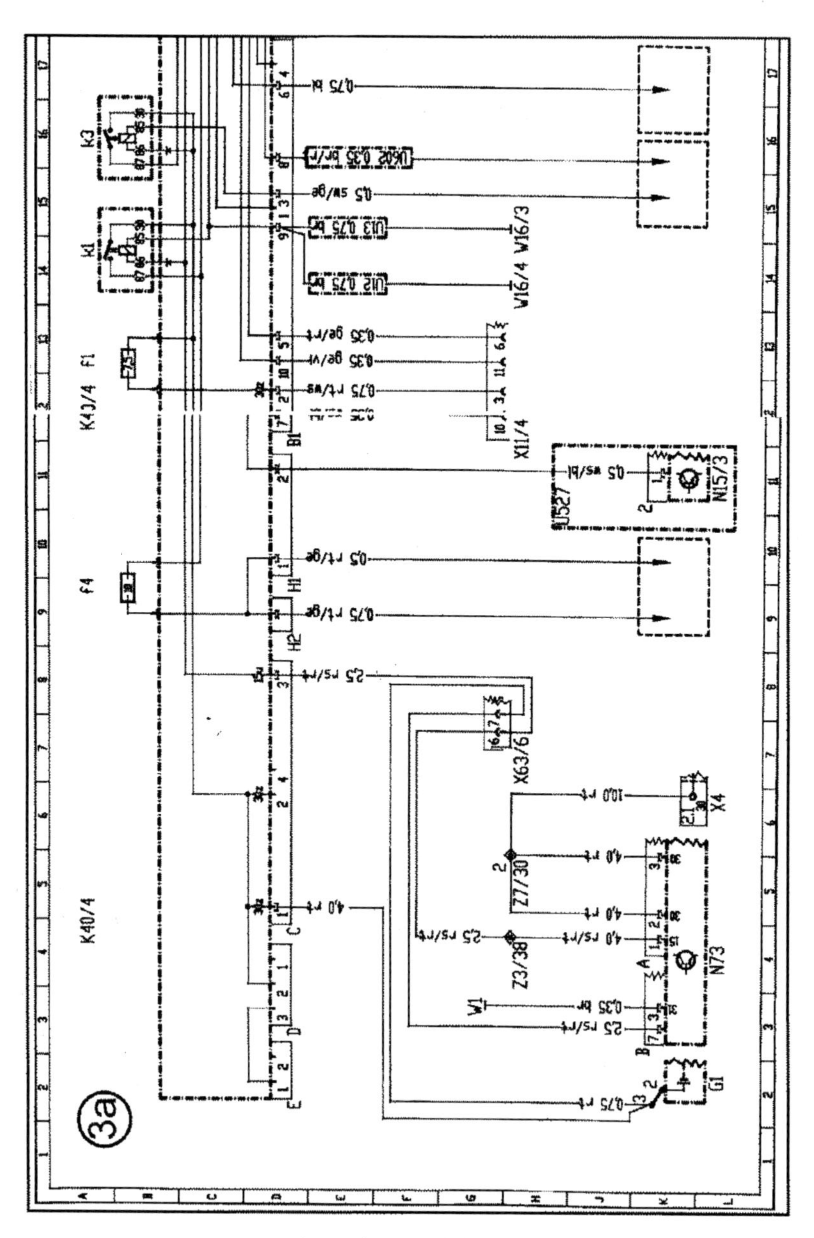

Legend, see end of manual

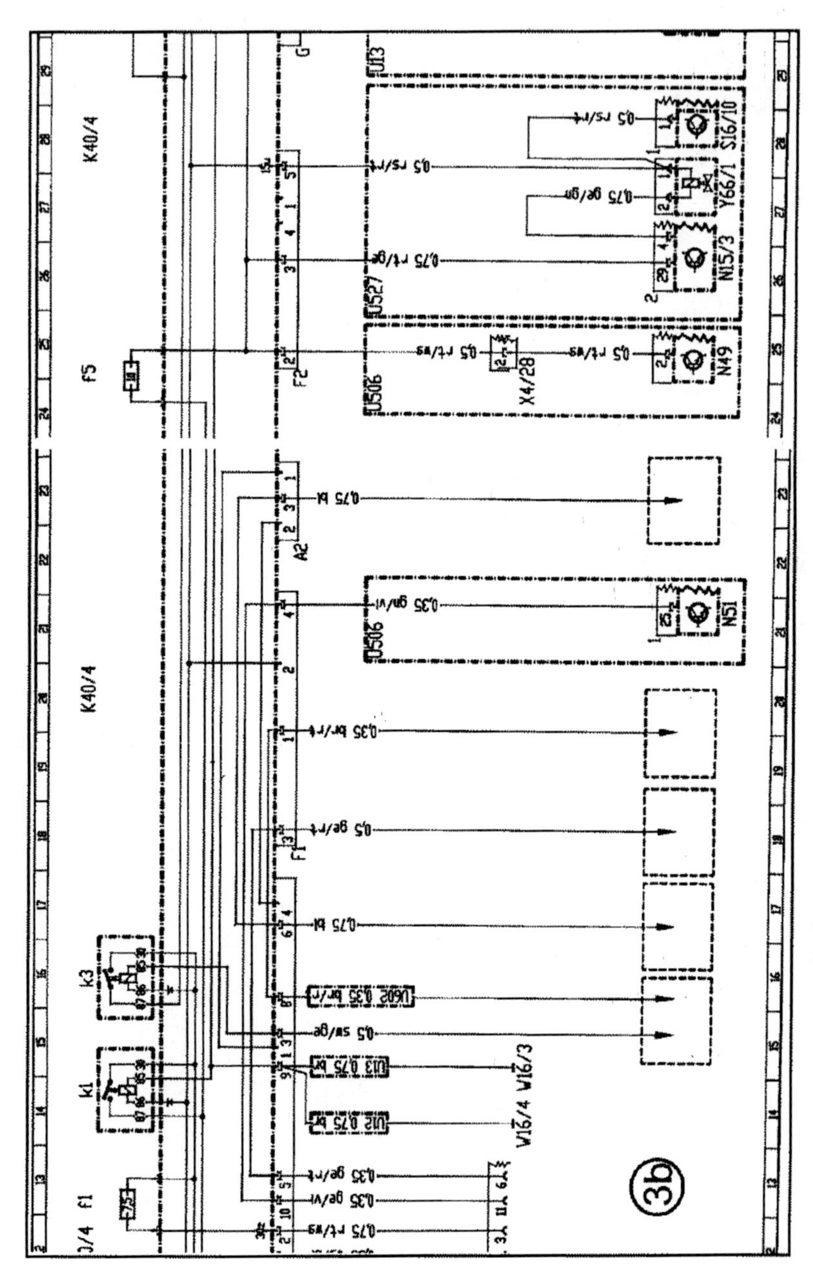

Legend, see end of manual

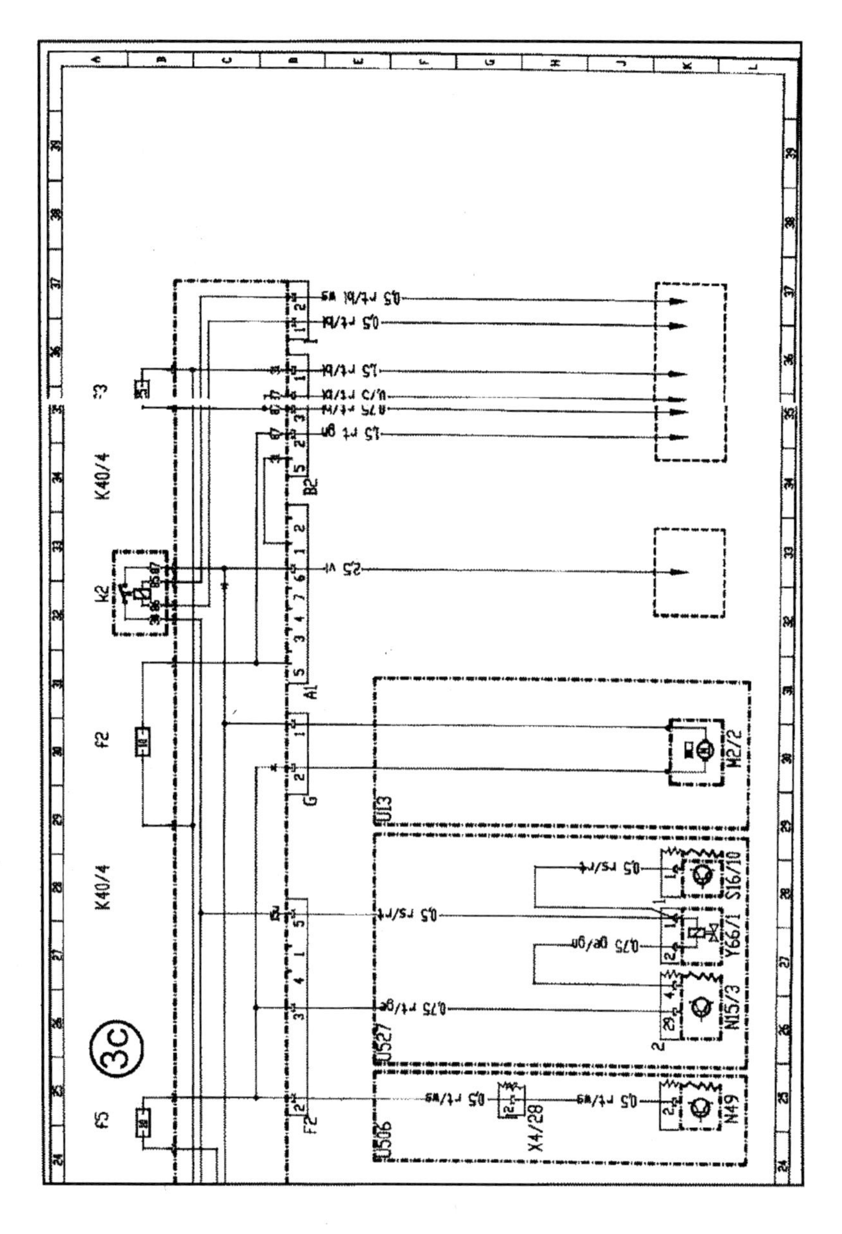

Legend, see end of manual

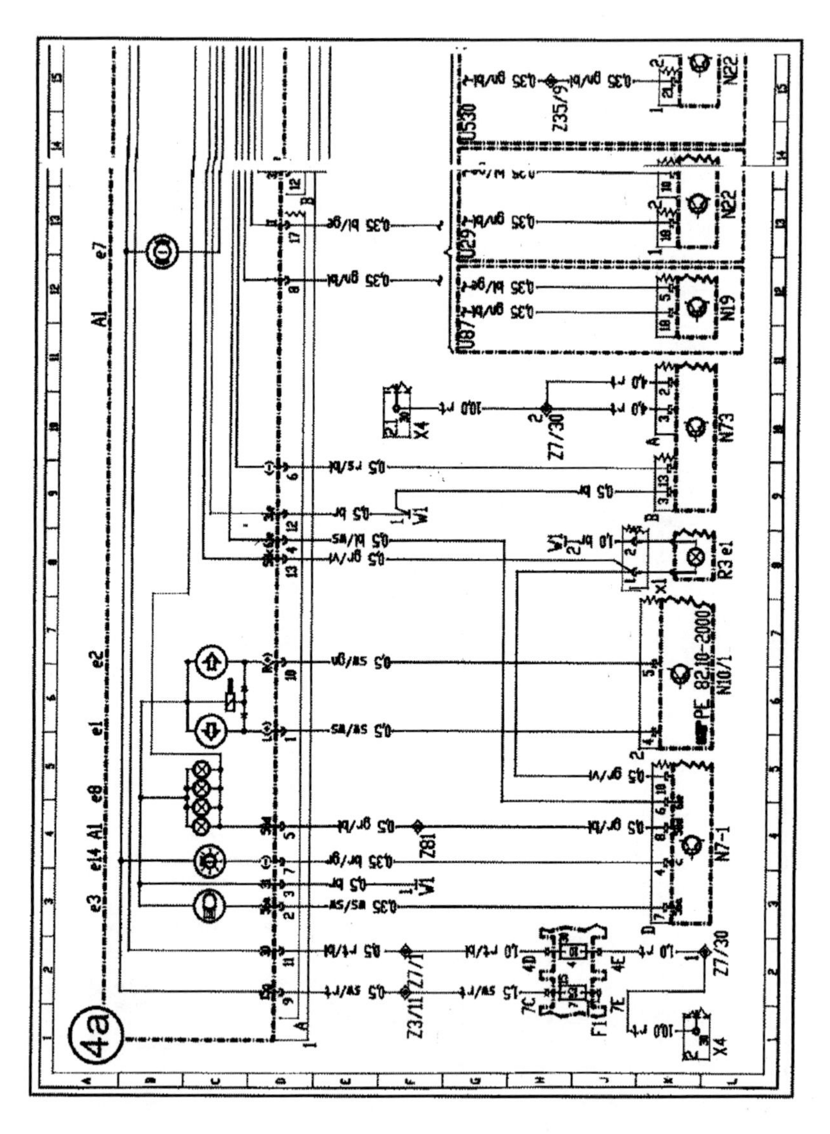

Legend, see end of manual

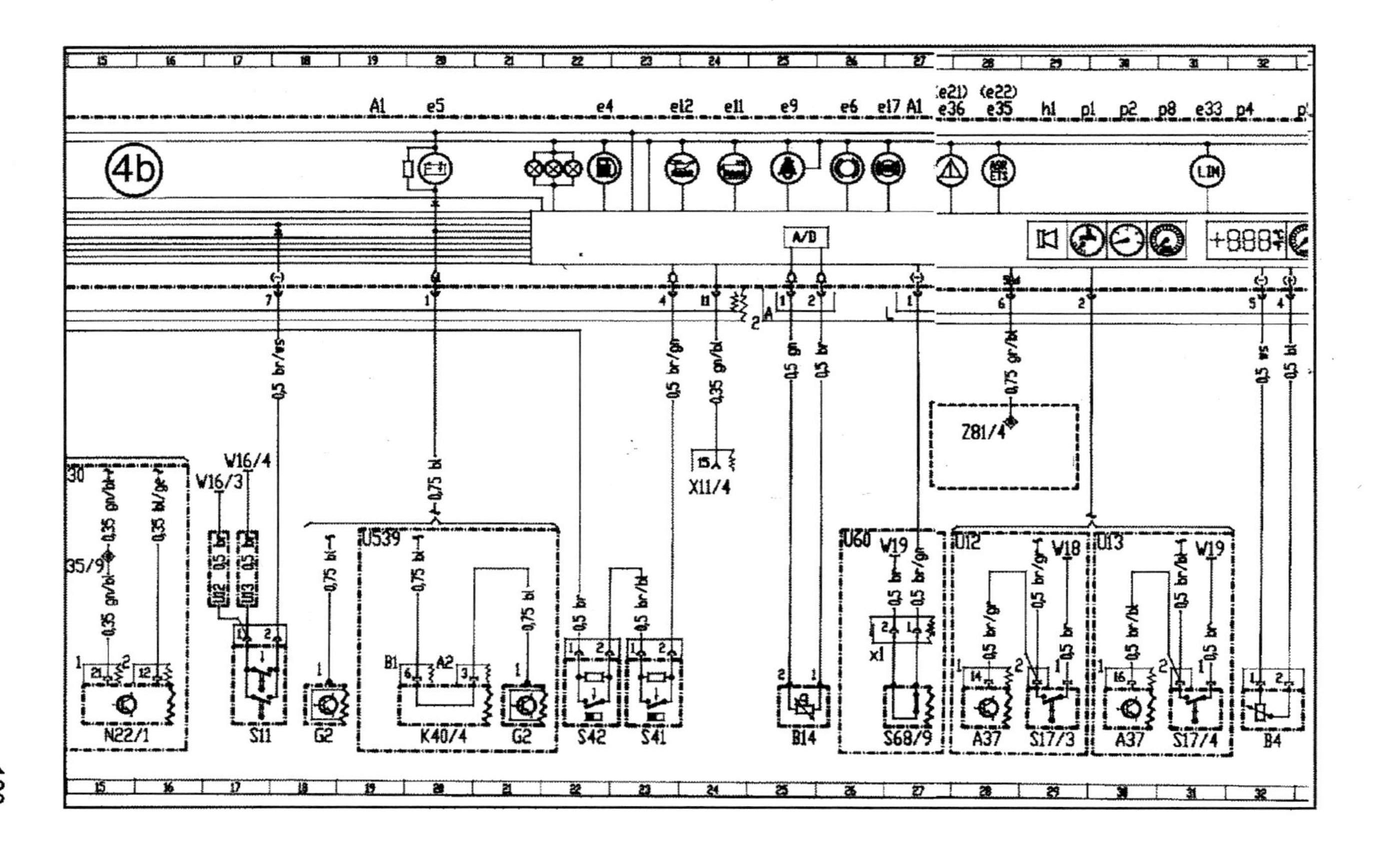

4b
A1
e5
e4
e12
e11
e9
e6
e17
(e21)
(e22)
e36
e35
h1
p1
p2
p8
e33
p4
A/D
N22/1
35/9
V16/3
V16/4
S11
G2
U539
K40/4
S42
S41
X11/4
B14
U60
W19
S68/9
Z81/4
U12
V18
A37
S17/3
U13
S17/4
B4

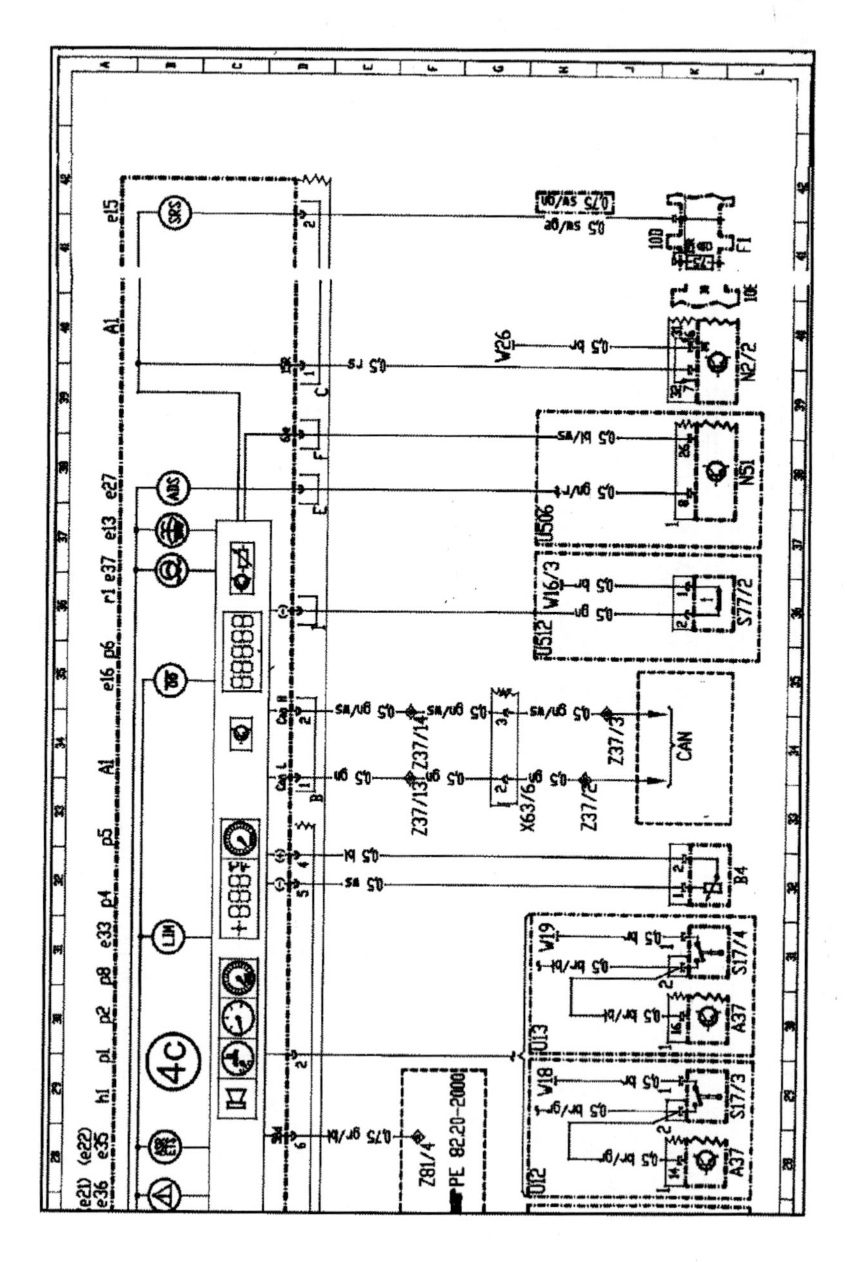

Legend, see end of manual

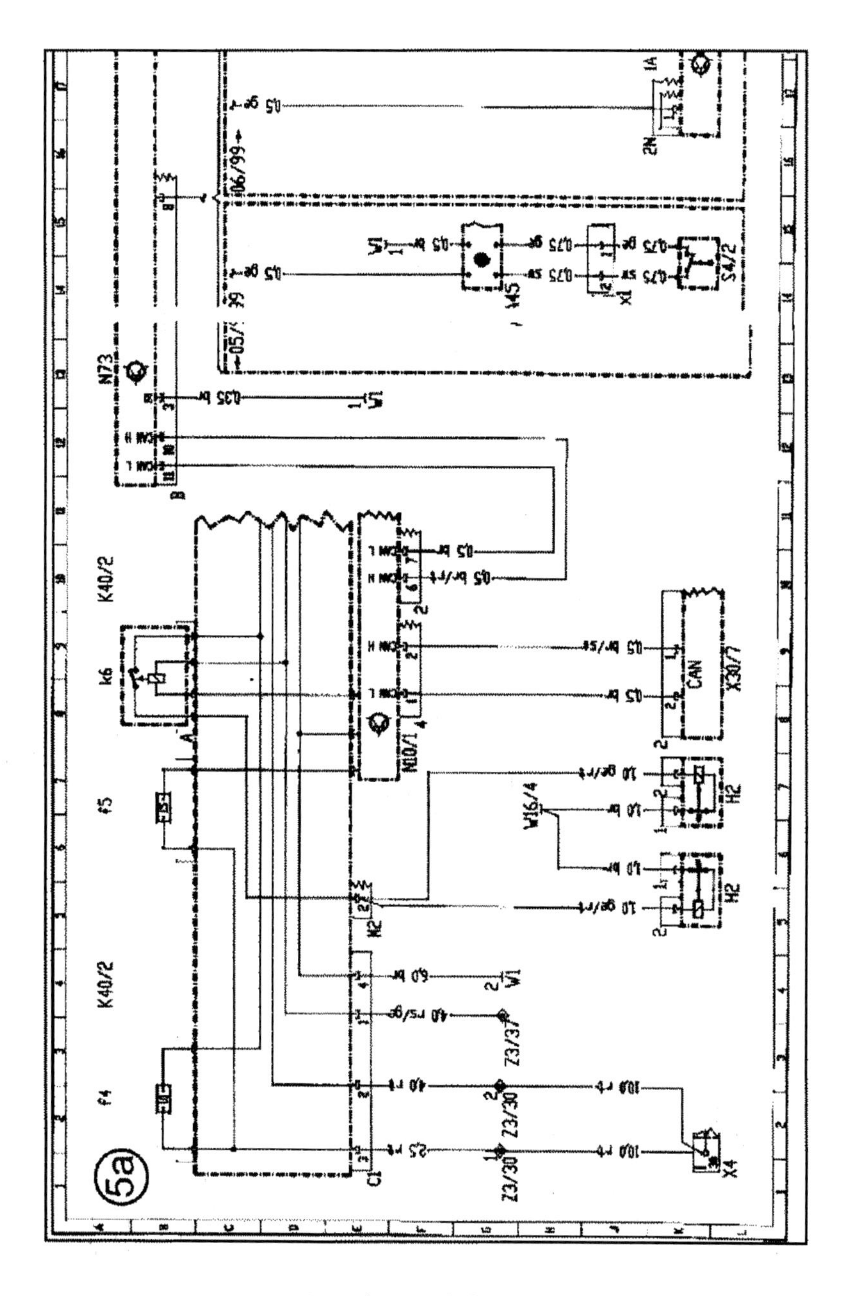

Legend, see end of manual

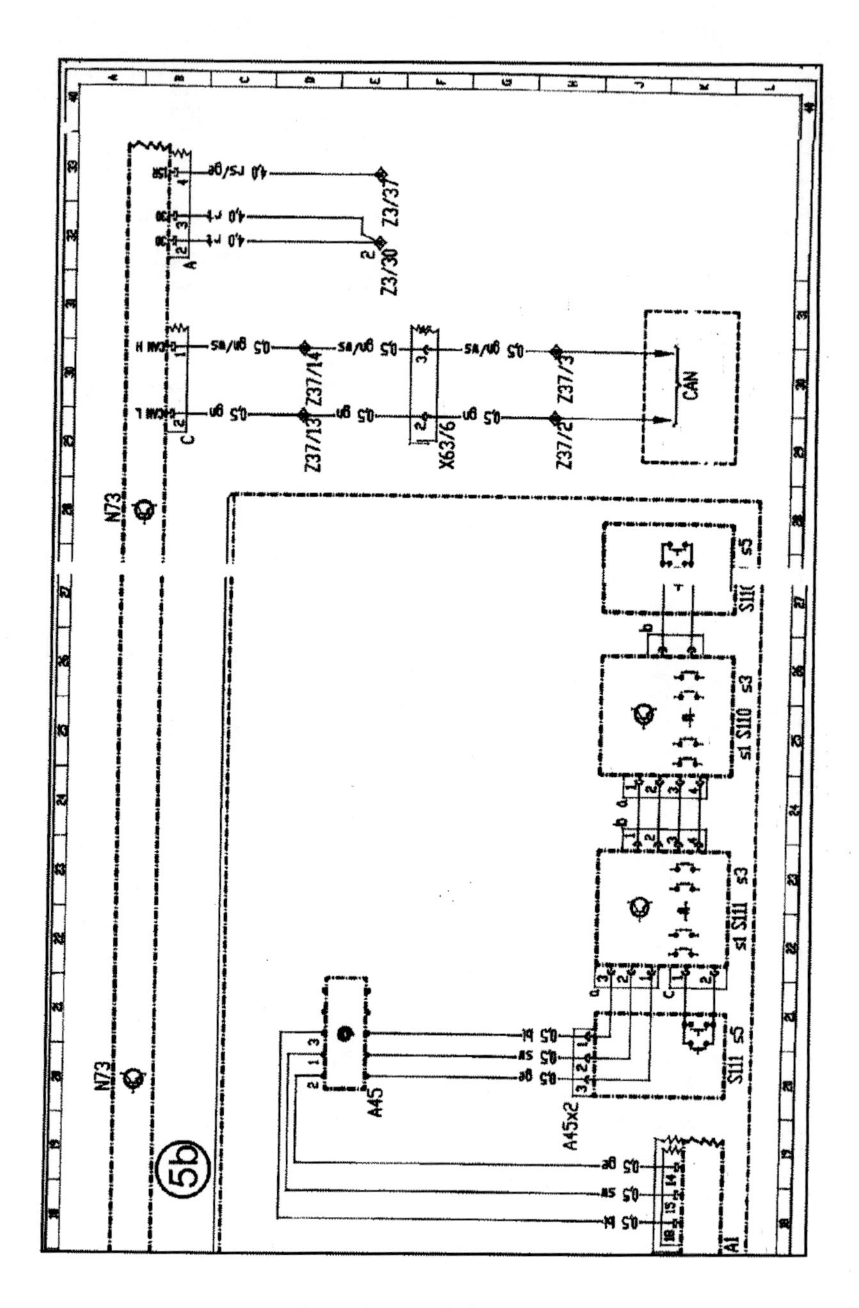

Legend, see end of manual

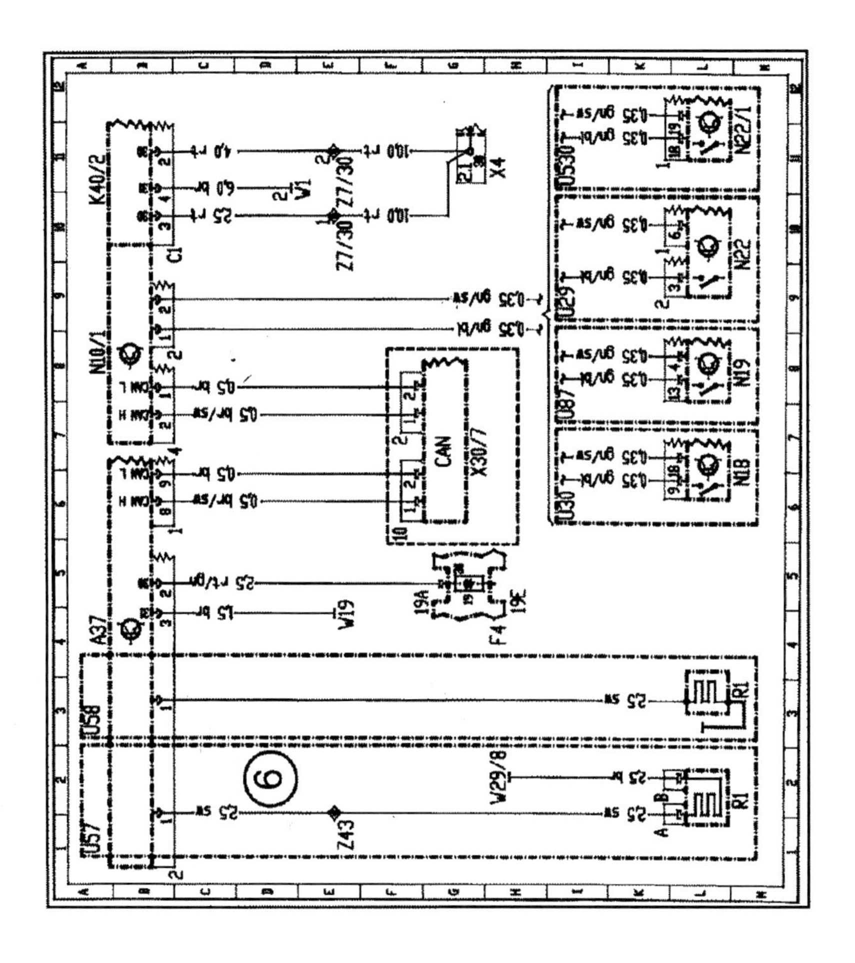

Legend, see end of manual

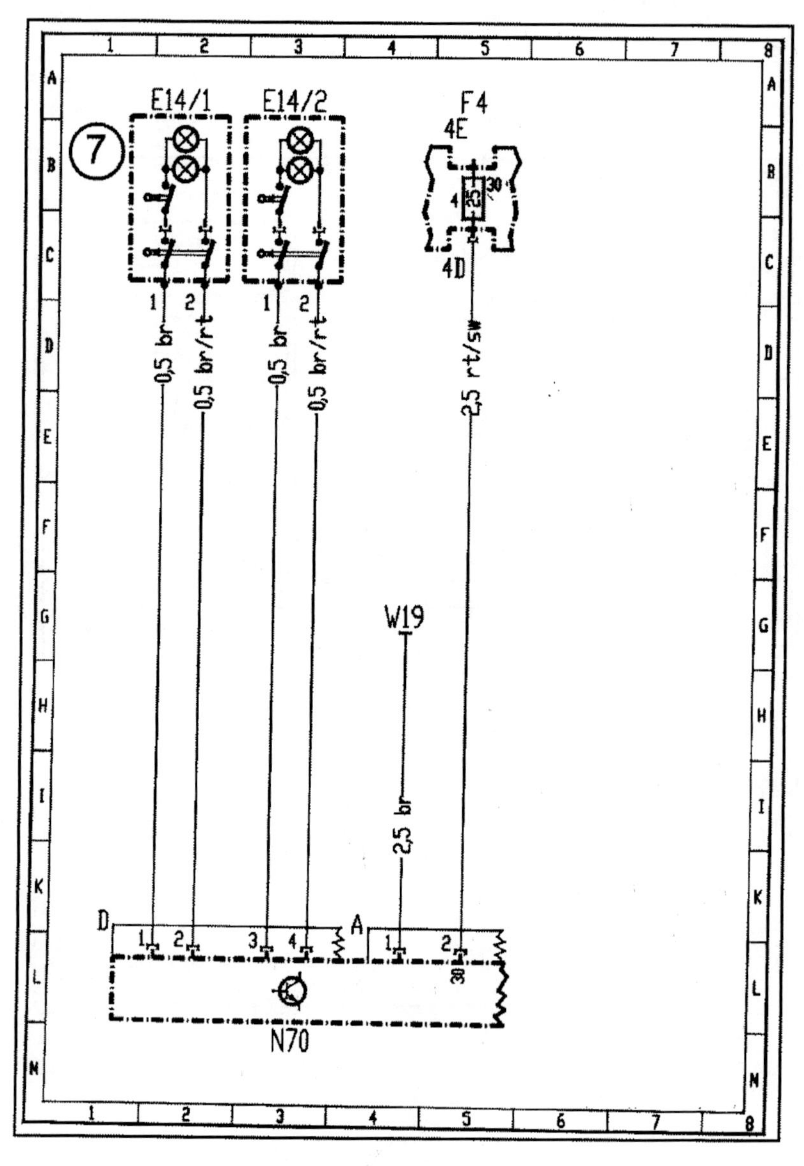

Legend, see end of manual

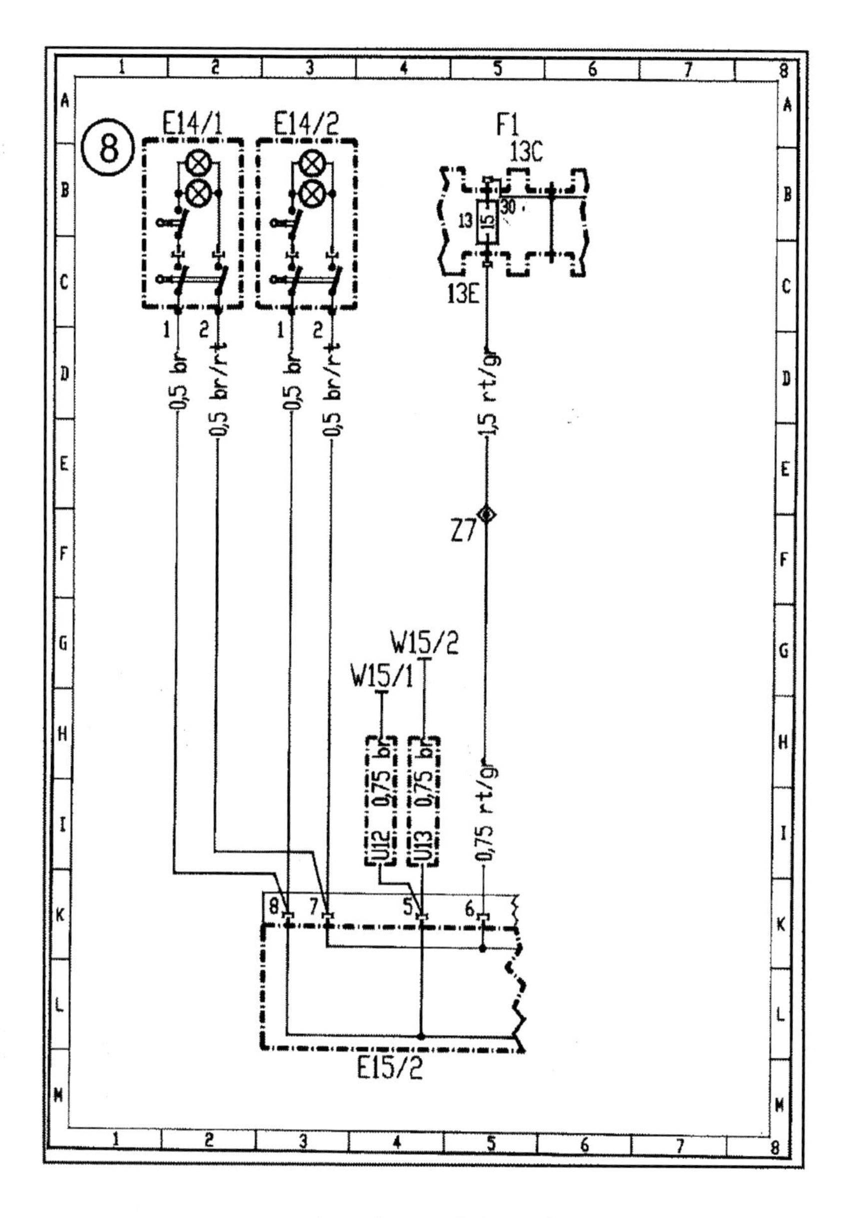

Legend, see end of manual

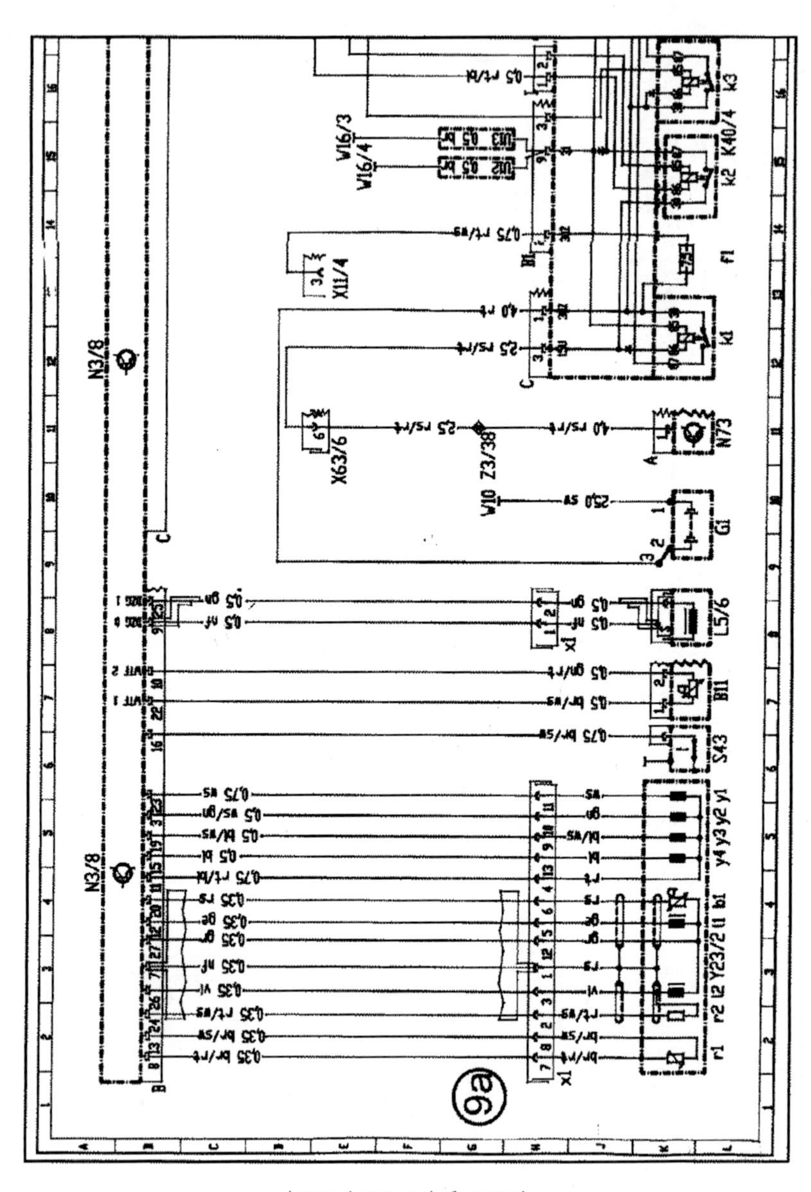

Legend, see end of manual

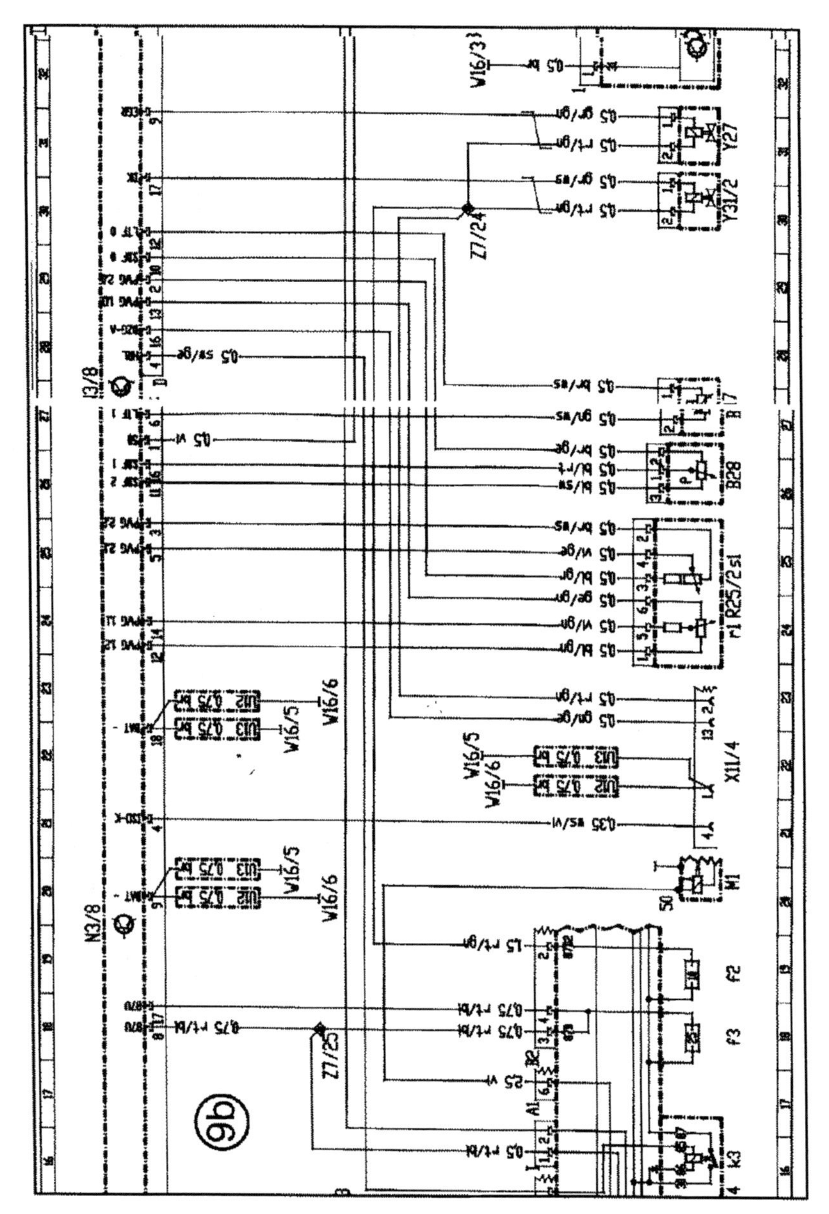

Legend, see end of manual

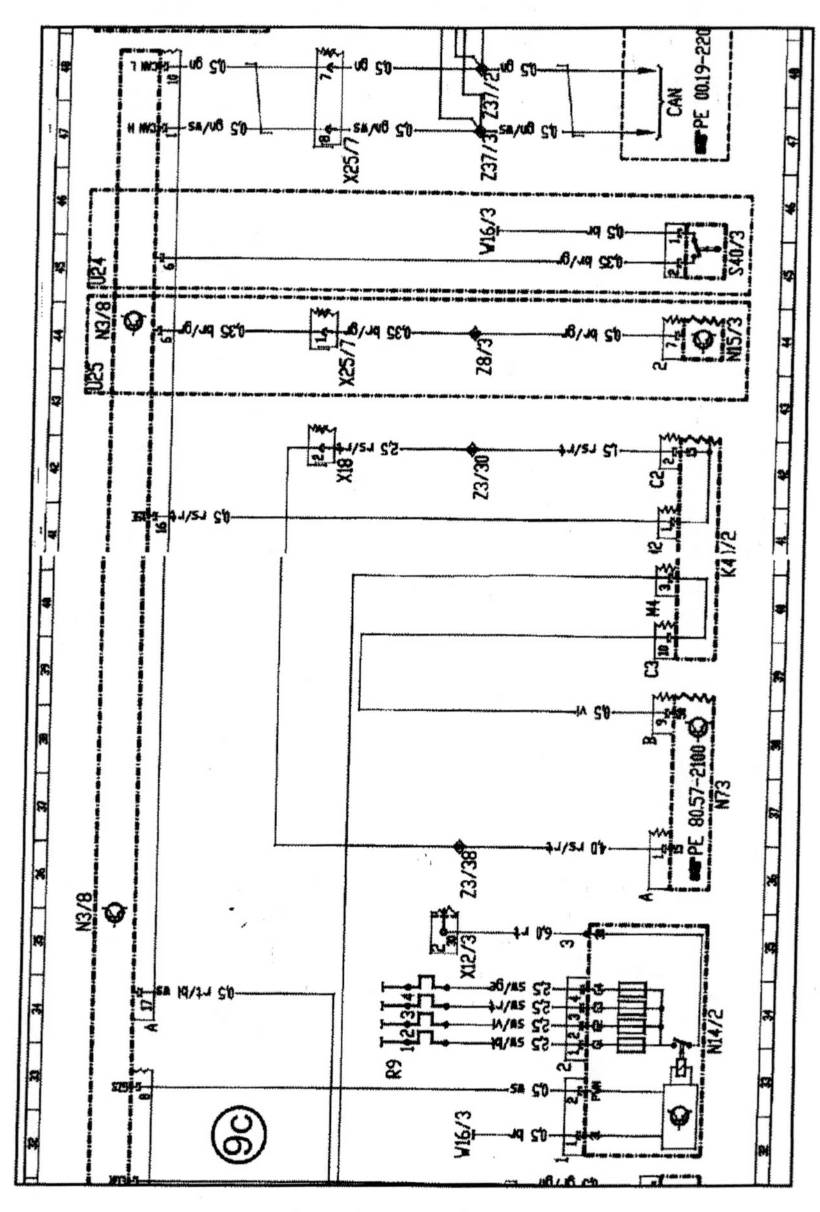

Legend, see end of manual

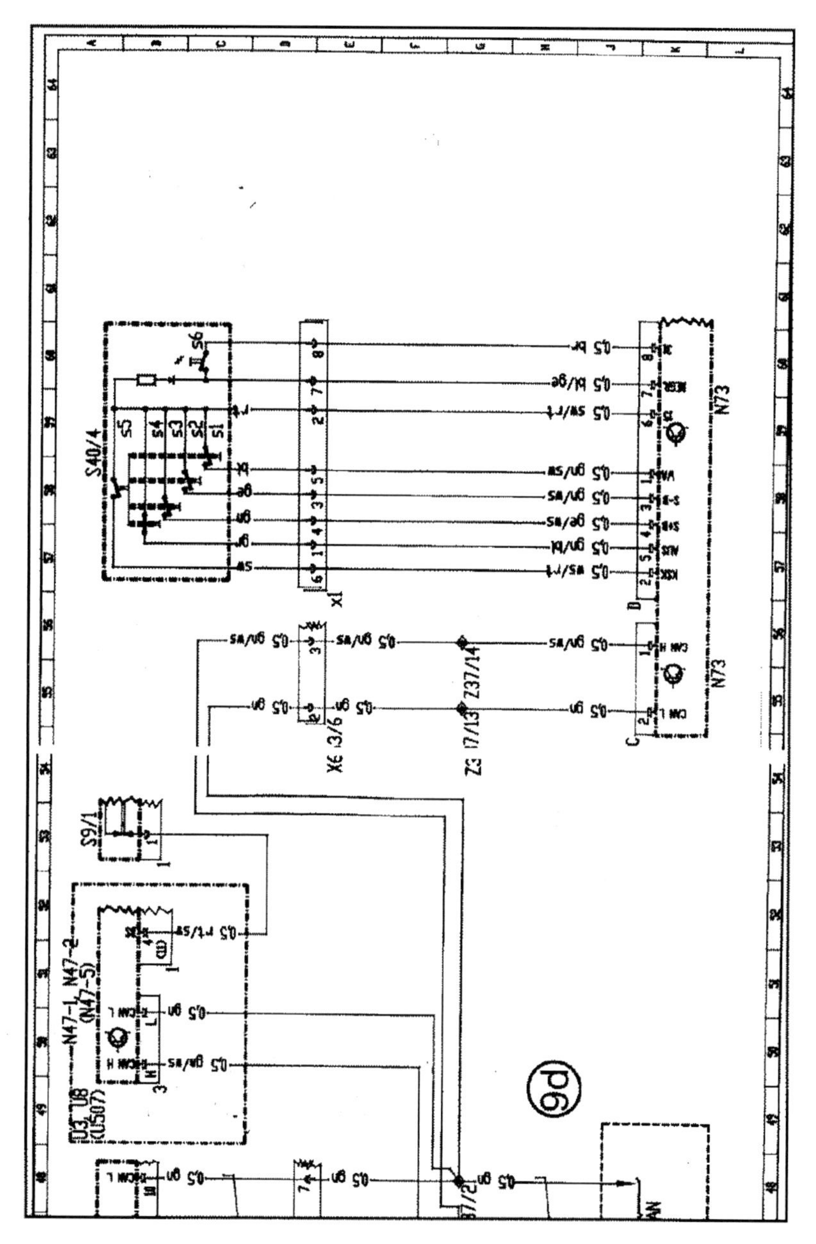

Legend, see end of manual

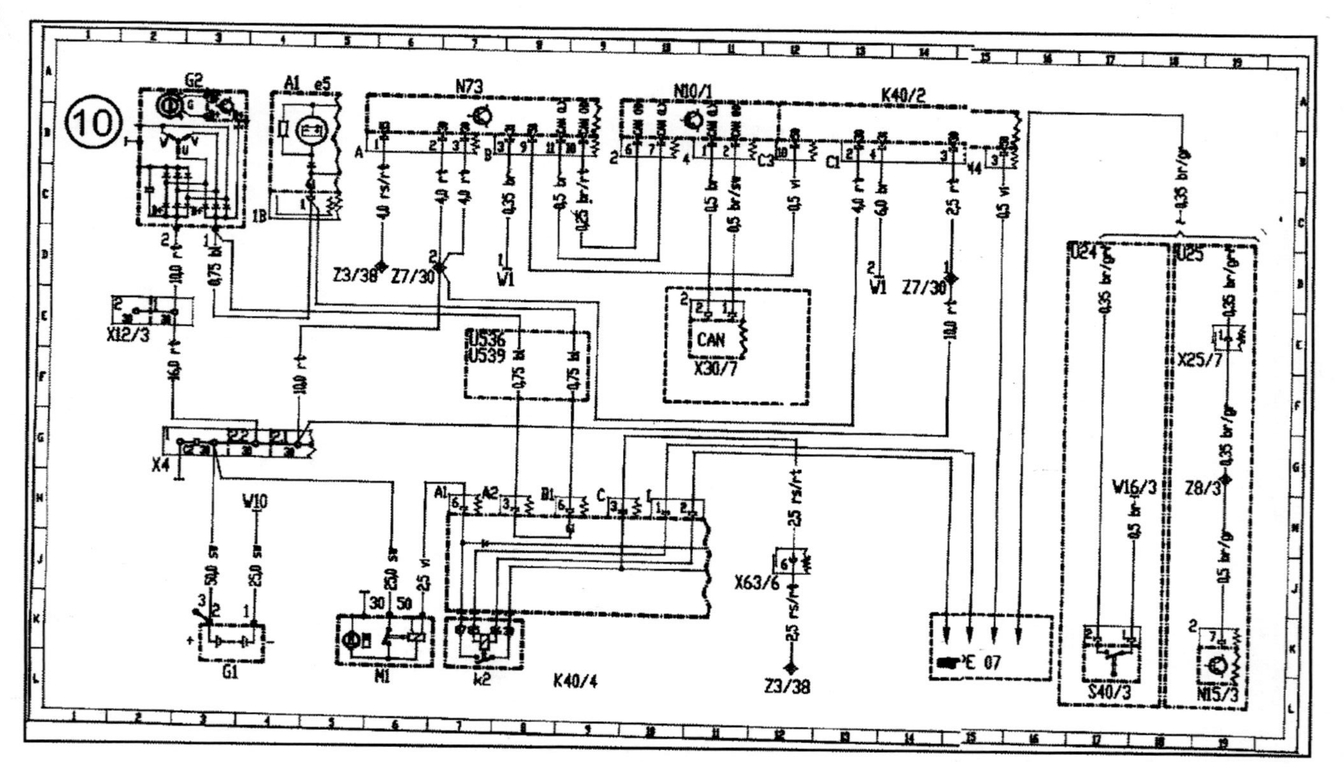
10
G2
A1 e5
1B
N73
N10/1
K40/2
C1
C3
X12/3
Z3/38
Z7/30
W1
U536
U539
CAN
X30/7
X4
W10
G1
M1
k2
K40/4
A1
A2
B1
X63/6
E 07
U24
U25
X25/7
W16/3
Z8/3
S40/3
N15/3

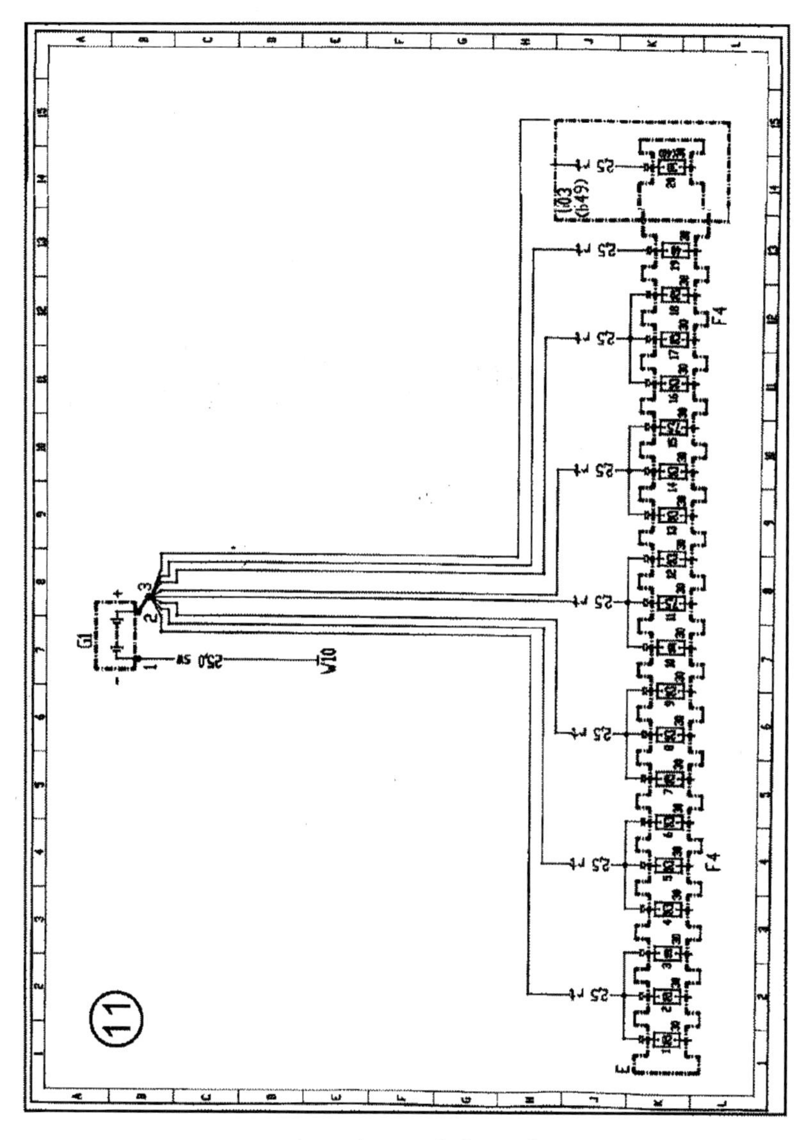

Legend, see end of manual

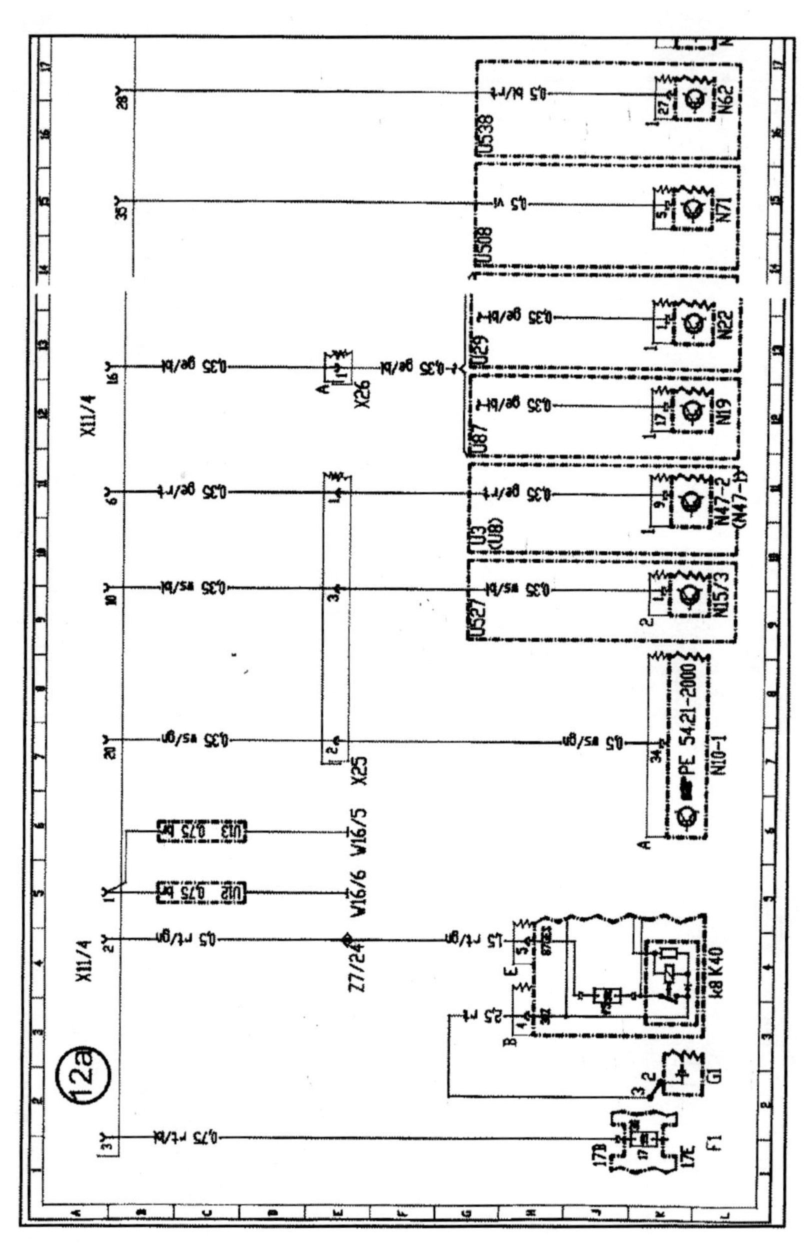

Legend, see end of manual

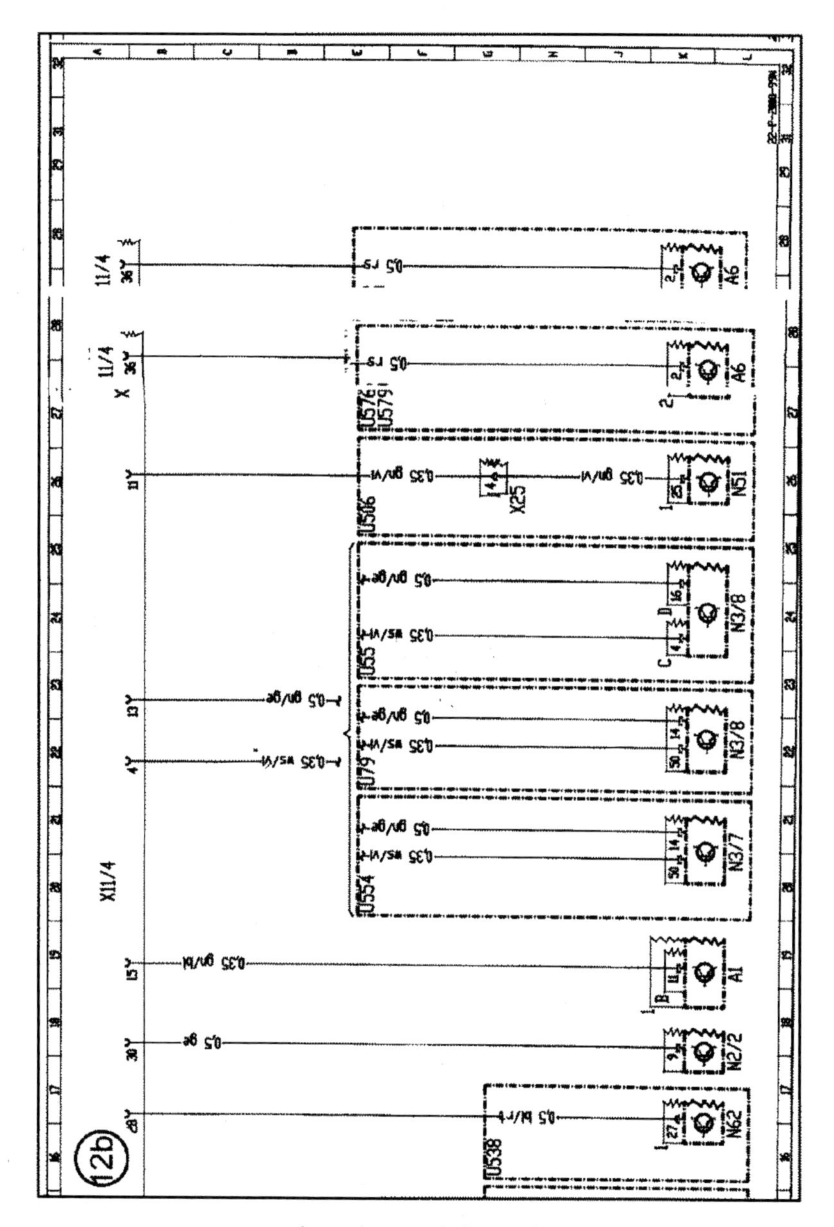

Legend, see end of manual

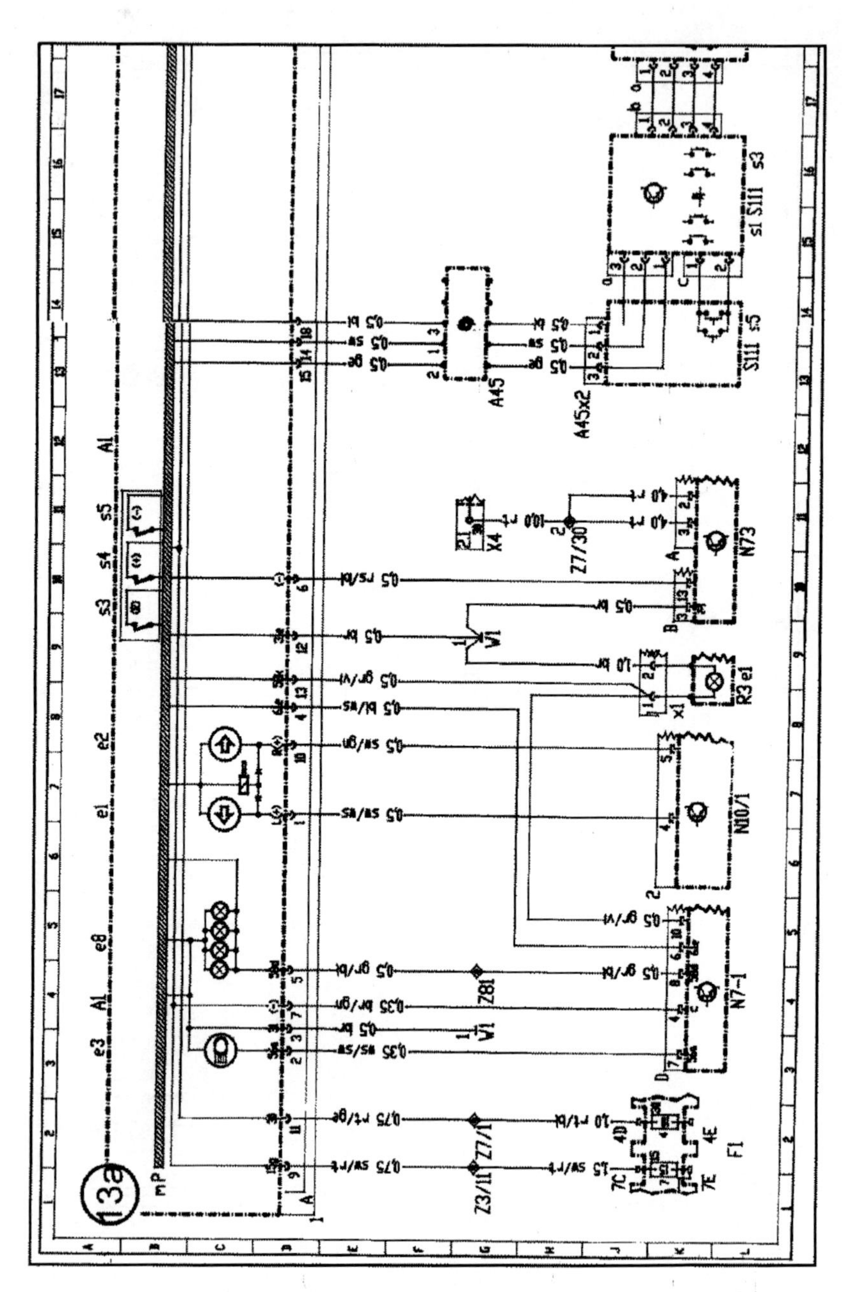

Legend, see end of manual

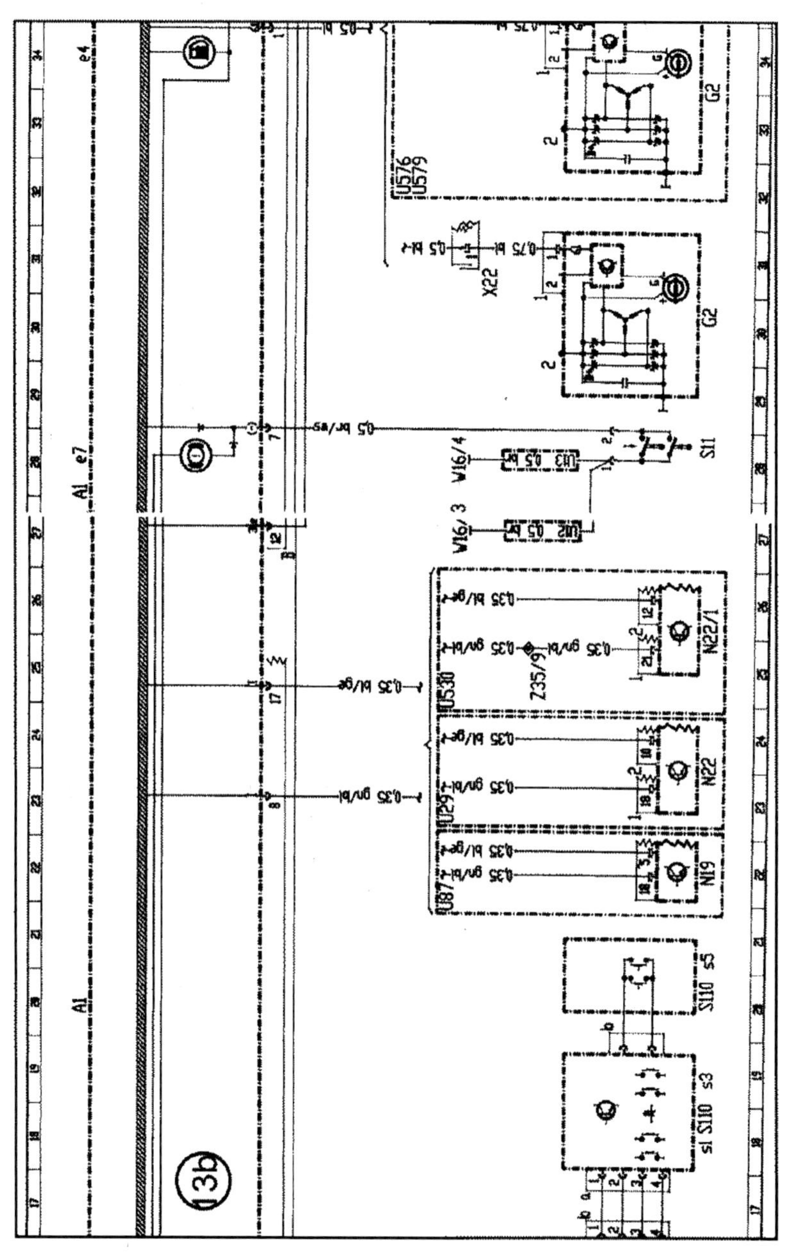

Legend, see end of manual

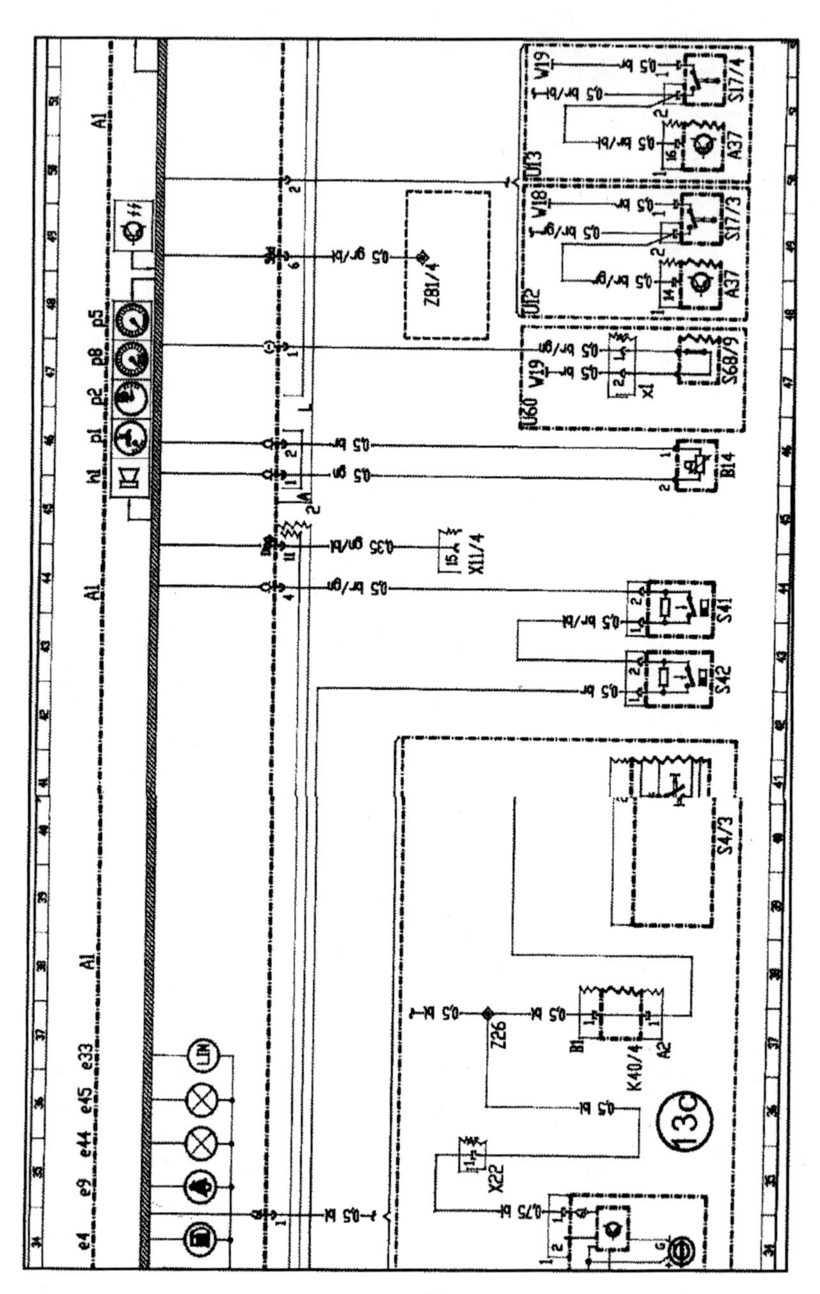

Legend, see end of manual

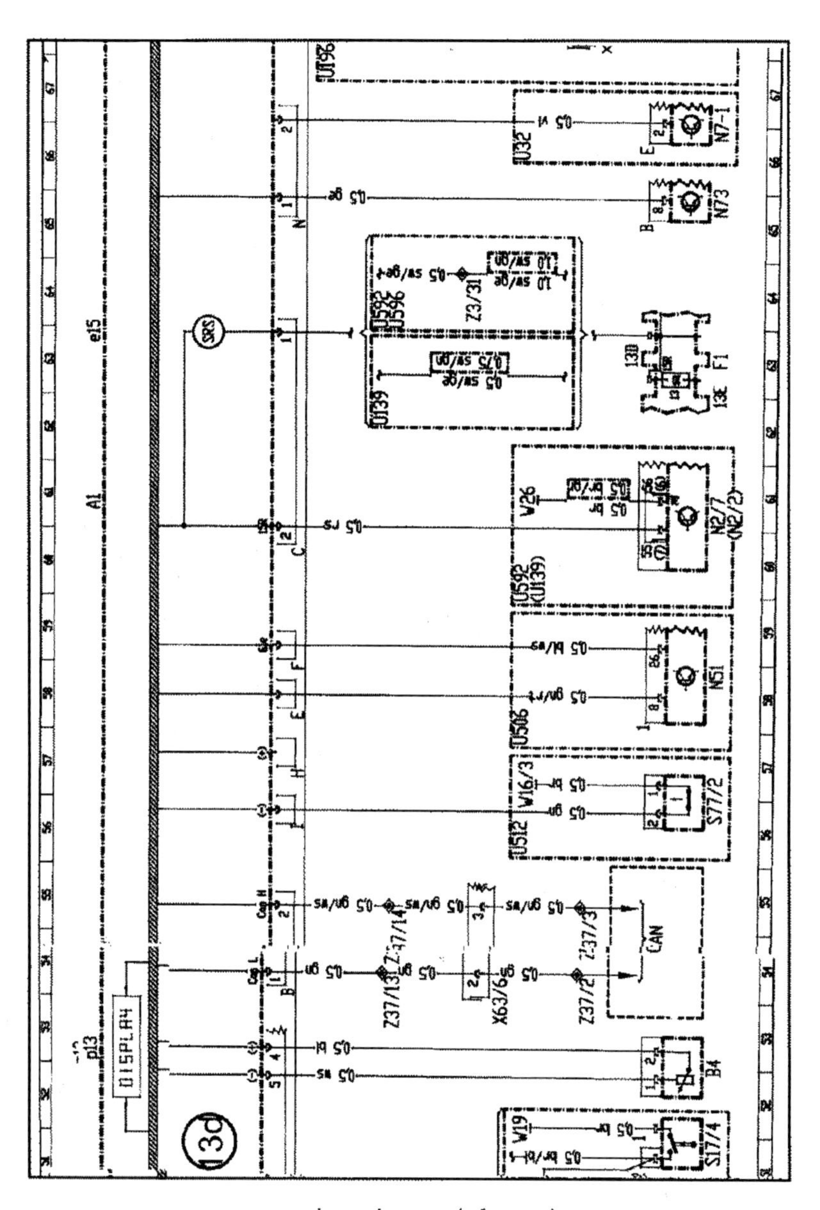

Legend, see end of manual

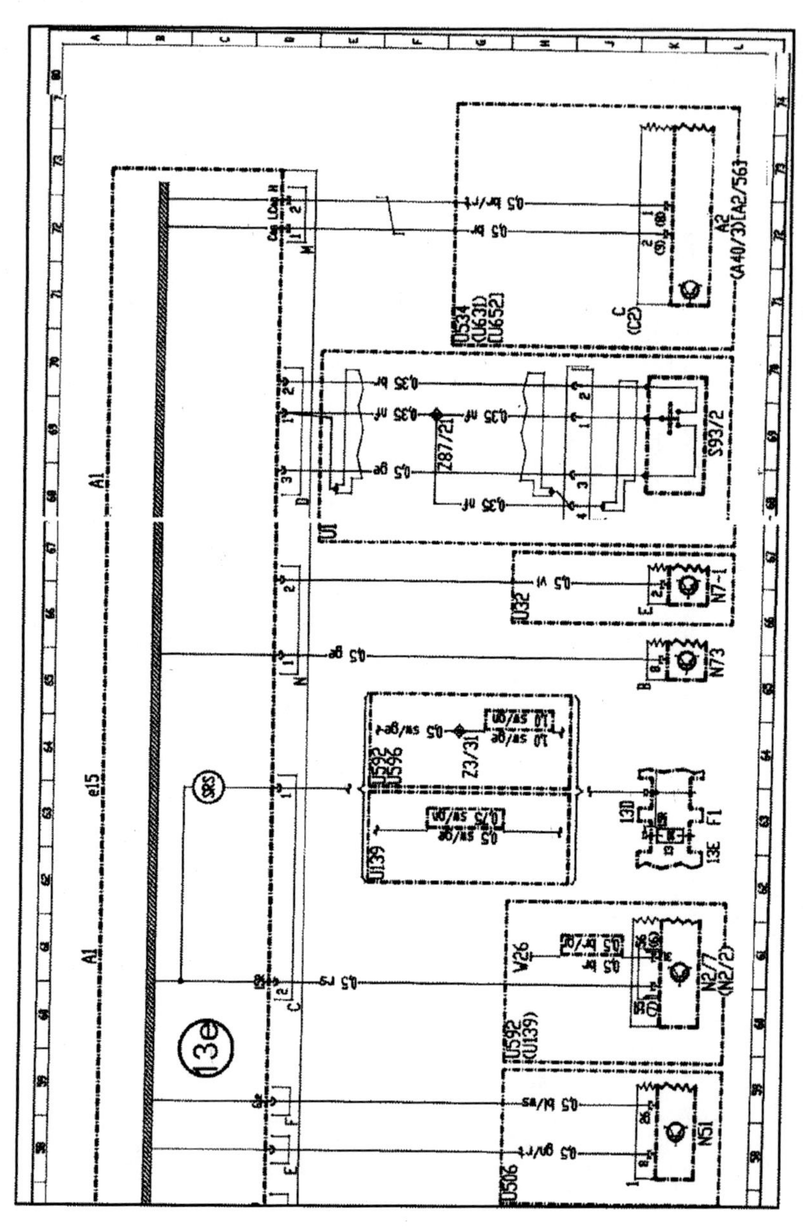

Legend, see end of manual

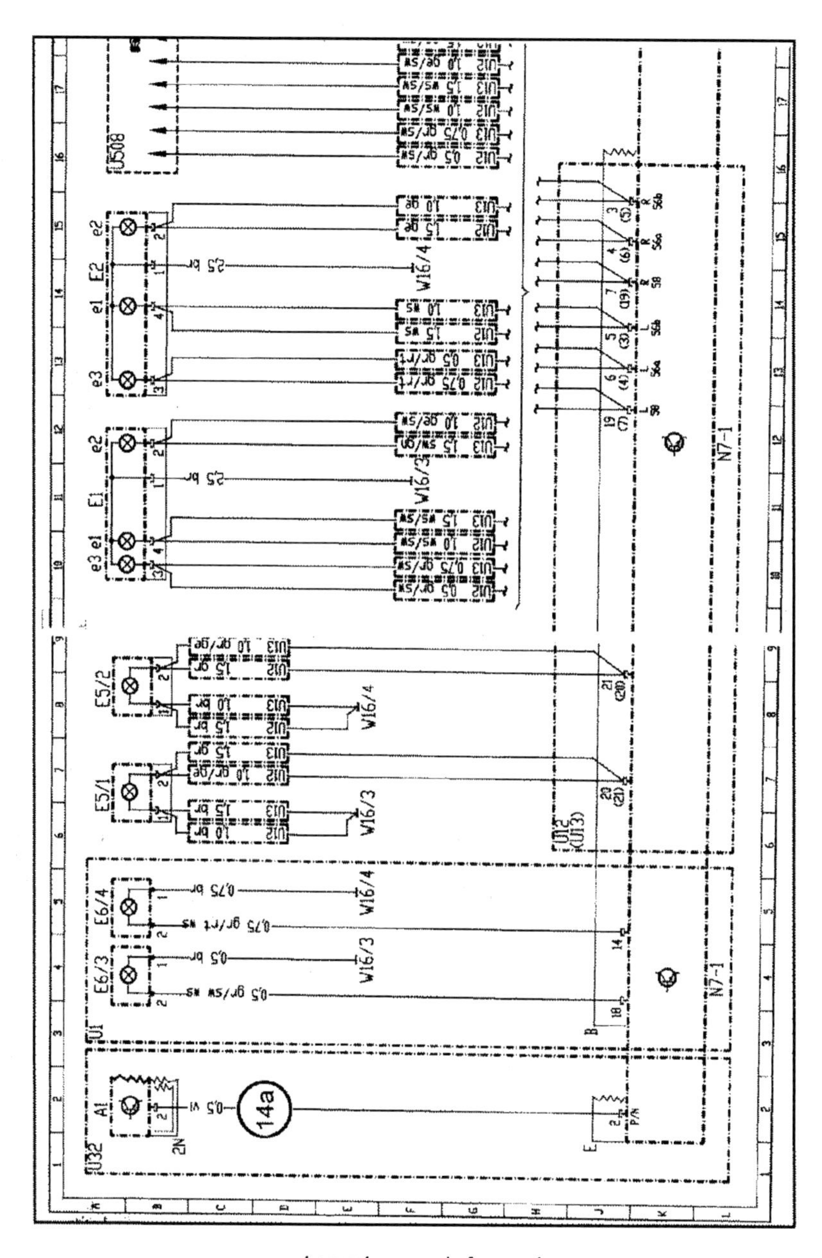

Legend, see end of manual

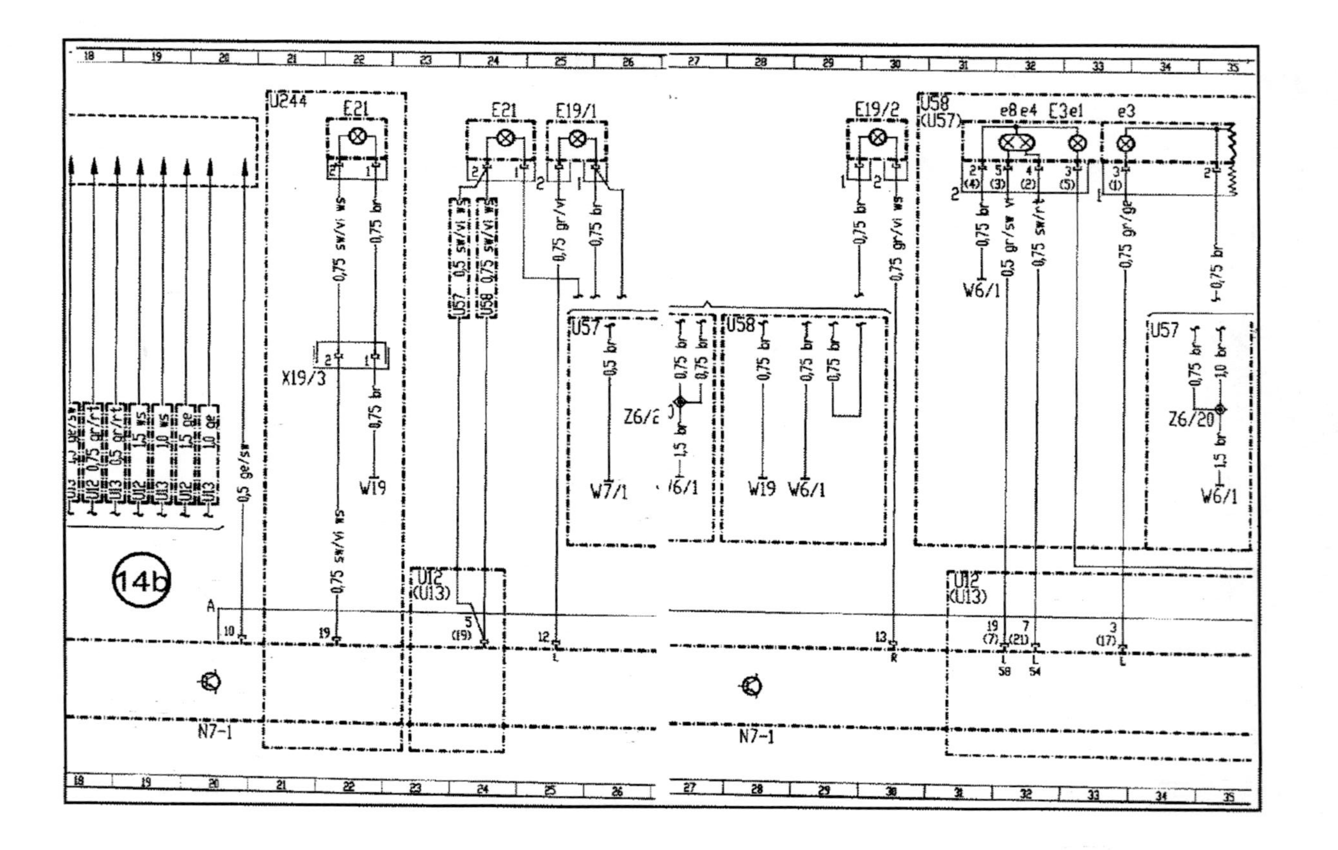
14b
U244
E21
X19/3
W19
E21
E19/1
U57
U58
W7/1
E19/2
W19 W6/1
U58
(U57)
e8 e4 E3e1
e3
W6/1
U57
Z6/20
W6/1
U12
(U13)
N7-1
N7-1

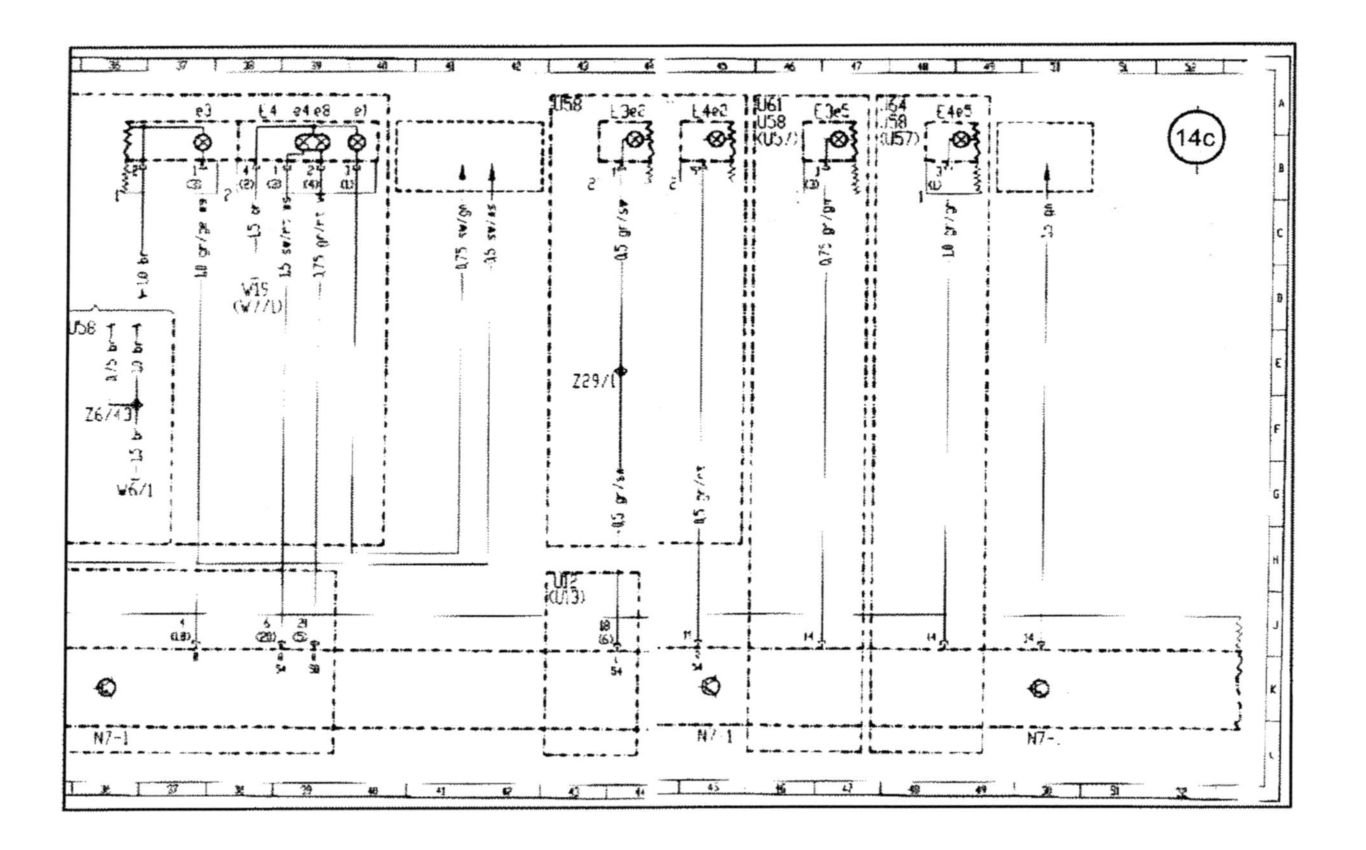
14c
e3
E4
e4 e8
e1
W15
(W7/1)
U58
Z6/43
W6/1
L3e2
L4e2
Z29/1
U61
U58
E3e5
E4e5
N7-1

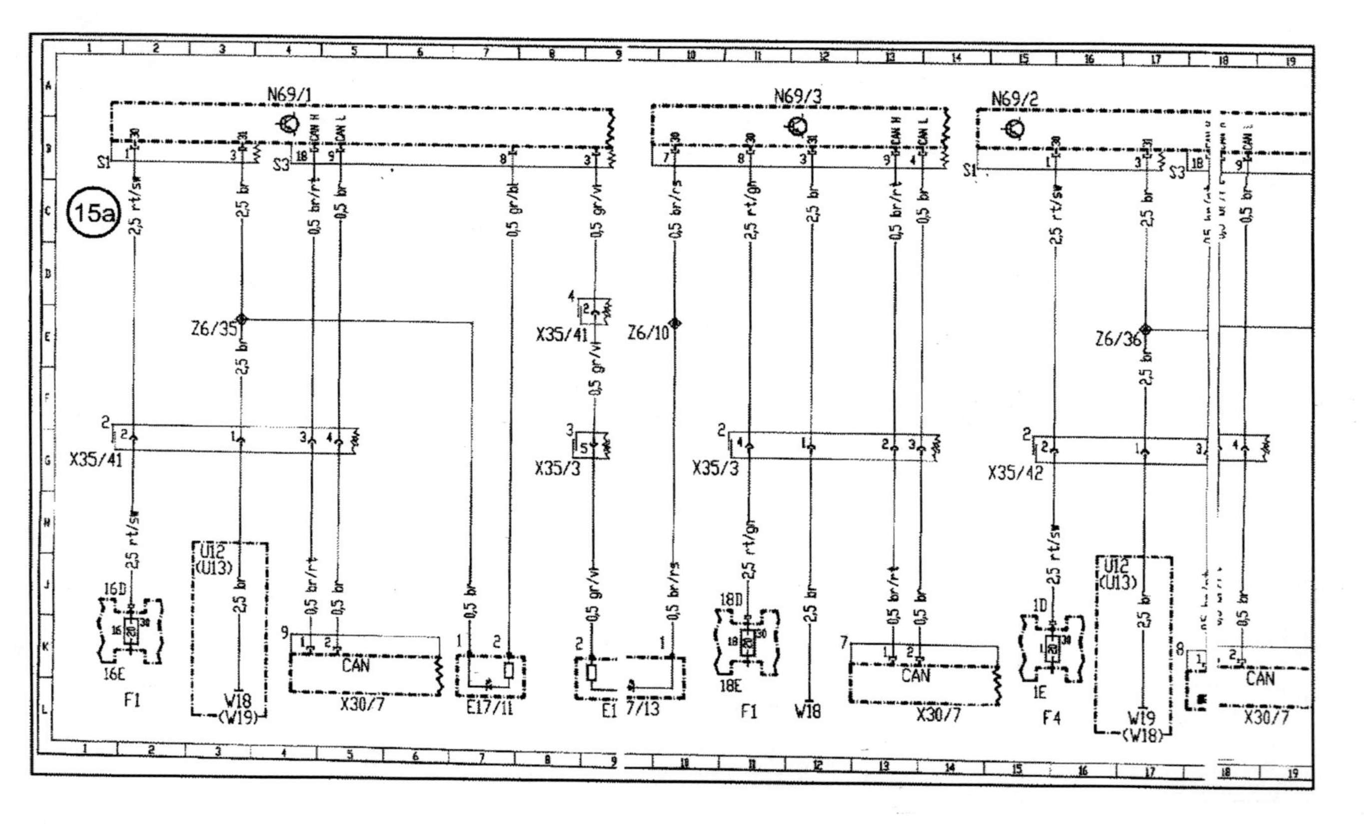
15a
N69/1
N69/3
N69/2
S1
S3
X35/41
X35/3
X35/42
Z6/35
Z6/10
Z6/36
U12
(U13)
W18
(W19)
W19
(W18)
CAN
X30/7
E17/11
E1 7/13
F1
F4
16D
16E
18D
18E
1D
1E
CAN H
CAN L
2,5 rt/sw
2,5 br
0,5 br/rt
0,5 br
0,5 gr/bl
0,5 gr/vi
0,5 br/rs
2,5 rt/gn

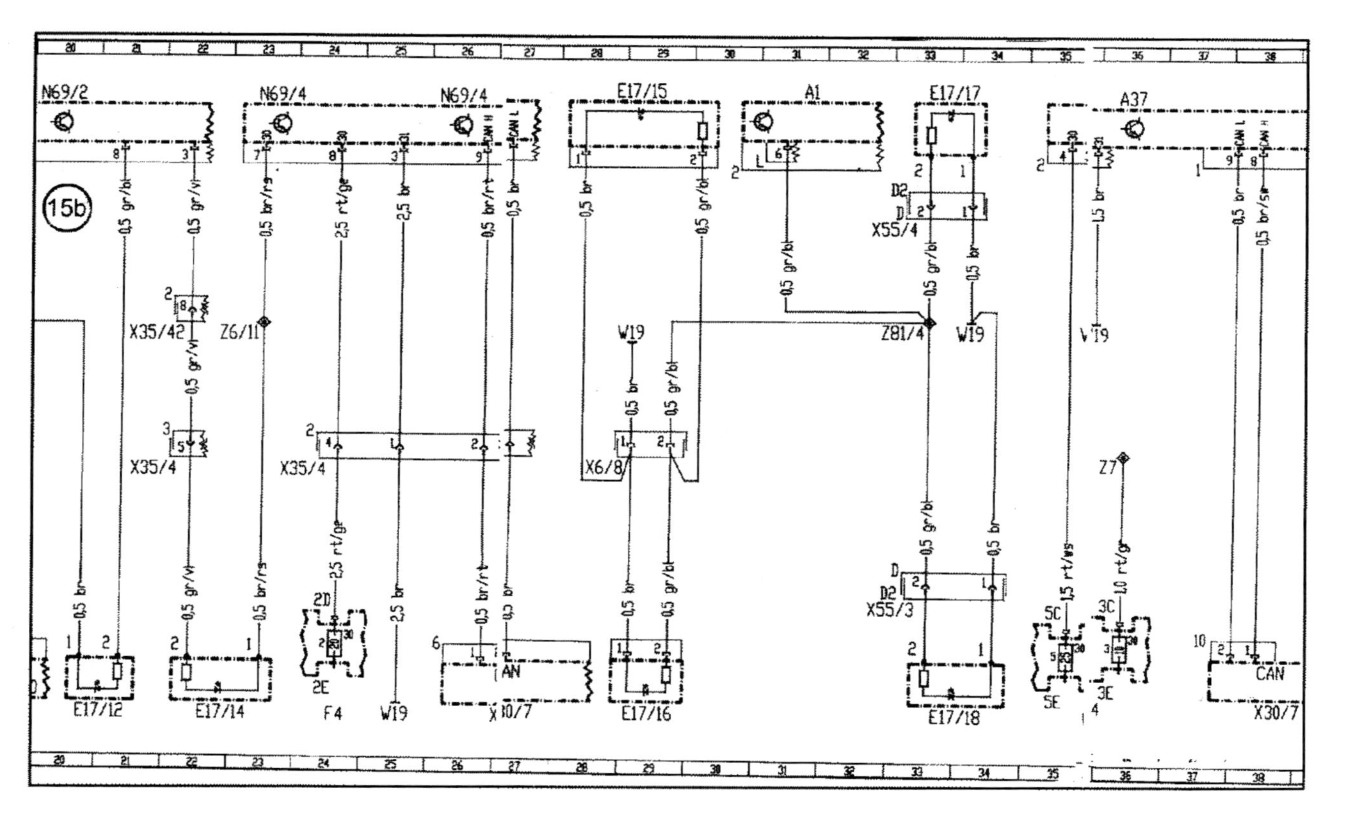
15b
N69/2
N69/4
N69/4
E17/15
A1
E17/17
A37
X35/42
Z6/11
X35/4
X35/4
X6/8
Z81/4
X55/4
X55/3
W19
Z7
E17/12
E17/14
F4
X10/7
E17/16
E17/18
X30/7
CAN
CAN H
CAN L
0,5 gr/bl
0,5 gr/vi
0,5 br/rs
2,5 rt/ge
2,5 br
0,5 br/rt
0,5 br
1,5 br
0,5 br/sw
1,5 rt/ws
1,0 rt/ge

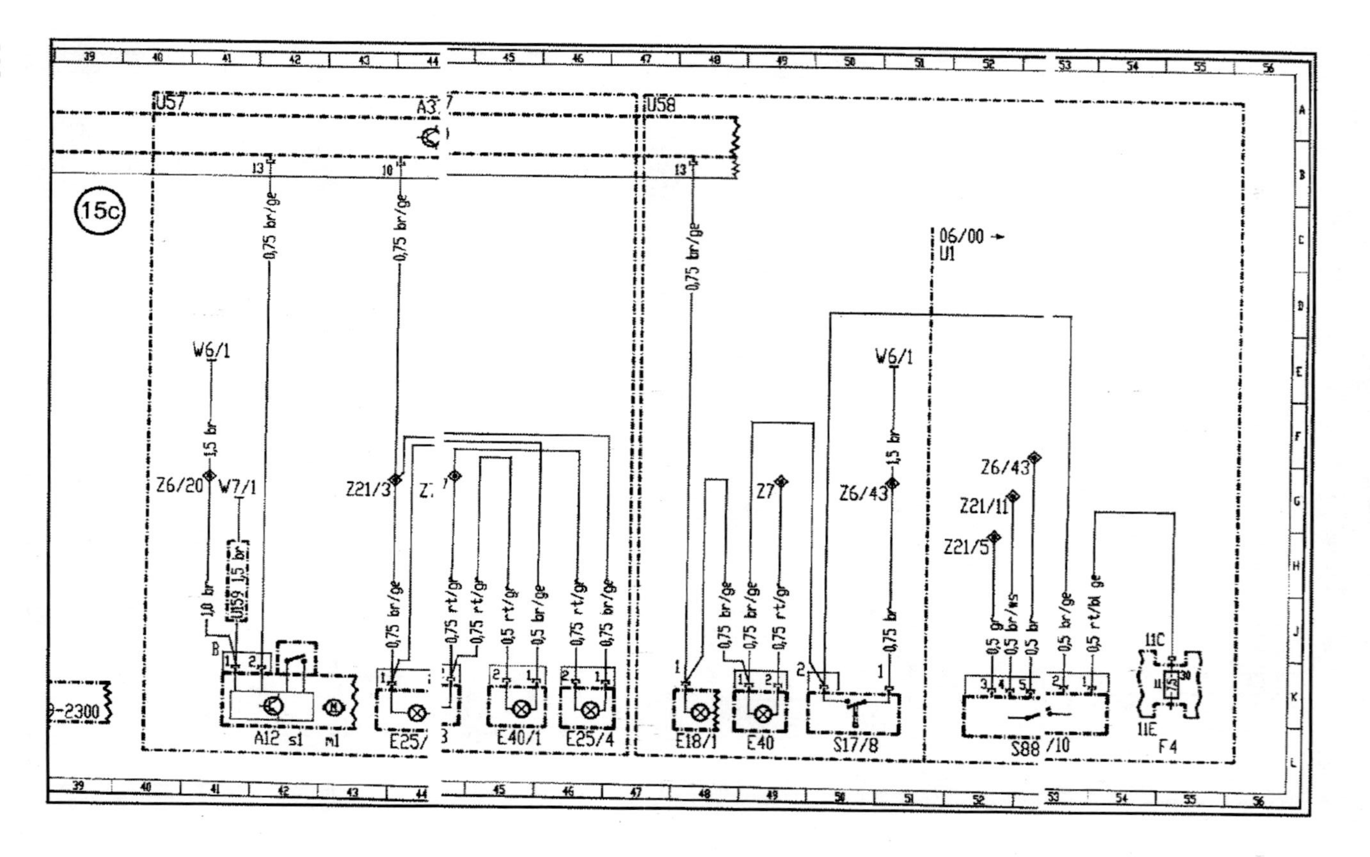
15c
U57
A3
U58
06/00 →
U1
W6/1
1,5 br
Z6/20
W7/1
U159 1,5 br
1,0 br
0,75 br/ge
Z21/3
0,75 rt/gr
0,5 rt/gr
0,5 br/ge
Z7
Z6/43
0,75 br
Z21/11
Z21/5
0,5 gr
0,5 br/ws
0,5 br
0,5 rt/bl ge
11C
11E
A12 s1 m1
E25/
E40/1
E25/4
E18/1
E40
S17/8
S88 /10
F4
9-2300